Rights, Resources, Culture, and Conservation in the Land of the Maya

The editors and authors dedicate this book to Mary Lindsay Elmendorf in appreciation of her encouragement, leadership, and example of working with Maya communities to improve living conditions—with dignity—for present and future generations.

Mayan woman showing Mary Elmendorf how to weave a hat from palm leaf strips (photo by Kappie Spencer).

Rights, Resources, Culture, and Conservation in the Land of the Maya

Edited by

Betty B. Faust,
E. N. Anderson,
and John G. Frazier

Foreword by June Nash

PRAEGER

Westport, Connecticut
London

Library of Congress Cataloging-in-Publication Data

Rights, resources, culture, and conservation in the land of the Maya / edited by Betty B. Faust, E.N. Anderson, and John G. Frazier ; foreword by June Nash.
p. cm.
Includes bibliographical references and index.
ISBN 0-89789-731-5 (alk. paper)
1. Mayas—Mexico—Yucatân (State)—Social conditions. 2. Maya women—Mexico—Yucatân (State)—Social conditions. 3. Indigenous peoples—Ecology—Mexico—Yucatân (State) 4. Conservation of natural resources—Mexico—Yucatân (State) 5. Biotic communities—Mexico—Yucatân (State) 6. Ecosystem management—Mexico—Yucatân (State) 7. Yucatân (Mexico : State)—Social conditions. 8. Yucatân (Mexico : State)—Economic conditions. 9. Yucatân (Mexico : State)—Environmental conditions. I. Faust, Betty Bernice. II. Anderson, Eugene N. (Eugene Newton), 1941– III. Frazier, John G.
F1435.1.Y89R54 2004
333.72′089′97427—dc22 2003058154

British Library Cataloguing in Publication Data is available.

Library of Congress Catalog Card Number: 2003058154
ISBN: 0-89789-731-5

First published in 2004

Praeger Publishers, 88 Post Road West, Westport, CT 06881
An imprint of Greenwood Publishing Group, Inc.
www.praeger.com

Printed in the United States of America

The paper used in this book complies with the Permanent Paper Standard issued by the National Information Standards Organization (Z39.48-1984).

10 9 8 7 6 5 4 3 2 1

CONTENTS

Photo essay follows chapter 6.

FOREWORD

Living in a Mayan World

This homage to Mary Elmendorf also pays tribute to the Mayan world about which she wrote and to which she contributed so much. The many contributors who have studied the past and present accommodation to the environment from the perspectives of archeology, biology, ethnography, political ecology, and social anthropology make us aware that the Mayan world still exists in the consciousness of the living heirs of Classic Maya civilization. This world is a reference point for many practices as well as memories. Mary Elmendorf devoted herself to both family planning and improvement of water quality in Mayan communities, as part of a development process that recognized limits and included planning for the future. Her insights were precursors of the population-environment policies currently influencing both conservation and development work.

In reading Mary's autobiographical reflections, I learned how much more we had in common besides working with Mayan people. I too had worked in the American Friends Service Committee in the spring of 1950 in Tepic. Villa Rojas introduced me to the people of Amatenango del Valle, proving to be as gracious and wise a host as he had been when introducing Mary Elmendorf to Chan Kom. Ivan Illich published Mary's first book in CIDOC (Center for Intercultural Documentation, Cuernavaca, Mexico), where my thesis on the social structure of Amatenango del Valle was also published. I too worked in the Peace Corps, but as an instructor for volunteers going to Guatemala, while Mary Elmendorf helped shape participatory action programs in the Peace Corps and the Overseas Education Fund. I share with her the desire to make the lives of the people with whom I did fieldwork more rewarding, but never accomplished in practice what Mary Elmendorf did throughout her working life. I admire, but could not possibly emulate, the way in which Mary could gain the attention of people in high policymaking positions to do what was right, at the same time cultivating their awareness of the needs and capabilities of community development agents.

At a time when anthropologists, especially those inclined to postmodern expressions, fail to orient the reader to the place of investigation or the chronology of events, it is rewarding to read a book that has a strong sense of place and time. E. N. Anderson and Betty Faust have succeeded in bringing together contributors with decades of research on the Yucatecan environment, both social and physical. Their personal experience of the changes they describe builds on the many archeological and ethnohistorical studies carried out in the area. Like Mary Elmendorf, many of the contributors came to the Yucatan when the peninsula was densely forested. They have heard the stories of old chicleros and seen the succession of henequen workers, loggers, cattle ranchers, Guatemalan exiles (in the 1980s), Chiapanecan self-exiles from the Lacandón (after the 1994 uprising), and nongovernmental organizations (NGOs) from many countries with their own agendas and constituents to serve. Revitalization of the economy sometimes means reverting to old crops, such as using henequen for making paper, a proposal Faust describes. But instead of the exploitative labor practices and the devastating clear-cutting of the forest, some of the new projects attempt to avoid the mistakes of the past, turning to small-plot, mixed cultivation practices managed by the indigenous people themselves.

Multiple studies in the same area by different investigators allow comparisons showing how different enterprises generate—or impede—the "collective identities" necessary for the survival of a culture. One of the most tempting—and potentially disruptive—intrusions is that of tourism, a disturbance that "makes Maya people themselves unsure what is real and what is theme park" (E. N. Anderson, chapter 6). Some practices endure and others are invented in response to new technologies. Slash-and-burn agricultural strategies continue, but vary in the degree to which careful firing techniques and managed regrowth maintain the forest.

One of the greatest challenges to survival for Mayan communities is the very success of past efforts leading to increased population growth in this peculiar environment where karst topography combines with deserts and jungles that alternate seasonally, from six months of the dry season to six months of the rainy season. Faust, Anderson, and Frazier remind us of the importance of addressing population pressures by interpreting them with reference to past events, including those that occurred from 200 C.E. to the Classic collapse, and comparing this with the growth correlated with increasing tourism between 1970 and 1995, when the population tripled despite a gradual decline in the relatively high birthrates. Elmendorf was highly conscious of how such an increase could defeat future development, regardless of how effective the conservation efforts were. Her work with family planning helped reduce birthrates, but uncontrolled migration policies resulted in overall population growth. What she stressed, and others attended to, was the need for a holistic approach that would conjugate the net impact of all these factors by using the Earth Charter to provide a framework for communities to work toward ecological integrity and social justice.

The kind of participatory research pioneered by Mary Elmendorf is precisely what is needed in bringing together indigenous peoples' strategies for coping with their local environments and the exogenous knowledge and resources of NGOs and government agencies, all of which are needed to address population pressures on available resources.

E. N. Anderson discusses how people coming from such different backgrounds can address common problems, such as technological transformations and commercial opportunities that introduce incentives for overcutting forests and depleting wildlife. Plan Forestal, initiated by the Mexican government with advice from Germany, has succeeded in many communities in Quintana Roo because it works with local authorities, cultivating their goodwill, while introducing rigorous accountability for expenditures.

In my own experience, doing fieldwork in an indigenous community of Chiapas, intermittently from 1957–1958 and during summers and vacation periods from 1963–1967, returning again in 1988, I could see the great importance of accountability for public projects. In communities such as Amatenango del Valle, in which the population of the *cabecera* (county seat) was entirely indigenous except for the schoolteacher's family, the town operated beyond Mexican law and financial co-optation. In the 1950s, indigenous leaders operated within civil religious hierarchies without receiving compensation, under the titular head of a *ladino* (Spanish-speaking person in Chiapas, generally associated with the dominant culture) mayor and secretary who handled any money coming from outside. In my research projects during the period 1963–1967, indigenous leaders held salaried posts of mayor, police chief, and secretary in the township center, and the town received limited funds for government-specified projects. Conflict was incipient, as young literate leaders challenged the authority of age-ordered hierarchies, but the limited funds had not yet led to wholesale co-optation. This visit came in the interim between the earlier field sessions and my return in 1988, when large grants for communal projects were handed over to local leaders with no attention to accountability except for the assurance that the leaders would deliver the vote to the ruling party. The corruption of local political powers grew enormously, exceeding traditional restraints on behavior.

In Amatenango in the 1960s and 1970s, I saw a preexisting, shared morality based on mutual dependence between different social levels that was being undermined. This process began somewhat later in the Yucatan. Anderson underlines the importance of restoring the moral and motivational basis among all parties involved in conservation programs by ensuring accountability and by promoting coparticipation in management. Policing to ensure that the conservation practices are followed is now required in the highly commercialized exploitation of forests and fauna in the Yucatan; however, it is important that this be part of a joint policy, developed and implemented with local communities.

These lessons were not followed in the town of Pich, Campeche, studied by Betty Faust. This proud community, which had served as an administrative center provisioning agricultural products during the rise of Preclassic Mayan cities until Postclassic times, was on the verge of selling a large part of its communal lands (ejido) in the last half of 2002, after struggling with the possibility since 1995. The government did not begin to seriously regulate extraction of timber until the 1980s, when most of the valuable wood had already been sold. Reforestation was not attempted until 1995, when it was mandated by the federal agency responsible for supervising community forests. However, these trees will not be ready to sell for lumber until approximately 2025. Thus, the future is dim for young people who are forbidden to hunt or to cut wood in the forest, and who feel that they cannot wait until forest regrowth can ensure their livelihood. Their dilemma reflects the growing divide between them and the few wealthy Mayan families who enjoy investments in cattle and university or technical school education for their children, bought with the money they earned by selling logs from the community's forest.

Despite many disruptions throughout the peninsula, the landscape still provides a matrix on which people map local concepts of the past and of the cosmos. David Forrest indicates the potential value of these continuities in "constructing community" in the context of conflict over the use of resources on rural estates. The political importance of these links in claiming and reinforcing their hold on resources, especially water, is not lost to the Maya of Maní. Ellen Kintz and Amanda Ritchie also stress the "calculated interface with their natural ecosystem" in which Mayan people reveal the continuity with a cyclical worldview manifested in art and archeology. This continuity with the past, the authors show, provides Mayan communities with a field of intrinsic relations that preserve and maintain local autonomy. Yet the authors also recognize that the intensified stress on the environment, and the breakup of families and eventually communities by forced migration, is threatening the linkages that have survived generations.

John Frazier provides a welcome intervention from the perspective of his field of biological conservation, urging anthropologists to bring their message regarding the social dimension to the field of practicing ecologists. The need for communication between the biologists, the ecologists, the NGOs, and the anthropologists grows with the increasing complexity of the problems. Here again, Mary Elmendorf's multiple careers, as an activist deeply concerned with ecological issues and as a mediator and manager in projects seeking to better the relations among people and between human populations and the landscape, enabled her to cultivate the human communication needed to fulfill the multiple tasks required to achieve such goals.

While reading about the succession of land uses and fading prospects for a future that had once promised continuity for the Maya as a people, I recalled my return trip to the Mayan region after a twenty-year absence. I was shepherding a group of eight students and a faculty couple, visiting the archeological sites of Chichén Itzá and Uxmal, when we decided to go to lunch in one of Redfield's communities of study for his classic comparative work, *Folk Culture of Yucatan.* "We'll go to Dzitas," I said with authority, "which is a bustling industrial city and sure to have many restaurants." When we arrived at two in the afternoon, we found a deserted town center. The railroad station that once serviced henequen production was abandoned, and the only traffic impediment was stray dogs sleeping in the streets. When we approached someone to ask directions to the nearest restaurant, we were advised to go toward Chan Kom. During Redfield and Villa Rojas's study in the 1930s, the people of Chan Kom had begun to build a road to connect with people and activities in Chichén Itzá. Their road was connected with a new highway in 1971, one that now provides access to a rapid toll road going beyond Chichén Itzá to the new tourist city of Cancún. Now many of the sons and daughters of Chan Kom travel this road to work their "new milpa" in Cancún, perhaps frequenting the *típica* (a popular diner with regional fare) that we finally found. Even at midafternoon it was so busy that we had to wait in line for our "traditional food," served by waitresses wearing costumes inspired by tourist brochures. We wondered as we ate if some of the young people of Chan Kom now dream of being the owners of such a restaurant, just as their grandparents once dreamed into reality a future with a school, a road, electricity, and access to a wider world.

Reflecting on the changes experienced by both traditional Mayan and modern capitalist communities, I wondered whether an archeologist in the next millennium would be able to decode what was occurring at the turn of this second millennium of the current era. Would he or she perceive that Chan Kom, introduced to the anthropological world in the 1930s as the village "that chose progress," had become another example of capitalist overkill? Perhaps it would surpass the neighboring city of Dzitas and choose a path of "sustainable development" in the interim between our trip and the archeologist's arrival? Would the archeologist find any traces of Chan Kom's Mayan roots, or would this community be seen merely as another "Disneyland virtual community," serving as a mecca for tourism? And would there still have been tourists, whose remains would show up in the most recent strata excavated by those archeologists of the fourth millennium, or would they simply find scattered evidence of scavengers in a barren landscape? Mary Elmendorf, and those who come together to give evidence of the scholarly tradition she pioneered, call upon the readers of this inspired anthology to help make possible the best of future scenarios.

June Nash
Professor Emeritus
City University of New York

PREFACE

Perspectives Evolving from and with Mary Lindsay Elmendorf's Work in Maya Communities of Mexico's Yucatan Peninsula

This volume was conceived after a session titled Yucatec Maya Political and Cultural Ecology: Papers in Honor of Mary Lindsay Elmendorf, held at the Ninety-Sixth Annual Meeting of the American Anthropological Association in Washington, DC, November 19–23, 1997. At the end of a dinner in celebration of Elmendorf's eightieth birthday, the session speakers encouraged us to develop an edited volume based on the papers we had all presented earlier that day.

The eight chapters by session speakers have been thoroughly revised and updated, and we have included three additional chapters. The first is about Maya women, volunteered by Mary Elmendorf, who also suggested the inclusion of a chapter by Joann Andrews, a longtime friend of hers and founder of Pronatura Península de Yucatán, a local conservation organization. The third addition is by Jenny Ericson, who has done applied work in the Calakmul Biosphere Reserve, an area of great significance not only to the ancient Maya but also to contemporary issues central to the topic of this book.

The introductory chapter sets the scene, introducing the reader to the Yucatan Peninsula and the Maya who have lived there for more than 3,000 years. It alerts the reader to some of the outstanding issues that need to be considered if we are to understand culture and conservation in Maya (or any other) communities—particularly human rights, community, and the complexities of political ecology.

This is followed by Alicia Re Cruz's chapter on the background, contributions, and leadership of Mary Elmendorf, whose tireless work and unshakable principles have inspired not only those of us who have worked in the Mayab,[1] but also other academics and practitioners working in many nations. Elmendorf's pioneering contributions in the integration of community development, health, environmental issues, and women's empowerment have set precedents influencing academic and applied work in anthropology and in related fields. Today, participatory research, as it has come to be called, is a well-respected investigative tool, not only in the land of the Maya but around the world. In recognition of these substantive contributions, Elmendorf has received many awards, among them the 1982 Margaret Mead Award, as well as honorary recognition from the University of North Carolina in 1993. As Re Cruz explains, exploring the rich and productive career of Mary Elmendorf provides many important lessons for research and applied activities.

The third chapter is Elmendorf's own, on the lives of Maya women and their contributions to the environment of their families and communities. This is not only a synthesis of decades of her work, but it also provides an update on her latest innovative efforts, with a personal history that puts many of the developments and concepts into a human context. In 1976, her first study was published in the United States as *Nine Mayan Women*; it had previously been published in Mexico in both English and Spanish. (It was first published in 1972 in Mexico in English by the Center for Intercultural Documentation (CIDOC) and later in Spanish by the Secretaría de Educación Pública (Secretary of Public Education [a federal agency], SEP.) This classic monograph shows the importance of individual women to processes of change in the community of Chan Kom, Yucatan. This same community was originally studied by Redfield and Villa Rojas (1934) as part of the Carnegie Institution's project in the Yucatan Peninsula, and was later restudied by Redfield (1950). Elmendorf's 1972 work, in turn, provided a base for understanding the more recent changes reported in Re Cruz's 1996 book, *The Two Milpas of Chan Kom*. These four studies of the same community over six decades provide valuable information and insights concerning processes of cultural and social change in adaptation to new technologies, markets, and environmental conditions.

The next chapter is a history of the one of the first conservation organizations in the Yucatan, by its founder, Joann Andrews. This personalized history interprets the development of a local branch of Pronatura within the context of institutional, political, and conservation struggles in Mexico. Pronatura's primary focus is the establishment of protected natural areas and programs for working with communities living in or near protected areas. Andrews describes the challenges—and frustrations—of working with rural communities in an effort to promote biological conservation. An area of major importance to Pronatura Península de Yucatán, as well as to Andrews, is the Calakmul Biosphere Reserve. Jenny Ericson's chapter follows, describing population fluctuations and land use within communities neighboring the Calakmul Biosphere Reserve. The majority of people in these communities are "colonists" (recent immigrants to the area), originally from Chiapas and other Mexican states to the north. Many of them belong to Maya groups that are related to the Yucatecs, such as the Chol and Tzotzil Maya from Chiapas. They were settled in this area by the national government before the present reserve was established. The first of six case studies presented in this book, Ericson's results show the tensions between community development, modernization, and biological conservation in communities of resettled people. It is interesting to compare these communities with those of indigenous Yucatec Maya reported in the other five case studies, presented later. The descriptions of contrasting types of rural communities provide information concerning the range of variation in strategies and illustrate the necessity of understanding this diversity in relation to both ethnic identities and specific political histories, including migration.

E. N. Anderson's chapter follows, with a discussion of the dual problem of valuing the Maya forests and interpreting local perceptions of the forest and its resources. The case study he presents is based on research begun in 1990 in the community of Chunhuhub, in the interior of the state of Quintana Roo. Anderson reports a remarkable range of uses for native plants; he also relates the opinions and actions of the community and its history of resistance to destructive forms of development. Chunhuhub is a relatively traditional community, yet is subject to impacts of nontraditional economic pressures and opportunities. The adaptability of the Yucatec Maya people and their culture has led to modest prosperity (for many) in recent years, but the future is uncertain.

Betty Faust discusses changes observed over eighteen years of recurrent research in the community of Pich, in the interior of the state of Campeche, near the ruins of Edzná. Her chapter focuses on shifting perceptions of the values of timber cutting and wildlife hunting, analyzed in light of a community decision taken in October 2001 to sell 22,000 hectares of the forest commons to an agribusiness corporation that plans to invest $23 million in a local project for dairy production. Once again, the community easily qualifies as traditional, yet its modernization includes the conscious and willing relinquishing of a major resource base, despite previous indications that a transition to sustainable usage was possible. These earlier indications were described by the author in her 1998 book, *Mexican Rural Development and the Plumed Serpent*.

Grace Bascopé's article is based on twelve years of research in the community of Yaxunah, near the center of the state of Yucatan, southeast of the archeological site Chichén Itzá. The community had been regarded as being "in balance with its environment," but this chapter describes recent changes that are resulting in accelerating rates of environmental and social degradation. Bascopé describes the new strategies adopted to maintain livelihoods and the effects of these activities on the community's forest. She also makes clear that many residents are aware of their predicament and are concerned about the future. Once again, the impacts of external forces on a traditional Maya community are apparent.

David Forrest looks at the role of metaphors in conceptualizing territory, space, and resources on a Yucatecan hacienda, Santa Rita Komchen close to the archaeological site and town of Maní (south of the city of Mérida). Before the Spanish conquest, the area was part of the region sustaining the Postclassic Maya city of Maní, which (like Cobá; see chapter 10) has retained its identity as an urban site into the present. He examines metaphors as tools that local people use to create—and destroy—paradigms that legitimize the control of resources, tracing through several centuries the process by which these metaphors have interacted with political structures, affecting land tenure, access, and ownership. Forrest discusses the concepts of usufruct and land ownership, water management and rights in relation to the metaphors employed in the cultural construction of the landscape. These concepts are found to be central to historical changes in the relationship between the town of Maní and the hacienda of Santa Rita Komchen.

In the last of the case studies, Ellen Kintz and Amanda Ritchie investigate ecological concepts and practices in Cobá, Quintana Roo, southwest of Cancún. Working from a conceptual foundation in "deep ecology," they resort to yet another conceptual entry point into the human-environment paradigm. Interestingly, Kintz began work at Cobá in an archaeological project in the 1970s and has continued with ethnographic investigations during subsequent decades. A description of change over time in Maya adaptation to the tropical forest was provided in her 1990 book *Life Under the Tropical Canopy*. In the present chapter, she and coauthor Ritchie use this foundation of long-term investigation to expand the question of how the Maya value nature and understand its limits. Cobá, like Pich, Yaxunah, and Maní, is associated with an archaeological site and historical documents indicating continuity of cultural traditions and local environmental knowledge since the Maya Classic period.

The following two chapters grapple with theoretical and practical issues underlying the particularities of the preceding case studies. They are contributions to contemporary struggles to construct a theoretical structure to guide community conservation work. Engaged in an emerging synthesis, they reflect and express some of the tensions and conceptual conflicts involved in working with communities for biological conservation, including the management of resources and the protection of rights to those resources. Susannah Glusker's chapter on the history of human rights issues in Mexico is highly germane to complex issues of land tenure, livelihood, and control of resources. She puts community conservation issues into a political context, summarizing the Mexican history of human rights, particularly with respect to indigenous women. As Elmendorf explained in 1972, women are fundamental decision makers as well as agents of change, so the human-environment interaction depends intimately on their situation. A central aspect in Glusker's chapter is the description of the academic, administrative, and political life of Mariclare Acosta, whose career has interwoven with Elmendorf's over decades and who was named by Mexico's President Vicente Fox to be Undersecretary for Human Rights within the Ministry of Foreign Affairs. Acosta previously ran a nongovernmental human rights organization, after working on the staff of Amnesty International in Mexico. While the specifics vary by country and time period, the problems of Mexico in this arena are germane at a worldwide level: protecting the rights of local peoples whose livelihood depends largely on the use of "natural" resources. These peoples often find themselves squeezed between powerful corporations seeking to use their resources for the global economy and conservationists seeking to protect their resources and local ecosystems for the future of humanity.

The penultimate chapter is by John (Jack) Frazier, who explores the complexities of political and economic forces affecting biodiversity conservation. He documents the need for more involvement of social scientists in biological conservation, as well as the need for social scientists concerned with communities to be better informed concerning—and more involved with the formation of—higher level policy decisions. He delves into the problems of cross-cultural misunderstandings, but especially differences in power and how they impact organizational development and behavior. Included is an analysis of the nongovernmental organizations working with local communities and the need to grapple with this issue for more effective community participation in biological conservation. The discussion is grounded in his eleven years of experience working as a conservation biologist in the peninsula, plus more than three decades of work on the conservation of marine turtles, reptiles that migrate across oceans, often crossing between the coastal areas of different countries. Although trained as a biologist, he analyzes powerful economic and cultural factors that drive the human-environment interaction in our industrial, and postindustrial, civilization. This brings the book full circle, for in this treatise are the elements to understand why and how traditional communities described earlier in the case studies are being drawn into what Frazier calls "the postmodern, neoliberal Age of Globalization." In begging and challenging social scientists to widen their field of operation, he promotes some basic changes needed for interdisciplinary work on problems affecting the biosphere.

The book ends with the reflections of Mary Elmendorf on the previous chapters, relating them to her experiences knowing and working with communities in the peninsula for nearly forty years. It ends with her present efforts on behalf of the Earth Charter and perceptions of its potential for assisting biological conservation work with communities in the Yucatan Peninsula. The idea for this charter was conceived during the 1992 United Nations Conference on Environment and Development at Rio de Janeiro, Brazil. This charter is now emerging as a global mechanism to promote democratic process in decision making concerning the global environmental crisis, and it formed part of preparations for the World Summit in Johannesburg.

Although these chapters cover a diverse range of issues, they are united by one central theme: the effort to protect biological diversity and ecosystem services in the local resource bases of diverse human communities. Many Mayas and other peoples without much formal education, as well as scholars of indigenous and other ethnicities, have come to realize that the long-term survival and development of diverse human communities depends intimately on such protection. These resources are critical to their struggles for self-determination in defining their own cultural values and identities, as well as essential for the future of the biosphere and humanity at large. We hope that our reflections on years of work with Maya and other communities in the Yucatan Peninsula will be of practical use not only to others working in this region, but also to anthropologists, biologists, conservationists, agronomists, ecologists, and development agents working in other regions. Comparisons of this volume with other regional studies may contribute to the development of respectful—and effective—procedures for working in biological conservation with diverse communities, while improving their quality of life within a framework that honors cultural values, human rights, self-determination, and collaboration with others for the sake of all our children and the other species with whom we share and co-create our biosphere.

NOTES ON USAGE

1. Usage of the terms *Maya*, *Mayan*, and *Mayas* varies considerably by author. *Mayan* is universally used for the indigenous language of the Yucatan Peninsula. It is also frequently used as an adjective in references to Mayan literature, chronicles, values, and the like. However, among most contemporary specialists in Maya culture, the term *Maya* is currently used as both an adjective and a noun referring to the people. Those referred to sometimes include peoples indigenous to the entire larger Maya region (e.g., Chol, Tzotzil, Kanjobal, Tzeltzal, Quiché, and many others). At other times the term is used more narrowly to refer to the speakers of Yucatec Mayan who inhabit the Yucatan Peninsula of Mexico. Use of the plurals, Mayas or Mayans, is found more often in contexts emphasizing the variations between individuals rather than those learned patterns that are broadly shared in this culture area. Ethnographers such as June Nash (Foreword) and Mary Elmendorf (chapter 3 and concluding chapter 13) frequently use Mayan as an adjective, and sometimes as a noun as well. We have respected differences in usage between the authors of this volume since they reflect differences found in the existing scholarly literature.
2. The word *Yucatan* in colonial times referred to one politically defined unit that corresponded approximately to the geographical boundaries of the peninsula. Eventually British Honduras (now Belize) and Guatemala became independent, while the states of Quintana Roo and Campeche became politically separate from the state that continues to be called Yucatan (see map). When referring only to the State of Yucatan we specify that it is the state. Otherwise we use "the Yucatan" to refer to all three of the Mexican states in the Peninsula of Yucatan.

3. In this book, for reasons of convenience, we adhere to one system of writing Yucatec Maya. The written form developed by Catholic religious workers in the sixteenth century has been modified by linguistic research and standardized with the twentieth century publication (1980) of "the Cordemex," a dictionary compiled under the direction of Alfredo Barrera Vásquez. It has become the system of choice and normal use in Mexico (although place names maintain the original colonial orthography). We recognize that a substantially different system, originating in Guatemala in 1988, is being accepted in some of the ethnographic literature. However, after some debate among the editors, we have decided to use the Barrera Vásquez system in an effort to communicate efficiently with a wide range of readers already familiar with existing literature concerning the Peninsula of Yucatan.
4. Hispanic last names are traditionally double, with the father's last name followed by the mother's family name. English speakers commonly ignore the first Hispanic last name given in a set of two, treating it as a middle name, and then mistakenly use what is the author's mother's family name as the last name. We have cited authors with Hispanic last names as they appear in Spanish citations; some authors use both last names, some hyphenate the two last names, and some married women use their father's last name followed by "de" and their husband's father's last name. Our references are alphabetized according to the Spanish rules, under the paternal surname.
5. Following what is increasingly the international usage, we use B.C.E. (before the common era) as a nonreligious substitute for B.C. (before Christ) and C.E. (common era) for A.D. (anno Domini).

NOTE

1. In the local, indigenous language, Yucatec Maya, the term *Mayab* means literally "the land of the Maya," and is used to refer to the Yucatan Peninsula.

REFERENCES

Barrera Vásquez, Alfredo. 1991. *Diccionario Maya: Maya-Español, Español-Maya*, 2nd Edition (first edition 1980). Mexico City: Editorial Porrúa.

Elmendorf, Mary. 1972. *Mayan Women and Change*. Mexico: CIDOC Cuaderno No. 81.

———. 1973. *La Mujer Maya y el Cambio*. Mexico City: Secretaría de Educación Pública, SEP/70.

———. 1976. *Nine Maya Women: A Village Faces Change*. New York: Schenkman.

Faust, Betty B. 1998. *Mexican Rural Development and the Plumed Serpent*. Westport, CT: Bergin and Garvey Press.

Kintz, Ellen. 1990. *Life Under the Tropical Canopy: Tradition and Change Among the Yucatec Maya*. New York: Holt, Rinehart, and Winston.

Re Cruz, Alicia. 1996. *The Two Milpas of Chan Kom*. Austin, TX: University of Texas Press.

Redfield, Robert. 1950. *A Village That Chose Progress: Chan Kom Revisited*. Chicago: University of Chicago Press.

Redfield, Robert, and Alfonso Villa Rojas. 1934. *Chan Kom, a Maya Village*. Publication 448. Washington, DC: Carnegie Institute of Washington.

ACKNOWLEDGMENTS

We wish to thank those communities that have generously accepted us, participating in our research and development work. We are grateful for their collaboration and their willingness to teach and learn with us. Moreover, we wish to express a deep appreciation for the beauty of the rural and coastal landscapes and for the many interesting and beautiful species of flora and fauna that can still be enjoyed in the Mayab, the ancestral territory of the Yucatec Maya.

The editors thank their respective institutions: the Centro de Investigación y Estudios Avanzados del IPN in Mérida, Mexico; the University of California in Riverside; and the Conservation and Research Center of the Smithsonian Institution in Front Royal, Virginia. We appreciate the patience and assistance of Jane Garry and Jim Lance of Greenwood Publishing Group during the preparation of this book. We also thank Calvin Lindsay Elmendorf for his valuable and patient assistance in bringing his mother's chapters to press.

The editors' share of proceeds from the sale of this book will be donated to biological conservation activities that both benefit Maya communities of the Yucatan Peninsula and involve them as decision makers in planning, implementation, and evaluation.

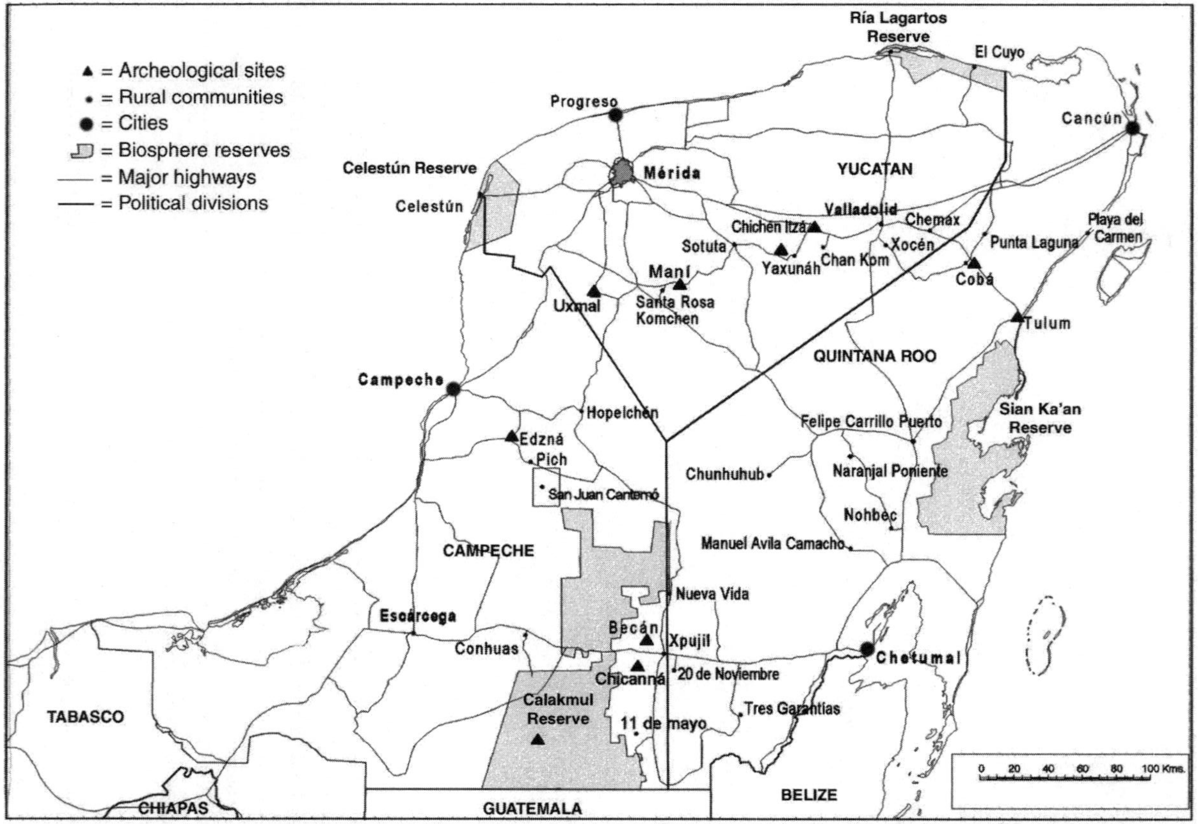

Sites in the Yucatan Peninsula mentioned in this book (*Source:* Laboratorío de Cartográfico, Departamento de Ecologia Humana, CINVESTAV, Mérida, Yucatan, Mexico).

1

Introduction: An Environmental and Cultural History of Maya Communities in the Yucatan Peninsula

E. N. Anderson
Betty B. Faust
John G. Frazier

THE YUCATEC MAYA PEOPLE

Today, over 700,000 Yucatec Mayas live in the southern Mexican states of Yucatan, Campeche, and Quintana Roo (see Pfeiler 1997), while some thousands more live in the neighboring countries of Belize and Guatemala. In addition to the Yucatec Maya (on whom this book is focused), there are many other peoples of Maya heritage in this region, including those of the two other southern Mexican states of Chiapas and Tabasco, as well as the countries of El Salvador and Honduras. These include the Quiché, Tzeltal, Tzotzil, Chol, Chontal, Kekchí, and Mam. All of these peoples share a heritage known to every reader of *National Geographic* or *México Desconocido*. There is much more to their heritage, however, than ancient temples. Especially important for this book is the Mayas' interaction with their environment—agriculture, forestry, and other forms of landscape and resource use. Their ancestors developed a great civilization whose city-states shared trade routes, architectural styles, ceramics, stone tool technologies, a calendar, and a system of writing. The remains of these city-states and their supporting agricultural villages are found throughout the Maya region, from the northern metropolis of Dzibilchaltún in Yucatan to the southern ones of Copán in Honduras and Usulután in El Salvador, from the western centers of Palenque in Chiapas and Comalcalco in Tabasco to the eastern ones of Cozumel in Quintana Roo and Altun Ha in Belize (Weaver 1993).

The Yucatec are the only ones among these who historically called themselves Maya (A. Barrera Vásquez 1991). The term has now been adopted by speakers of related languages. The Yucatecs' land is known in their own language as *Mayab*, referring to the area we know today as the Yucatan Peninsula. *Yucatec* is not a Maya word. It is an English translation of the Spanish *Yucateco*, which is based on the Spanish name for the region, *Yucatan*; this term is derived in turn from a misunderstanding dating to early sixteenth-century contact between the Spanish conquistadors and the Maya (see Frazier, chapter 12, this volume). Even as a geographical term, *Yucatan* continues to be confusing, as it refers to either the entire peninsula or to just the triangular-shaped Mexican state in the north that reaches south to a point near the center of the peninsula (see map at end of frontmatter). We use *Yucatan* to refer to the Mexican state and *the Yucatan* to refer to the entire peninsula. The chapters in this volume refer mainly to research done in rural communities of the three states of the Yucatan Peninsula: Yucatan, Quintana Roo, and Campeche. Most, but not all, of these communities could be described as culturally Maya. However, over the past twenty or more years there has been migration of peoples from northern Mexico into this peninsula, chiefly through government-sponsored projects, as well as short-term, but massive, influxes of tourists into the area. Immigrants have also come from Asian and Mediterranean regions seeking employment and economic opportunity; and for decades national and foreign oil crews have worked in Campeche.

While they are only part of the larger Maya world, the Yucatec Maya still display considerable diversity. They have, among other things, assimilated large numbers of Spanish immigrants, as well as smaller numbers of Chinese, Koreans, Lebanese Arabs, Africans, northern Mexicans, and others; the descendents of mixed marriages are considered to be Maya, if they remain in Maya towns and speak some of the language. There are also differences between various Yucatec Maya populations in dialect, local history, and degree of acculturation to Hispanic Mexican life (Brown 1988; Hervik 1999).

Yucatec communities differ among themselves in prosperity and in local culture: this is true today just as it was at the time of the Conquest. For example, towns of Quintana Roo's Zona Maya (the central part of the state) are intensely independent and often relatively well-off. In the state of Yucatan, the collapse of a once-booming industry based on the fiber of henequen (also called sisal, *Agave* spp.) has devastated and traumatized many communities (Benítez 1986). Jorge Flores Torres's (1977) and Alan Burns's (1983) collections of statements and stories from Yucatec Maya include comments on credit availability, fertilizer policy, and bureaucratic hindrance to farming. Politics at the "cornroots" has affected the structuring of daily work, and the government appears to the Maya (among many other rural Mexicans) to be more bureaucratic than supportive. There are also communities that live so much on tourism that they have become rather surrealist, the people not quite sure what is real and what is theme park (Castañeda 1996; Primack et al. 1998). Other towns within tourist country have been bypassed by the tourist dollar, left in poverty within the shadow of pyramids that earn Mexico millions of dollars annually in hard currency (Kintz and Ritchie, chapter 10, this volume). Some towns, notably Oxkutzcab in southern Yucatan, flourish on orchard agriculture (Morales 1987; Rosales 1988). Others adapt by becoming specialized crafts suppliers, like Tixkokob, also in southern Yucatan, with a sizable hammock industry, and Yaxunah with its wooden carvings (Bascopé, chapter 8, this volume). There are urban centers such as Mérida, Valladolid, Cancún, and Campeche, where most of the "big-city folk" are Yucatec Maya, at least by ancestry. While Redfield's (1941) model of the "folk-urban continuum" had heuristic value in his time, it is certainly too simple to give us much help in understanding today's complex and fast-shifting social and economic situations, many of which are driven by factors well outside the region.

In short, generalizations are difficult to make in Yucatec country (Brown 1988; Hervik 1999). Yet the Yucatec are still one people, united not only by common language and common heritage but also by a real and vibrantly living sense of identity and common destiny. They are aware of their complex identity and their historical ties to other Maya groups and to other indigenous peoples. In January 1996, Faust served as Spanish-English translator for one *h-men* (Maya healer and ritual expert Don Fidel of Kantunilkin, Quintana Roo) while attending the North American Indigenous Peoples' Human Values of Biodiversity Meeting (sponsored by the United Nations Environmental Program Global Biodiversity Assessment, Section 8, Tucson, Arizona, January 26–27, 1996, with representatives from Canada, the United States, and Mexico). In November 2001, she and Frazier attended the Foro del Plan Puebla Panamá (a forum concerning President Fox's proposed plan to coordinate economic development from the Mexican city of Puebla south to Panama) held in Quetzaltenango (Xela), Guatemala. In both of these meetings, leaders of various indigenous groups discussed common concerns regarding environmental degradation, the global economy, and the effects of these on their traditional ways, their natural resources, and the health of their people. In the second meeting, they were communicating by cellular telephones, the Internet, and fax machines, and most of the participants seemed to take for granted many other aspects of modern society. Within this emerging international, technofied context, the meanings of such words as *traditional* and *indigenous* are elusive and enigmatic.

What does one call a person of Maya descent who is proud of her or his Maya heritage, but lives in a city, works at a white-collar job, and speaks English and Spanish daily, in addition to the native Maya? Many such persons call themselves Mayas, but often do so only in restricted contexts. How traditional is a university-educated medical doctor or university president who still maintains his or her language and ethnic identification? What does it mean for a woman to wear the traditional *huipil* (a dress without a waistline that has flowers embroidered around the neck and the hemline), so common in some towns, so rare in others (Brown 1988)? When she wears the *huipil*, uses a cellular telephone, has an e-mail address, and participates in national conferences, is she traditional? Urban, educated Mayas can and do refer back to a world still present in many rural areas, which includes many traditional practices. These villages are linked to the outside world by close economic and cultural ties, but so were they during the Classic Maya period (Kintz 1990). Contrary to present-day interpretations of some of Redfield's statements, the Maya did not live in isolated communities before European contact nor during the Colonial period. Their economic and political interface with wider world systems is as traditional as their cornfields and gardens. They have used their traditions to generate "collective identities for surviving" under various conditions (Farriss 1984; see also Bartolomé 1988; Gossen 1994; Re Cruz 1996; Restall 1996, 1998). Maya culture persists, managing local resources for subsistence needs and for exchanges with wider economic spheres. New institutions evolve as Maya responses to national politics and global forces, in a process Esteban Krotz and his group have described as "re-socialization" (Krotz 1996). The risk is that in changing to meet new challenges, they may lose some of the local environmental knowledge accumulated over millennia of observations and experimentation, including techniques for long-term management of critical resources and of species that are key to ecosystem resiliency. Adoption of new techniques and products often carries long-term risks that are not immediately obvious. The changing attitudes regarding agrochemicals in the Yucatan Peninsula are a case in point: at first these modern agricultural aids were enthusiastically accepted, but today there are Yucatec farmers who are having second thoughts and even moving away from agrochemicals (as Anderson and Faust have both found in fieldwork). In attempting to meet today's new economic and political challenges, the Yucatec Maya may well make choices that undermine their children's resource base, but they also have the "cultural resilience" to reevaluate and later rectify mistakes.

ANCIENT MAYA HISTORY

Mayas have been farming the Yucatan Peninsula for at least 3,000 years. Long before 1492, they introduced some useful species from neighboring regions and constructed protected areas to increase the prevalence of both these and some native species they found desirable. Such constructions included terracing, ditched fields, enclosed household gardens, and *pet kot* (stone-circled, small areas of fruit trees planted and protected within the forest [see Gómez-Pompa, Flores, and Sosa 1987]). Discovery of these relic constructions as well as relative densities of useful tree species in formerly inhabited areas led various researchers to believe that local vegetation is anthropogenic. However, Rico-Gray and García-Franco (1991) presented convincing evidence that this was an unwarranted extrapolation. The survival of any species of plant over centuries depends on local microenvironments, including soil characteristics and the structure of the local plant community. Thus the present distribution of species in various parts of the Yucatan Peninsula is not a reliable indication of what was planted where by the ancient Maya. Species introduced or maintained in increased frequencies will have to compete with other species in the plant community once their human protectors are gone. Thus, it cannot be said that those areas not continuously inhabited by Maya agriculturalists still present anthropogenic plant communities, despite the findings of certain species in microhabitats such as relic stone circles that maintain soil moisture and microorganisms that favor these species. Nonetheless, the composition and structure of many contemporary plant communities on the Yucatan Peninsula have likely been affected by land use practices of Maya communities in the past and present, including the introduction of exotic species. Living Maya communities continue to protect useful trees and bushes in their household gardens, as well as planting them in their cornfields, to be harvested years later when that land is fallow.

Maya agriculture includes both gardens and shifting cultivation (swidden). This form of agriculture is known locally as *milpa*. It includes the polycropping of multiple varieties of maize, beans, and squash, as well as other cultigens (manioc, sweet potatoes, chilies, tomatoes, fruit trees, etc.). It consists of cutting the woody vegetation in a plot of land, setting it on fire, and then farming the plot for a year or two before moving on to another plot that has regrown for ten to thirty years. It is thus a recycling system of agroforestry that makes use of long-term, natural cycles for producing biomass and decomposing it to enrich the soil. It is not simply a process of randomly cutting and torching the forest. Certain trees, such as palmettos (*Sabal* spp.) and chicozapotes (*Manilkara zapota*)[1], are saved. Others are coppiced. Only a few large trees are killed outright. Timing and conditions of succession are managed to create desired types of forest regrowth (Benedict and Steggerda 1937; Várguez Pasos 1981; Jones 1982; Gómez-Pompa, Flores, and Sosa 1987; Gómez-Pompa and Kaus 1990; Terán and Rasmussen 1994). Forests on hilltops and strips of forest between agricultural fields are often maintained as hedgerows, providing seed reserves for forest regrowth, habitat for many kinds of animals, and shade for farmers. Maya management strategies reduce risk, but the system still depends on soil, climate, populations of insects, birds and other fauna interacting with vegetation, and much more. No part of the system exists in isolation or can be understood without at least some reference to the rest. Although most of the Yucatan Peninsula was relatively densely populated during the Classic period, large areas were sparsely inhabited in Colonial times, and even at the end of the twentieth century counties[2] with an average of fewer than 10 persons per square kilometer comprised 68 percent of the peninsula's land surface (Batllori et al. 2000). The number of persons whose subsistence needs can be supported by milpa agriculture is 35 per square kilometer in the peninsula. This indicates that in most of the Yucatan carrying capacity has not been exceeded, assuming traditional farming practices; rather, access to land has diminished.

Over the past two decades, the Mexican government has officially designated 27,000 square kilometers (18 percent) of the peninsula's land area[3] as Areas Naturales Protegidos (ANPs) (SEMARNAT 2002; see Andrews, chapter 4, this volume, for examples). Some of these protected areas lie within counties of relatively high density; however, we can calculate that 57 percent of the Yucatan is in counties that both have fewer than ten persons per kilometer and are *without* government protection. Such areas are rapidly becoming the targets of cattle ranchers and agribusiness corporations (see Faust, chapter 7, this volume, for an example).

The Yucatec Maya who have been least affected by Spanish colonialism are probably the Itzá Maya, who live in Guatemala's El Petén around Lake Flores, and the Lacandón, who live in Chiapas, near Palenque. Linguistic analysis has indicated that both of these indigenous peoples are of Yucatec Maya descent, despite being geographically separated from the Yucatan Peninsula. Scott Atran and his co-workers have described the first with attention to their fine-tuned, relatively sustainable relationship with the local forest (Atran 1999; Atran et al. 2001). The Lacandón are a much smaller group that has recently seen much of its area invaded by outsiders, but their earlier relationship with the forest was well documented by Alfred Tozzer (1907) and later by Gertrude Duby (1944) and others. The knowledge that both these groups have of their respective local environments, and their management concerns, are comparable to accounts of many elders among Yucatec Mayas living in the peninsula, and known to several authors of this volume, indicating the similarity of traditions among Yucatec peoples, despite geographic separation (Villagutierre 1933 [original 1701]; Rice 1987).

By 1000 B.C.E., the Maya were moving toward cultural complexity, based on widespread use of swidden agriculture. By the Middle Preclassic (1000–300 B.C.E.), they were building canals and water reservoirs at Edzná, Campeche (Matheny et al. 1983), and by the Late Preclassic similar enterprises had developed in the urban centers of Cerros, Belize, and El Mirador, Guatemala (Scarborough 1996). Ridged fields have been reported along the Río Candelaria of Campeche (Siemens and Puleston 1972) and the Río Hondo of Belize (Cliff 1988), and raised fields are documented in some of the seasonal swamps in Belize and elsewhere (Culbert 1996; Culbert and Rice 1990). By the Classic period (400–900 C.E.) the Maya were building terraces in Río Bec, Campeche (Turner 1979).

Such intensive agriculture fed urban populations, including full-time specialists whose artistic and scientific accomplishments are famous (Coe 1984; Sharer 1994). The Maya perfected a calendar system capable of predicting eclipses and movements of planets, a hieroglyphic writing system based on a combination of syllables and ideographs, a sophisticated religious system expressed in icons and painted images accompanying written texts (codices), a complex mathematical system that included a concept of zero, and monumental architecture oriented to astronomical events (Coe 1984; Weaver 1993; Sharer 1994).

This high culture had what might be described as a cultural dialogue with the neighboring Olmecs, whose management of river water with levees for maize agriculture has been estimated to have begun as early as 2250 B.C.E. By the Early Preclassic (1150 B.C.E.), these Maya neighbors had already achieved complex social organization, with cities and monumental sculptures along the southern coast of the Gulf of Mexico. By the Middle Preclassic (800 B.C.E.), the Maya had begun to develop their own cities with monumental architecture and sculpture along the Pacific coast of Guatemala and Chiapas (the Soconusco), proceeding into the highlands of Guatemala and across into the lowlands of the Petén and the Yucatan Peninsula to its north. The early sites show Olmec influence but also the beginnings of a distinctive Maya style (Weaver 1993, 52–76).

The cities of the Yucatan and neighboring Guatemalan Petén faced a water management problem due to the paucity of groundwater, for in those areas rainwater seeps downward through the limestone subsoil and does not form surface rivers or lakes. This problem combined with a six-month dry season to create a challenge for Maya engineering and administrative talents; the response was the development of complex systems including large artificial reservoirs, underground cisterns, wells, and canal networks (see Scarborough 1996; Scarborough and Gallopin 1994). An example is the capacity of reservoirs in Tikal, Guatemala, that were constructed to hold sufficient water to supply this ancient metropolis for eighteen months (Gallopin 1990, 85–87).

The decline of the Maya Preclassic cities around 200 C.E. correlates with recently discovered evidence of a regional, extended drought. Similar evidence applies to the declines of the Classic Hiatus (535–593 C.E.), when Maya civilization seems to have taken a downturn, and the Classic Collapse (760–910 C.E.), when the great cities of the southern lowlands were gradually abandoned and left to ruin (Sharer 1994; Gill 2000). In between these decades of severe drought, the Maya were able to build their great cities. However, the extreme drought at the end of the Classic correlates with a massive abandonment of cities in the southern lowlands and a decline of the Maya population. Some northern cities initially grew from the refugee influx, but later they also declined. During the Postclassic, some cities eventually recovered, but at a lower level of population than they had during the Classic. These were the cities that had reliable water supplies: deep lakes, large rivers, and/or a water table shallow enough to be accessed through hand-dug wells or sinkholes in the limestone rock. Notable among these sites are the Petén lakes of Guatemala; the Mexican sites of Maní, Mayapán, and Cobá in Yucatan; and the city of Lamanai in Belize (Gill 2000, 314–325; Hodell et al. 2001).

During the Postclassic, cities grew along the coast, from Champotón in the west to Cozumel in the east, eventually to be reported by Spanish explorers in the sixteenth century. However, the cultural accomplishments of these smaller, Postclassic populations are of lesser renown than those of their Classic predecessors. The overall population of the Maya lowlands—excepting some areas with plentiful water sources—shrank, according to some estimates, by more than 80 percent during the period 800–1000 C.E. (Santley 1990, 342–343; Adams et al. 1999, 196).

The drought hypothesis is not new among the many proposed causes of the Classic Collapse (e.g., Culbert 1974; Coe 1984; Lowe 1985; Sharer 1994; Hodell, Curtis, and Brenner 1995;). Evidence of soil erosion had previously led to charges of environmental abuse by the Maya (notably in Copán [Fash 1991]).[4] This seems well supported for Copán, but elsewhere the recurrent droughts discussed by Gill were severe enough to have disrupted even the least damaging of Maya agriculture. In Pich, Campeche, contemporary farmers report planting larger plots in response to traditional signs that it will be a dry year. Such increases in areas marginal for agriculture could well have resulted in drastically increased erosion (Faust, field notes 1999). Collapse of population would also have led to abandonment of labor-intensive terraces and canal systems. In the short term, such abandonment also results in increased erosion, as both these modifications require continual human labor to repair the effects of gravity and rainfall. Once cleared of forest for agriculture, the soil on slopes will erode unless terraces are continually repaired to prevent leakage. In earthen canal systems, rainwater also erodes the sides of the ditches, the soil eventually settles in the bottom and, if not recovered to use as soil topping, it moves downstream to settle in lakes (Culbert 1977, 518; Enge and Whiteford 1989, 66–67).

Some have argued that the Collapse was most dramatic in the wet areas of the Maya lowlands, specifically the central Petén, and least so in the driest, and that therefore it could not have resulted from drought. The north, while drier in terms of annual rainfall, is lower in altitude than the Petén; its ground surface is closer to the phreatic water table, which is less directly dependent on the rains than the artificial reservoirs used in the south (Scarborough and Gallopin 1994). Another factor to consider is that the winds that bring the rain shift their paths in response to global temperature. The easterly winds that bring the rains during the agricultural season may well have moved farther north during the drought centuries associated with global cooling (Folan et al. 1983; Folan et al. 2000; Gunn and Folan 2000). In short, there is much to learn, but drought appears at present to be the most likely candidate as the trigger for decline.

SPANISH EXPLORATION AND COLONIZATION

The populations of the Classic period city-states did not disappear completely. Even long after the Spanish conquest, when John Stephens and Frederick Catherwood explored Maya ruins in the 1800s (Stephens 1841, 1843), they found that local people often knew where the ancient sites were. However, there were no dense populations in the southern area, except around permanent lakes (Rice 1987). Northern and coastal cities had been smaller during the Classic than those of the south, but they were still functioning as living cities when the

Spanish arrived. A 1517 voyage captained by Hernández de Córdoba encountered various towns along the coast of the Yucatan Peninsula. Chronicler Díaz del Castillo (1942, original n.d., 1500s) reported the cities now known as Campeche and Champotón as having 2,000 and 8,000 houses, respectively (or approximately 10,000 and 40,000 persons, given the conservative estimate of five persons per house normally used in such estimates). It appears safe to assume that these communities were cities, in contrast to the much smaller villages and hamlets documented for the southern, interior lowlands during this initial period of Spanish exploration (see also Villagutierre 1933 [original 1701]; Jones 1982; Roys 1957; Scholes and Roys 1968).

Although the Yucatan was not conquered until 1545, the earlier exploratory expedition captained by Hernández de Córdoba initiated a peninsula-wide smallpox epidemic, according to reports given to later visitors. The native population at the time of first contact (1519) has been estimated at 800,000 for the entire peninsula, and by the time the Conquest was over (around 1550), it was estimated at about 250,000 (Clendinnen 1987, 36). These estimated losses have been variously ascribed to warfare, virgin-soil epidemics, and disruptions of the local ecosystem through ecological imperialism (Crosby 1972, 36–38, 48–51, 98–99; 1986, 199–202; Clendinnen 1987, 36–37). Religious repression was serious, especially under the regime of Fray Diego de Landa, who burned many Maya texts and had many Mayas tortured (to death or permanent disability) on charges of practicing the old religion (Landa 1937 [original 1566]).

The Spanish appeared at a time when local political entities (the Itzá, the Chi, the Cocom, and others) were contending for power in the north and central parts of the Mayab. The Spanish had the advantages of "guns, germs, and steel," (Diamond 1998) as well as horses, large-scale support by ships, and much else. Locally, they played their familiar game of setting groups against each other, allying themselves with one faction (often a weaker one) in order to smash the strong and set up a hegemonic order supported by initially friendly natives, though the Maya too could play politics and often resisted conquest for substantial periods or aligned with the Spanish to be able to destroy their compatriots from competing groups (Roys 1957; Clendinnen 1987; Patch 1993; Restall 1996, 1998). Once the Spanish had consolidated their power, there was time to impose stiff taxes and other fees, and to practice savage religious repression. The effects of disease on populations with no prior history of exposure were catastrophic to social (not only immunological) resistance efforts. Like other Native Americans, the Maya saw their population decline precipitously, to a small fraction of its former size. This may have relieved pressure on the land, if indeed the population had rebounded from the ninth-century Collapse sufficiently to have been straining its resource base prior to the Conquest. There are no clear references to intensive systems of cultivation in the Yucatan Peninsula from the Spanish chronicles, although there are reports for Chiapas (Villagutierre 1933 [original 1701]), and Hellmuth (1977) has referred to documents in the Archivo General de las Indias (General Archives of the Indies) in Seville (Spain) that report such systems in both Chiapas and Guatemala.

During the Colonial period, the demands from church and ranch owners were the subject of numerous complaints by the Maya regarding the need to invest time and energy in food production for family subsistence and to pay taxes (Restall 1998). An uneasy, but relatively stable, accommodation was made to the domination of these European outsiders, who in turn were required to make substantial adjustments of their own in order to minimize risks from hurricanes, droughts, floods, ticks, malaria, and pirates, as well as uprisings of local Mayas (Clendinnen 1987). A world of subsistence agriculture emerged, in which even the hacienda owners displayed genteel poverty more often than great opulence (Restall 1996, 1998; Rugeley 2001). The population was so low that the lack of a labor force was a serious problem, and laborers could escape to live in forested areas (Contreras-Sánchez 1990), where they occasionally encountered mission priests who referred to them as "apostate Indians" (Scholes and Roys 1968, 251–290).

The Maya never quite accepted defeat, and resisted Spanish attempts to tighten economic control (Bricker 1981; Clendinnen 1987; Bartolomé 1988; Dumond 1998; Patch 2002). The resistance was rarely if ever "pan-Maya" (Dumond 1998)—the Spanish made sure of that—but revolts did succeed in keeping much of the interior peninsula (particularly the southern and eastern areas) out of Spanish hands for years at a time. The last independent Yucatec Maya state, that of the Itzá Maya in what is now called "the Guatemalan Petén," did not fall until 1697, nearly 200 years after the Conquest began (Villagutierre 1933 [original 1701]; Rice 1987). Large areas of the Yucatan Peninsula had de facto autonomy into the twentieth century; efforts to establish tighter control by the Spanish colonial administrators and later by the Mexican government have routinely met with resistance, and often with revolts (Reed 1964; Bricker 1981; Joseph 1986; Sullivan 1989).

Throughout the Colonial period, Maya water management strategies were generally replaced by European ones, while the catastrophic loss of population removed economic incentives for intensive methods of agriculture, not to mention the relative scarcity of an available workforce. Culbert (1977, 518) has analyzed productivity per unit of labor for both swidden and raised-field systems, finding that swidden is more efficient as long as adequate land is available for fallow. For nearly five centuries, there was insufficient population density in areas under Maya control to make intensive systems economically attractive (Jones 1982). The techniques were forgotten; neither Maya nor Spaniard needed them. The Maya continued to live by their traditional form of swidden, the form of agriculture that produces the most food for the least labor (though not on the smallest area of land). Maize was the main crop, providing about 75 percent of calories—as it still does today in the more remote, subsistence-based communities (see White 1999 for the Classic diet in general; for twentieth-century communities see Benedict and Steggerda 1937; Peraza López 1986). Beans, squash, chilies, fruits, and root crops provided additional nutrients and some proteins, while meat was probably supplied largely from game, though turkeys, muscovy ducks, and dogs were raised for food (Terán and Rasmussen 1994, 1995; White 1999). The colonial Spanish regime concentrated its production energies on sugar, cattle, logwood for black dye (Contreras-Sánchez 1990; McJunkin 1991), and cotton (Patch 1993; Restall 1996), most of which went for export. Of these, only cattle remain somewhat important locally, and even they are rarely exported nowadays, due to growing consumption by the upper and middle classes of Mérida and Campeche, as well as by the tourists in the Cancún area (Anderson, personal research). Throughout Colonial times, periodic droughts resulted in widespread deaths, many of which occurred rather quickly from dehydration (Gill 2000), but which continued to occur after the drought because of crop failures resulting in famine. The use of windmills and wells eventually reduced the impact of droughts on domestic water supplies, while the associated famine from crop losses was relieved through the transportation of foods by both railroad and ship (Gill 2000).

POSTCOLONIAL MAYA HISTORY

Independence from Spain in 1821 brought an almost immediate desire in the Yucatan Peninsula for independence from Mexico. By the 1840s, it seemed that there was a change of government almost every month in the regional capitol of Mérida. The Maya, disaffected, eventually took matters into their own hands. The enormous uprising widely known as the Caste War liberated the more remote and subsistence-oriented parts of the Mayab from the Mexican (or Yucatecan Creole) governments.[5] The huge area that is now Quintana Roo remained independent well into the twentieth century.

Final peace between the Mexican government and rebel Maya of Quintana Roo was not declared until 1901, and some areas remained free until the 1930s; the last battle was fought in Dzula (next door to Chunhuhub) in 1934. Villa Rojas, traveling in central Quintana Roo in the 1930s, found that he was in a hostile and thoroughly independent realm; he had to use deception to survive, and he was often importuned to get help from the United States in renewing the independence movement (Villa Rojas 1945; Sullivan 1989). Even in recent times, one of the authors (Anderson) has heard tales from rural villagers of the alleged plan by the great Mayanist Sylvanus Morley to help reliberate Mayan Quintana Roo with American help, though there is not the slightest evidence that Morley (who was active in the region from 1907 to 1948) ever planned any such thing (see Sullivan 1989).

Swidden agriculture still has a reputation in some quarters as being primitive and simple (Leach 1999); the burning in particular is often assumed to be destructive. It is well to point out, then, that Maya traditional agriculture is incredibly complex and sophisticated. It requires detailed knowledge of when to cut, when to burn, how to burn, how to plant, what to plant, what soils to use for what crops, how to control pests, which crops to interplant, and much more (Benedict and Steggerda 1937; Várguez Pasos 1981; Gómez-Pompa, Flores, and Sosa 1987; Gómez-Pompa and Kaus 1990; Terán and Rasmussen 1994, 1995; Faust 1998; Faust and Bilsborrow 2000).

In the late nineteenth and early twentieth centuries, haciendas in a large area east and south of Mérida, Yucatan, rose to vast wealth through the henequen industry. The best description of this is still Benitez's (1986) classic *Ki: El drama de un pueblo y una planta* (see also Wells 1985 and Joseph 1986). Yucatecan inventors found a cheap, simple way to separate the fiber from the irritating leaf tissue, launching henequen's rise to dominance in the world trade in hard fibers for twine and rope. The *auge* (the Gilded Age) lasted from the late nineteenth century well into the twentieth (Wells 1985). Although regarded with nostalgia by many in modern Yucatan, it was a hard time for the Maya in the henequen zone. Conditions on the plantations were appalling. Debts were inherited and manipulated through company stores, producing effects similar to slavery; Maya workers were literally bought and sold (Turner 1911; Joseph 1986). The period is still recalled as the "Age of Slavery" by the descendents of those affected (Burns 1983), in contrast to an earlier period when hegemony was exercised in a more genteel fashion (Bracamonte y Sosa 1993; Patch 1993). The period of so-called "free enterprise," the time of Porfirio Díaz, brought not only economic destitution but also loss of freedom to the Maya.

The Mexican Revolution ended forced labor by the 1920s, but prosperity never really returned to the henequen zone. Henequen began to decline in Mexico, due in part to the end of this form of slavery, but also because of competition with producers in other nations, particularly Brazil and Kenya. Then came synthetics; the Mexican government tried to keep the business going, pouring subsidies into a dying industry (Villanueva 1985; Benítez 1986) while also providing incentives for people to migrate to the coast and take up fishing (Fraga and Paré 1990; Hale 2000) or work in the tourist industry in and around Cancún. The final end of henequen did not come until the 1980s. By then, tourism had come to replace henequen as the golden goose, with fishing a secondary alternative. The children of henequen workers moved to Cancún, to work as waiters, filling station attendants, construction hands, and road builders, as well as to various coastal communities where they learned to fish, buying boats and outboard motors with government subsidies (Fraga and Paré 1990; Faust and Bilsborrow 2000; Hale 2000).

While henequen rose to prominence in a large part of the state of Yucatan, the rest of the peninsula continued to support itself through subsistence agriculture, logging of precious woods such as mahogany (*Swietenia macrophylla*) and cedro (*Cedrela odorata*), and collecting chicle latex for chewing gum from the chicozapote (*Manilkara zapota*). Cutting of dyewood (*Haematoxylon campechianum*) for textile dyes continued to support an industry for some decades (Contreras-Sánchez 1990; McJunkin 1991). Cattle and fruit were (and still are) locally important in the economy of the rural Yucatan Peninsula, cattle being especially common in the central north and southeast, and fruit in the center (Villanueva 1985). Brief booms of other agricultural commodities, such as rice and cotton, suffered the same fate as the henequen boom—though with a far more modest upward trajectory.

Introduced agricultural systems often have either accommodated to Maya ways or failed. Commercial fruit production is one example of successful accommodation (Morales Valderrama 1987). Failures include monocrop cultivation of henequen, cotton, rice, sesame, and vegetables (Sandoval-Palacios 1982; Gates 1993; Faust and Bilsborrow 2000). Fruit production in monocrop orchards has been threatened by the *tristeza* virus (*Closterovirus* spp.) and by a fruit fly (*Anastrepha* sp.). Coconut groves along much of the coast were devastated during the 1980s and 1990s by another disease that caused yellowing of the fronds and subsequent death to the palm trees; since then, a vaccine has been developed and resistant dwarf varieties are being planted to replace the susceptible larger trees (Faust, discussions with agricultural extension agents in Campeche and botanists in the Centro de Investigaciones Científicas de Yucatán). Increases in cattle production and fishing followed the decline of henequen, but fishing has begun to decline as a result of overuse (Hale 2000). Sugarcane production continues in the area around Champotón, Campeche. Watermelon, tomatoes, chilies, and some other vegetables like *calabazas* (the local varieties of squash) and eggplant are cultivated in small patches throughout Yucatan and Campeche, as cash crops and for domestic consumption (Prieto 2000, 228–231).

By the second half of the twentieth century, the national government began concerted efforts, first to transfer landless peasants from overpopulated areas in the north of Mexico to "unused" lands in the Yucatan Peninsula and later to turn the coast of Quintana Roo into the Mayan Riviera, connecting it with Chichén Itzá, Mérida, Uxmal, Campeche City, Palenque, Edzná, and Calakmul in the Ruta Maya to serve the tourist industry. The effects of these policies have severely reduced land available to Maya communities for hunting and agriculture, as well as reducing their political and economic independence. Large tracts of land have also been bought by both contemporary cattle ranchers and large agribusiness corporations growing cotton, citrus, papayas, and other commercial monocrops. Often these had previously been private lands held for investment purposes but freely available for use by neighboring Mayas. Sometimes they were officially national lands, also available for hunting, swidden agriculture, beekeeping, and extraction of wood and other forest products (see Emery and McLain 2001 for a comparison with the use of nontimber forest products in the United States).

The 1992 change in Article 27 of the Mexican Constitution allowed rural communities to legally sell their communal lands for the first time since these had been granted during the land reform that followed the Mexican Revolution. Lacking money for investing in irrigation and machinery, lacking good access to markets and information, and faced with disturbing shifts in weather patterns, many Mayas have voted to divide their ejido lands in order to sell them. Like small farmers fleeing the Oklahoma dustbowl in the 1930s and blacks pushed out of sharecropping by agricultural mechanization of Southern farms in the United States during the 1960s, today's Mayas generally take what little cash and belongings they have to the city, seeking jobs, education for their children, and a chance to participate more fully in the modern world.

Traditional village and town dwellers, especially in remote or conservative sections of the Yucatan Peninsula, speak Maya, and indeed women and children are still usually monolingual in the most remote forest settlements. When they migrate to the cities, most people gradually lose many of their rural customs, but many continue to speak Maya and maintain Maya identity. In the early twentieth century, Mérida was a truly bilingual metropolis, and even today Maya is still commonly heard, especially in the southern half of the city. Despite steady conversion of the young to monolingual Spanish speaking, Maya speakers are still plentiful. The census for the year 2000 found that there are nearly 76,000 Maya speakers in Campeche, over 163,000 in Quintana Roo, and over 547,000 in the state of Yucatan (INEGI 2000a, 270, 282, 287), out of total populations of 691,000, 875,000, and 1,658,000, respectively (INEGI 2000a).

Throughout this area, in spite of the ongoing demographic transition (see Preito 2000 for an analysis of this transition), the rate of population increase is high. In the state of Yucatan, the population nearly tripled from 1970 to 1995, with an average annual growth rate of 4 percent (Eastmond, García, and Córdoba 2000, 57–59). The national census for the year 2000 reported population increases of nearly 3 percent in Campeche, 2 percent in Yucatan, and a sobering 6 percent in Quintana Roo (INEGI 2000b, 3). The number of births per thousand was thirty for Quintana Roo, twenty-three for Yucatan, and twenty-eight for Campeche (INEGI 2000b, 4.) These rates are greatly reduced from the 1950s rates of forty-eight per thousand in Campeche, forty-five in Yucatan, and forty-four in Quintana Roo (INEGI 1996, 58, 66, 70). This reduction in birthrate has been complemented in recent years by medical attention provided by small local clinics. The infant mortality rate in central Quintana Roo (to take one example) has fallen from 50 percent in Alfonso Villa Rojas's time (Villa Rojas 1945) to very low figures today (Quintana Roo reports some 8.13 percent of children dying by the age of 12 [INEGI 2000c, 83]). Through her work with family planning, community development, water quality, and sanitation, Mary Elmendorf contributed to this impressive decline in both infant mortality and births (see Elmendorf 1972, 1976, 1980, and chapters 3 and 13 in this volume). It is important to note that the decline in birth rates has occurred despite simultaneous increases in bottle-feeding, which allows for resumption of ovulation that is otherwise generally suppressed by lactation (Howrigan 1988).

Malnutrition due to poverty and overcrowded agricultural lands remains the major problem—expressing itself indirectly, through deaths from minor infectious diseases that would not normally kill a well-nourished child or adult. Using UN standards, the national census for the year 2000 reported that of the children under five years of age, 33 percent are malnourished in the state of Yucatan, 28 percent in Quintana Roo, and 26 percent in Campeche; however, less than 5 percent of these children are severely malnourished (INEGI 2000d, 240). Malaria has been virtually eliminated from this region, and other serious infectious diseases are rare, though dengue has recently increased, with travel advisories posted.[6] Mary Elmendorf's research and development work has contributed to the reduction of population growth rates; however, transition to a stable population size is far from complete (Prieto 2000). Even the relatively slower rate of growth has created problems of resource overuse in many rural communities. Fallow times have been shortened in response to population pressure, but this tends to reduce harvests per hectare, increase the need for costly inputs, and require more work. Although there is relatively under-populated land in the southern area, people are reluctant to move away from the comforts of modernized communities and easy access to major cities. Would-be migrants are also in competition with both large-scale commercial agricultural projects (e.g., rice, sugar cane, cattle, citrus orchards) and the newly established Areas Naturales Prótegidos (ANPs), which exclude them from access to land. This situation is likely to get worse before it gets better; analysis of the demographic transition indicates that the population will keep growing for approximately another twenty years (Prieto 2000).

Formal education is advancing and is often bilingual: Maya and Spanish. Rural Maya of a couple of generations ago could expect perhaps two to three years of schooling; few indeed went beyond six years of primary school. Today, although many communities now have middle and high schools, education for the Maya still lags behind that of central and northern Mexico (INEGI 1993, 42). Approximately 93 percent of children in Mexico between the ages of 6 and 14 are currently in school; 48 percent of those from 15 to 17; and 20 percent of those between 18 and 23. While statistics for the Yucatan Peninsula are not presently available, the respective percentages are probably lower. Nevertheless, the majority of Maya children are now finishing at least middle school, and many go on to complete high school. Small, rural towns like Chunhuhub and Pich have even become "exporters" of teachers, nurses, veterinarians, computer programmers, and other white-collar workers to both cities and rural areas of southern Mexico. Highly educated professionals are still relatively few among the Maya, but there is a substantial, and rising, number of Maya medical doctors, lawyers, politicians, scientists, and writers making sure that the Maya will not get lost in the rush of modernization. Serious Maya literature, in virtual eclipse for a couple of centuries, has come back with a rush; poetry, short stories, and prose are flowering (Ligorred Perramon 1997). Some of the impetus for these developments came from Radio Peto, the Maya radio station begun by Mexico's Instituto Nacional Indigenista. President Fox has promised to make educational reform a priority of his administration.

Redfield and Villa Rojas (1934) began researching the process of change in the original study of Chan Kom. Both continued; Redfield (1941, 1950) analyzed change across various communities in Yucatan. Villa Rojas (1945) studied Maya communities in Quintana Roo. In the 1970s, both Elmendorf (1976) and Press (1975) published studies of change (in Chan Kom and Pustunich, respectively). In the following decade, Sandoval-Palacios (1982) presented a brilliant analysis of the environmental and social impacts of a rural development project in Campeche. Farriss (1984) presented a widely respected description of Maya colonial history as a collective struggle to maintain identity, while Gilbert Joseph (1986), Morales Valderrama (1987), and Rosales González (1988) published detailed analyses of agricultural change, and both Eric Villanueva (1985) and Allen Wells (1985) focused on economics and culture during the henequen era. During the 1990s, various of the authors in this volume, as well as Esteban Krotz (1996), Arzápalo and Gubler (1997), Peter Hervik (1999), and others have continued the focus on processes of change, continuing the work begun by Redfield, Villa Rojas, and Elmendorf.

Y2K MAYA: EXPERIMENTS, INITIATIVES, AND ENVIRONMENTAL IMPACTS

Environmental problems are beginning to be serious in the Yucatan Peninsula, but in comparison with many other indigenous peoples (in the United States as well as elsewhere in the Third World) the Yucatec Maya are coping relatively well with a difficult situation. Decades of research and applied work in agricultural and community development, water quality, sanitation, and population growth have gradually begun to include concerns with two new foci: (1) processes of global change, including the postindustrial economy and global warming; and (2) local environmental quality, affecting health, management of natural resources, and conservation of biodiversity. The population continues to increase, while agricultural development is very slow, especially in the interior of the Yucatan Peninsula, and biological conservation is only beginning to be understood as a goal.

The rapid and chaotic privatization of the early 1990s accentuated changes in processes that affect the whole fabric of rural society. Government statistics showing economic improvements often do not capture the complexity of the process of change. For example, commercial sales of agricultural products increased substantially according to the government statistics presented in a 1996 report by Cuanalo, Navarrete, and Gómez. From 1960 to 1994, honey production doubled in the state of Yucatan, while cattle production increased three times; pork, nine; poultry, fifteen; chicken eggs, seven; watermelon, seven; oranges, nine; and tomatoes, ten. However, these increases in commercial sales are not necessarily net increases in total production, but rather reflect the transition to a cash economy, and malnutrition is still commonly around 30 percent in children under five years of age (Faust, discussions with local doctors). Mayas increasingly buy products produced in commercial operations and trucked to their communities, including not only packaged snacks and drinks but also chickens, tomatoes, and bananas—items that used to be produced in home gardens (and thus are unrecorded in statistics). Overall, the change to greater dependence on commercial production appears to have primarily benefited non-Maya entrepreneurs, with Mayas working as inexpensive labor, although it is difficult to ascertain distribution of benefits.

Under shifting cultivation, agricultural plots must be moved every year or two. This requires large community landholdings and a governing body to reassign the lands and resolve conflicts. Changing weather and variations in regrowth of forest mean that fields have to be reassigned on a strategic basis, the strategy covering a very wide area. Local agricultural development for commercial sales is tightly connected to prices that are affected by international competition through the North American Free Trade Agreement (NAFTA). NAFTA has allowed heavily subsidized agricultural produce from the United States to flood Mexico. Formerly, the Maya combined subsistence agriculture with small-scale commercial sales, the money from which was used to buy tools, clothing, medicine, and other items important to their lifestyle. However, the majority of what they traded was extra corn and beans, sold in local cities. But these food staples are now being supplied more cheaply through international trade. Adoption of permanent, mechanized cultivation is difficult because in tropical areas the heat and humidity accelerate biodegradation of organic matter in the soil, while the diversity of insects and other plant pathogens threaten monocrops. Monocrops, in turn, are economically desirable in the world of commerce because the transportation industry can handle them efficiently, with economies of scale.

In 1990, the minimum wage in Mexico was the U.S. dollar equivalent of \$3.30 per day, and 34 percent of the working population of the Yucatan Peninsula earned less than that (Eastmond, García, and Córdoba 2000, 58). From 1970 to 1990, the average yearly income per capita in the peninsula went from \$539 U.S. dollars to \$1,604 (Garza and Rivera 1995, as cited in Baños Ramírez 2001, 62). The increase is correlated with a shift from 55 percent of economically active individuals being employed in primary production (e.g., farming, logging, fishing) to only 27 percent by 1995 (Baños Ramírez 2001, 69), as many found at least sporadic employment in commerce, industry, and service occupations. The rapid increase in assembly plants following NAFTA (North American Free Trade Agreement) resulted in employment of 28,000 people by 1999 in Yucatan State, at between \$3.00 and \$6.00 U.S. a day, while wages in the petroleum industry in Campeche were substantially higher (Baños Ramírez 2001, 168–169). Despite these sectors of improved income, however, Balam (1999) has determined that average per capita food consumption is measurably less nutritious than in 1959.

Access to steady employment outside of subsistence agriculture is still rare in most rural communities but is growing as government policy increasingly directs foreign investors to establish assembly plants and agribusiness firms near existing rural communities in order to reduce migration to urban centers. Still, the majority of jobs are in cities. Mechanization of agriculture continues to push many from the countryside. Urban employment is not always abundant, though assembly plants, oil production, and tourism development have absorbed many workers and brought capital to parts of the region. Although President Fox has promised democratic process, transparency, and rule by law, he is frequently opposed by a divided congress and hampered by lack of experience in government. He has strongly opposed the widespread illicit drug trade, which has been found to have connections to many government figures, including those at the top; the Yucatan Peninsula has been of considerable importance in drug trafficking. But some of Fox's proposals have threatened rural Mexicans, including many Maya. In December 2003 he requested that Congress tax food and medicine. This was not accepted, to the great relief of the rural poor who, compared to the more affluent, spend a very great proportion of their income on food and medicine.

Economic shifts are a great problem for the Maya and the environmental health of the Mayab. The term *neo-liberalismo* encompasses various political and economic changes intended to bring the benefits of the free market. Of course, no absolutely free market exists anywhere in the world (Hong Kong may once have been close), nor could a country with the political traditions of Mexico change overnight into a free-market economy. Change has been happening in a crazy-quilt pattern. Some public-sector firms are "privatized"—sometimes by being sold (cheaply) to relatives of leading politicians. Some subsidies have ended; others have not. Protection of biodiversity and ecosystem health are very difficult to establish as priorities in this rapidly shifting economic environment. Regulation to protect resources for future generations runs against the grain of free-enterprise enthusiasm. Many government programs benefiting the poor have been cut, while others continue and new ones begin. Through it all, the government has retained much of its power over the economy and over society as a whole—making *neoliberalism* a very confusing term.

Orchard cultivation has been a successful development project, one that has improved and expanded the traditional, small-scale orchards run by villagers. Beginning in the López Portillo administration of the 1960s, with a citrus project for exporting orange juice, the government has expanded its efforts to reach small farmers through various programs, including a webwork of *corredores frutícolas* (bands of fruit trees along main roads) that blend at their farther edges into the exquisitely lovely forests native to the area. Technical assistance and improved varieties of many species of fruit trees have been provided; however, recommendations for use of biocides are only now beginning to take into account their effects on biodiversity and water quality. Mixed plantings are now recommended by many technical advisers, as this reduces the risks of insect and disease loss compared to extensive monocropping with biocides. While not necessarily environmentally ideal, these orchards do not produce the rapid and catastrophic ecological deterioration seen with row cropping and cattle ranching. They replace the natural forest with a tree canopy, which provides some habitat for wildlife and results in less soil erosion than most monocrops. They are relatively "sustainable" so long as the trees are maintained properly and extensive monocropping is avoided. They produce a great deal of food, often throughout the year; fruit too small, misshapen, or plentiful to sell can be eaten at home, greatly contributing to rural nutrition. Such fruits as mangoes, mameys, and avocados are nutritionally superb—far higher in calories, minerals, and vitamins than most temperate-zone fruits (Anderson, unpublished research).

Modern cattle ranching, with "improved pasture grasses," has come to much of southern Mexico, including the Yucatan Peninsula, but its profitability is highly debatable. Especially in Quintana Roo, many tracts of extremely valuable forest have been wantonly cleared for cattle ranches that soon fail; the land reverts to worthless scrub—a fire hazard, returning very slowly to high forest, if ever. While Maya agriculture may once have covered a great deal of the Yucatan Peninsula, it was a type of agroforestry, clearing small plots (1–2 hectares) with maintenance of roots, stumps, and strips of high forest, in order to encourage rapid forest regrowth. A number of researchers analyzing both historical records and the dryland grass and shrub communities have concluded that the savannas found in the Maya area are the result of colonial cattle ranching (e.g., Wagner 1964; Vargas 1983), although others have attributed them to Maya practices (e.g., Lundell 1945; Turner 1974, 1976). Today, successful cattle ranching seems to depend on finding areas where grass flourishes naturally and then managing both native and nonnative forage grasses better than many local ranchers care to do (Anderson, reports by ranchers in central and western Quintana Roo). On the whole, the enormous problems of tropical cattle ranching (Painter and Durham 1995; Sponsel Headland, and Bailey 1996) indicate that this activity should not be regarded as a potential source of future wealth, although a new dairy cattle project involving supplemental feed that combines manioc and corn raised continually in irrigated fields is being tried in Pich, Campeche (Faust, chapter 7, this volume).

Other agricultural initiatives also remain rather experimental. Commercial production of vegetables (chilies being especially profitable) became a casualty of monocrop cultivation and especially of pesticide overuse, which has led to population explosions of whiteflies that carry viral diseases (Sally Humphries, personal communication to Anderson). New chili farms, using less pesticide, have arisen and may succeed. Intensive mixed vegetable farming, using French intensive systems, drip irrigation, complementary planting, and other traditional (yet sophisticated) techniques from around the world, has been far more successful but is so labor intensive that it has not been adopted by many rural Maya (Anderson, field data).

Forestry provides another option, but the temptation to overcut is almost irresistible. In both Pich, Campeche, and Chunhuhub, Quintana Roo, commercial exploitation of precious woods has ended because of overcutting (as other chapters recount). However, some large forest ejidos in Quintana Roo have done better with long-term, intensive, technical assistance in forest management, product design, and marketing provided by international projects, particularly the Plan Piloto Forestal (Primack et al. 1998).

Despite the fact that some resources have been locally overexploited, including fisheries, forests, and soil (Faust and Bilsborrow 2000, 91–95, 97; Hale 2000, 214–219), commercial exploitation of some species has barely begun. For example, native fruit, medicinal herbs, ornamental plants, native stingless bees, and even peccaries have been farmed successfully in several experimental projects (Anderson, field data), but commercial success has been slow in coming, and several of these enterprises have failed. Sustainable forestry has been introduced quite widely in Quintana Roo, specifically for mahogany; there is a need to introduce it for all commercial species over the entire region. Related to agroforestry is the development of high-quality woodworking, including furniture and door manufacturing, in Quintana Roo, using native woods; such projects have generally failed but could succeed with better marketing (Anderson, field data). Controlled hunting combined with ecotourism is working as a subsistence strategy for some communities in Quintana Roo (Anderson, chapter 6, this volume) and could be much more widely attempted.

The success of marine fisheries was short lived, providing for an escape from the collapsed henequen industry, but now showing all the signs of overcapitalization and overexploitation (Hale 2000). Competition for nearshore costal fisheries resources has reached such levels that fishermen from some villages are simply not allowed into the waters used by other fishing villages. Fishermen of Celestún, Yucatan, reported to Faust (interviews on December 13, 2003) that in 2002 when there were no fish in their area, they went to neighboring Isla Arena, on the Campeche side of the border. The fishermen of that community did not welcome them, as they had in previous periods of scarcity. This time their boats were burned, their nets and motors stolen, and they themselves were beaten, arrested, and taken to jail in the City of Campeche. The governors of the two states had to work out a solution, allowing the fishermen to return to their families. While some blamed the protected area of the Celestún Reserve for preventing them from fishing in their own area, others indicated that there are simply fewer fish and more people trying to fish than was the case a few decades ago. A complicating factor is that, where fisheries are still lucrative, the products are often exported to Mexico City or out of the country, where the acquisitive power is far greater than in the coastal villages. Acquisition of freezing facilities by fishing cooperatives has enabled such exportation, along with commercial operations in Progreso, Lerma, and Chetumal.

Exportation of other products of Maya communities has suffered from the poor development of marketing. Small, often isolated Maya communities have no ready access to any markets—let alone the specialized and profitable market niches that would allow sales of such items as handicrafts and high-quality fruit. Otherwise, the superb fruit, beautiful hardwoods, exquisite traditional crafts, flavorful vegetables, and incomparable honey of the Maya villages languish unsold, or command a tiny fraction of their world market value. We have often seen pigs eating fruit that would be worth five dollars apiece in U.S. markets, such as anonas (or cherimoyas, *Annona squamosa*, *A. reticulata*, and *A. purpurea*) and mameys (the sweet potato–flavored fruits of *Pouteria mammosa*), while high-quality bananas (*Musa sapientum*) and chicozapotes (*Manilkara zapota*) are not far behind.

While various opportunities await development, a general slide downward into rural poverty and environmental stress is already well under way in parts of the Yucatan Peninsula, as in other parts of the Third World. The September 2002 World Summit on Sustainable Development in Johannesburg, South Africa, focused on increases in both the distance between rich and poor and the environmental degradation that threatens many species, including our own in the long term (see Frazier, chapter 12, this volume, for an analysis of the economic and political forces involved, and Frazier 1997 for a critique of the concept of sustainable development). Elmendorf has again read the signs and come out in the lead. On October 16, 2001, she called on colleagues to introduce in that year's November meeting of the American Anthropological Association in Washington, DC, a resolution in support of the Earth Charter (Earth Charter Commission 2000). This document describes a program through which communities can work toward ecological integrity combined with social and economic justice. The objectives derive from studies of consumption patterns, namely by industrialized societies, and how these affect the world community through interdependence, driving ecological and social problems globally. The detailed analyses used in this international effort show the need to base both domestic and foreign policies on Earth Charter principles, particularly fairness in conservation efforts. The present volume contributes to the effort by documenting the complex interactions among issues such as rural poverty, economic exploitation, human rights, cultural survival, resource degradation, and environmental protection in the Yucatan Peninsula.

Actions such as those recommended in the Earth Charter can help save the Yucatan Peninsula's rural society and its fragile, troubled environment from short-term exploitation, which is likely to result in the local extinctions of already endangered species of jaguars, ocelots, anteaters, tapirs, howler and spider monkeys, several varieties of parrots, and many more species of orchids and other endangered plants (CONABIO 2002), as well as causing irreversible degradation of thin, tropical soils on a fractured karst topography.

The Maya are currently squeezed between various pressures of a changing climate, growing population, changing—and more competitive—markets, desires for more participation in the modern world, and lack of knowledge concerning the environmental effects of the various alternatives becoming available. This book reports studies of accommodation to the countless pressures and problems of change. The Maya, like the rest of humanity, enter the new millennium in a troubled time, but they have survived troubled times before, and their adaptability and solidarity have not yet failed them. We hope that the following chapters will contribute to the collaboration of anthropologists, conservation biologists, development agents, and others in assisting Maya and other communities struggling with the various challenges and potentials emerging in this new millennium.

NOTES

1. In the Yucatan Peninsula the fruit is also known as just *zapote*. The history of the scientific name of this tree is somewhat involved. First named *Achras zapota* in 1752 by Linnaeus, other names that have since been applied to it include *Achras sapota*, *Sapota achras*, and *Manilkara achras*. Here, we follow the usage proposed by Pennington (1990) in his monograph on the family Sapotaceae: *Manikara zapota*.

2. Statistics on human population density are published by the national census bureau of Mexico by county; hence inferences cannot be made at a finer scale.

3. In addition to the land areas that are protected, there are also protected marine areas, totaling 5,479 square kilometers, according to the same source.

4. The Copán valley is a confined space of fertile soil surrounded by mountains. Such a geography invites overuse; indeed, it is showing signs of deforestation and erosion today. Most of the lowland Maya world lacks such rigid geographic limits. Maya farmers and city-states could spread out till they came up against farmers expanding outward from other centers. At that point, a military strategist would expect war rather than peaceful suicide by overcultivation. It is no surprise, therefore, to find much evidence for exactly this state of affairs in Maya inscriptions, archaeological data, and contact-era histories (cf. Gill 2000). The claim by Redman (1999, 145) that the Maya collapse was "primarily due to the extended period of intense human exploitation" simply cannot be sustained; the evidence, so far, is against it.

5. The rebellion was not at first limited to the Maya (Dumond 1998), and it never involved all the Maya. The Maya of the western and central Yucatan Peninsula remained, on the whole, peaceful. Many took to calling themselves *mestizos* to distinguish themselves from the rebel Maya, and in some parts of the Yucatan Peninsula, especially the old towns of the Puuc, *mestizo* still means a Maya rather than a person of mixed ancestry and culture.

6. For example, the Travel Medicine Program of Canada gives the Secretary of Health in Mexico as the source of information, citing 8,363 laboratory confirmed cases of dengue in Mexico for the year 2002, as of November 14, 2002 (www.TravelHealth.gc.ca, consulted on December 17, 2003).

REFERENCES

Adams, R. E. W., H. R. Robichaux, F. Valdéz Jr., B. Houk, and R. Mathews. 1999. Transformations, periodicity, and urban development in the Río Azul/La Milpa region. Appendix 2 in *Río Azul, an ancient Maya city*, edited by R. E. W. Adams. Norman: University of Oklahoma Press, 190–207.

Arzápalo Marín, Ramón, and Ruth Gubler, eds. 1997. *Persistencia cultural entre los mayas frente al cambio y la modernidad*. Mérida, Mexico: Universidad Autónoma de Yucatán.

Atran, Scott. 1999. Itzaj Maya folkbiological taxonomy: Cognitive universals and cultural particulars. In *Folkbiology*, edited by Douglas Medin and Scott Atran. Cambridge, MA: Massachusetts Institute of Technology Press, 119–204.

———, Douglas Medin, Elizabeth Lynch, Valentina Vapnarsky, Edilberto Ucan, and Paulo Sousa. 2001. Folkbiology doesn't come from folkpsychology: Evidence from Yukatek Maya in cross-cultural perspective. *Journal of Cognition and Culture* 1: 3–42.

Balam, Gilberto. 1999. *Un fantasmo recorre el mundo. El depredador más voraz de la historia*. Mérida, Mexico: Maldonado Editores.

Baños Ramírez, Othon. 2001. *La modernidad rural mexicana a fines de milenio: El caso de Yucatán*. Mérida, Mexico: Universidad Autónoma de Yucatán.

Bartolomé, Miguel Alberto. 1988. *La dinámica social de los Mayas de Yucatan: Pasado y presente de la situación colonial*. Mexico City, Mexico: Instituto Nacional Indigenista.

Batllori, Eduardo, Federico Dickinson, Ana García, Manuel Martín, Ivan González, Miguel Villasuso, and José Luis Febles. 2000. Socioecological regions of the Yucatan Peninsula. In *Population, development and environment in the Yucatan Peninsula: From ancient Maya to 2030*, edited by Wolfgang Lutz, Leonel Prieto and Warren Sanderson. Laxenburg, Austria: International Institute for Applied Systems Analysis, 33–53.

Benedict, Francis G., and Morris Steggerda. 1937. *The food of the present-day Maya Indians of Yucatan*. Washington, DC: Carnegie Institute of Washington.

Benítez, Fernando. 1986. *Ki: El drama de un pueblo y de una planta*. 3rd ed. Mexico City, Mexico: Fondo de Cultura Económica.

Bracamonte y Sosa, P. 1993. *Amos y sirvientes: Las haciendas de Yucatán, 1789–1860*. Mérida, Mexico: Universidad Autónoma de Yucatán.

Bricker, Victoria. 1981. *The Indian Christ, the Indian king*. Austin: University of Texas Press.

Brown, Denise. 1988. Catálogo de comunidades Mayas en tres zonas de la peninsula de Yucatan. Research report for Sostenibilidad Maya, available at MEXUS, University of California at Riverside. (Project funded by MacArthur Foundation.)

Burns, Alan. 1983. *An epoch of miracles: Oral literature of the Yucatec Maya*. Austin: University of Texas Press.

Castañeda, Quetzil. 1996. *In the museum of Maya culture*. Minneapolis: University of Minnesota Press.

Clendinnen, Inga. 1987. *Ambivalent conquests*. Cambridge, UK: Cambridge University Press.

Cliff, M. A. 1988. Domestic architecture and origins of complex society at Cerros. In *Household and community in the Mesoamerican Past*, edited by R. R. Wilk and W. Ashmore. Albuquerque: University of New Mexico Press, 199–225.

Coe, Michael. 1984. *The Maya*. 3rd ed. London: Thames and Hudson.

CONABIO. 2002. *Comisión Nacional para el Conocimiento y Uso de la Biodiversidad* (National Commission for Knowledge and Use of Biodiversity, in Mexico). Available at: www.conabio.gob.mx. Accessed August 30, 2002.

Contreras-Sánchez, Alicia del C. 1990. *Historia de una tintorea olvidada: El proceso de explotación y circulación del palo de tinte, 1750–1807*. Mérida: Universidad Autónoma de Yucatán.

Crosby, Alfred W. 1972. *The Columbian exchange: Biological and cultural consequences of 1492*. Westport, CT: Greenwood Press.

———. 1986. *Ecological imperialism: The biological expansion of Europe 900–1900*. Cambridge, UK: Cambridge University Press.

Cuanalo, H., R. Navarrete, and A. Gómez. 1996. Producción Agrícola. Research Report for Proyecto Población y Desarrollo Sustentable en Yucatán (Population and Sustainable Development in Yucatán), a project jointly undertaken by CINVESTAV (Centro de Investigación y de Estudios Avanzados [national government's Center for Research and Advanced Studies]), CONAPO (Comisión Nacional de Población [national government agency in Mexico]), and IIASA (International Institute for Applied Systems Analysis [Austria]).

Culbert, T. Patrick. 1974. *The lost civilization: The story of the Classic Maya*. New York: Harper and Row.

———. 1977. Maya development and collapse: An economic perspective. In *Social process in Maya prehistory: Essays in honor of Sir Eric Thompson*, edited by Norman Hammond. London: Academic Press.

———. 1996. Agricultura maya en los humedales de las tierras bajas mayas. Primera conferencia magistral. In *VI Encuentro de los Investigadores de la Cultura Maya*, no. 5. Campeche, Mexico: Universidad Autónoma de Campeche, 14–19.

———, and D. Rice, eds. 1990. *Pre-Columbian population history in the Maya lowlands*. Albuquerque: University of New Mexico Press.

Diamond, Jared. 1998. *Guns, germs and steel*. New York: W. W. Norton.

Díaz del Castillo, Bernal. (1942, original, n.d., 1500s). Account of Díaz del Castillo, archival document, reproduced in *Historia Verdaera de la Conquista de la Nueva España*. México, D. F., vol. 1, chapters II–VI, 1904. Translated by Henry R. Wagner. In *The Discovery of Yucatán by Francisco Hernández de Córdoba* (Whittier, CA: The Cortez Society, 1942), p. 63. [A more easily accessible volume is Díaz del Castillo, Bernal. 1928. *The Discovery and Conquest of Mexico, 1517–1521*. Translated by A. P. Maudslay. New York: Harper.]

Dumond, Don. 1998. *The machete and the cross*. Lincoln: University of Nebraska Press.

Duby, Gertrude. 1944. *Los lacandones, su pasado y su presente*. Mexico City, Mexico: Biblioteca Enciclopedia Popular.

Earth Charter Commission. 2000. *The Earth Charter: Values and principles for a sustainable future*. San José, Costa Rica: Earth Council.

Eastmond, Amarella, Ana García, and Juan Córdoba. 2000. Recent population and education trends on the Yucatan Peninsula. In *Population, development, and environment on the Yucatan Peninsula from Ancient Maya to 2030*, edited by Wolfgang Lutz, Leonel Prieto, and Warren Sanderson. Laxenburg, Austria: International Institute for Applied Systems Analysis, 54–72.

Elmendorf, Mary. 1972. *La mujer Maya y el cambio*. Mexico City, Mexico: Centro Intercultural de Documentacíon (Center for Intercultural Documentation, CIDOC) Cuaderno No. 81.

———. 1976. *Nine Maya women: A village faces change*. New York: Schenkman.

———. 1980. *Changing patterns of fertility: The impact of contraceptive technology on a Maya village*. Research Report for Research Institute for the Study of Man, A.I.D. Project no. 9323-0616. Washington, DC: U.S. AID.

Emery, Marla, and Rebecca J. McLain, eds. 2001. *Non-timber forest products: Medicinal herbs, fungi, edible fruits and nuts, and other natural products from the forest*. New York: Food Products (Haworth).

Enge, Kjell I., and Scott Whiteford. 1989. *The keepers of water and earth: Mexican rural social organization and irrigation*. Austin: University of Texas Press.

Farriss, Nancy. 1984. *Maya society under colonial rule: The collective enterprise of survival*. Princeton, NJ: Princeton University Press.

Fash, William. 1991. *Scribes, warriors and kings: The city of Copán and the ancient Maya*. London: Thames and Hudson.

Faust, Betty B. 1998. *Mexican rural development and the plumed serpent*. Westport, CT: Bergin and Garvey Press.

———, and Richard Bilsborrow. 2000. Maya culture, population and the environment in the Yucatan Peninsula. In *Population, development and environment in the Yucatán Peninsula: From ancient Maya to 2030*, edited by Wolfgang Lutz, Leonel Prieto, and Warren Sanderson. Laxenburg, Austria: International Institute for Applied Systems Analysis, 77–107.

Folan, William J., B. Faust, W. Lutz, and J. Gunn. 2000. Social and environmental factors in the Classic Maya Collapse. In *Population, development, and environment on the Yucantan Peninsula: From ancient Maya to 2030*, edited by Wolfgang Lutz, Leonel Prieto, and Warren Sanderson. Laxenburg, Austria: International Institute for Applied Systems Analysis, 2–32.

———, J. Gunn, J. D. Eaton, and R. W. Patch. 1983. Paleoclimatological patterning in southern Mesoamerica. *Journal of Field Archeology* 10: 453–468.

Flores Torres, Jorge. 1997. *Los mayas yucatecos y el control cultural*. Mexico City and Mérida, Mexico: Universidad Autónoma Chapingo and Universidad Autónoma de Yucatán.

Fraga, Julia, and Luisa Paré. 1990. *La costa de Yucatán*. Mérida, Mexico: Universidad Autónoma de Yucatán.

Frazier, J. 1997. Sustainable development: Modern elixir or sack dress? *Environmental Conservation* 24 (2): 182–193.

Gallopin, G. 1990. *Water storage technology at Tikal, Guatemala*. Cincinnati, OH: University of Cincinnati.

Gates, M. 1993. In *default: Peasants, the debt crisis, and the agricultural challenge in Mexico*. Boulder, CO: Westview Press.

Garza, Gustavo, and S. Rivera. 1995. *Dinámica macroeconómica de las ciudades en México*. Vol. I. Mexico City: INEGI.

Gill, Richardson B. 2000. *The great Maya droughts: Water, life, and death*. Austin: University of Texas Press.

Gómez-Pompa, Arturo, J. S. Flores, and V. Sosa. 1987. The "pet kot," a man-made tropical forest of the Maya. *Interciencia* 12: 10–15.

———, and Andrea Kaus. 1990. Traditional management of tropical forests in Mexico. In *Alternatives to deforestation*, edited by Anthony Anderson. New York: Columbia University Press, 45–64.

Gossen, Gary. 1994. From Olmecs to Zapatistas: A once and future history of souls. *American Anthropologist* 6 (3): 553–570.

Gunn, J. D., and W. J. Folan. 2000. Three rivers: Subregion global climate response surface for the southwestern Maya lowlands (Candelaria, Usumacinta, and Champotón watersheds). In *Global climate change and cultural memory*, edited by R. McIntosh, S. McIntosh, and J. Tainter. New York: Columbia University Press.

Hale, Lauren. 2000. A dynamic simulation model of population impacts on the environment: A fisheries model. In *Population, development and environment in the Yucatan Peninsula: From ancient Maya to 2030*, edited by Wolfgang Lutz, Leonel Prieto, and Warren Sanderson. Laxenburg, Austria: International Institute for Applied Systems Analysis, 204–227.

Hellmuth, Nicholas. 1977. Cholti-Lacandon (Chiapas) and Petén-Ytzá agriculture, settlement pattern and population. In *Social process in Maya prehistory: Studies in honour of Sir Eric Thompson*, edited by Norman Hammond. New York: Academic Press, 421–445.

Hervik, Peter. 1999. *Mayan people within and beyond boundaries*. Amsterdam: Harwood Academic.

Hodell, David A., Mark Brenner, Jason H. Curtis, and Thomas Guilderson. 2001. Solar forcing of drought frequency in the Maya lowlands. *Science* 292: 1367–1370.

———, J. H. Curtis, and M. Brenner. 1995. Possible role of climate in the collapse of Classic Maya civilization. *Nature* 375: 391–394.

Howrigan, G. A. 1988. Fertility, infant feeding, and change in Yucatan. In *Parental behavior in diverse societies: New directions for child development*, no. 40 edited by R. A. Le Vine and P. M. Miller. San Francisco, CA: Jossey-Bass, 37–50.

INEGI (Instituto de Estadística, Geografía, e Informática). 1991. *Censo general de población y vivienda de 1990*. Aguascalientes, Mexico: INEGI.

———. 1993. *Felipe Carrillo Puerto, Estado de Quintana Roo: Cuaderno estadístico municipal*. Aguascalientes, Mexico: INEGI.

———. 1996. *INEGI estadísticas históricas de México*. Aguascalientes, Mexico: INEGI.

———. 2000a. *Censo general de población y vivienda*. Aguascalientes, Mexico: INEGI.

———. 2000b. *Cuaderno de información oportuna regional*, no. 67. Aguascalientes, Mexico: INEGI.

———. 2000c. *Tabulados básicos: Quintana Roo*. Aguascalientes, Mexico: INEGI.

———. 2000d. *Sector alimentario en México, edición 2000*. Aguascalientes, Mexico: INEGI.

Jones, Grant D. 1982. Agriculture and trade in the Colonial period, southern Maya lowlands. In *Maya subsistence: Studies in memory of Dennis E. Puleston*, edited by Kent V. Flannery. New York: Academic Press, 275–293.

Joseph, Gilbert M. 1986. *Rediscovering the past at Mexico's periphery: Essays on the modern Yucatan*. University City: University of Alabama Press.

Kintz, Ellen. 1990. *Life under the tropical canopy: Tradition and change among the Yucatec Maya*. New York: Holt, Rinehart, and Winston.

Krotz, Esteban, ed. 1996. *Cambio cultural y resocialización en Yucatan*. Mérida, Mexico: Universidad Autónoma de Yucatán.

Landa, Fray Diego de. 1937 (original 1566). *Yucatan before and after the conquest*. Translated by William Gates. New York: Dover.

Leach, Helen M. 1999. Intensification in the Pacific: A critique of the archaeological criteria and their application. *Current Anthropology* 40: 311–340.

Ligorred Perramon, Francesc. 1997. *U Mayathanoob Ti Dzib/Las voces de la escritura*. Mérida, Mexico: Universidad Autónoma de Yucatán.

Lowe, John W. G. 1985. *The dynamics of apocalypse: A systems simulation of the Classic Maya Collapse*. Albuquerque: University of New Mexico Press.

Lundell, C. L. 1945. The vegetation and natural resources of British Honduras. In *Plants and plant science in Latin America*, edited by F. Verdoorn. New Bedford, MA: Chronica Botanica.

Matheny, Ray T., Deanne L. Gurr, Donald W. Forsyth, and F. Richard Hauck. 1983. *Investigations at Edzná, Campeche, Mexico: The hydraulic system*. Papers of the New World Archeological Foundation, No. 46. Provo, UT: Brigham Young University Press.

McJunkin, David. 1991. Logwood: An inquiry into the historical biogeography of *Haematoxylum campechianum* L. and related dyewoods of the neotropics. Ph.D. thesis, Deptartment of Geography, University of California at Los Angeles.

Morales Valderrama, Carmen. 1987. *Ocupación y sobrevivencia campesina en la zona citrícola de Yucatan*. Mexico City, Mexico: Instituto Nacional de Antropología e Historia-Secretaría de Educación Pública.

Painter, Michael, and William Durham, eds. 1995. *The social causes of environmental destruction in Latin America*. Ann Arbor: University of Michigan Press.

Patch, Robert. 1993. *Maya and Spaniard in Yucatan, 1648–1812*. Stanford, CA: Stanford University Press.

———. 2002. *Maya revolt and revolution in the eighteenth century*. Boulder, CO: M. E. Sharpe.

Pennington, T. D. 1990. Sapotaceae. *Flora Neotropica*, vol. 52. New York: New York Botanic Garden.

Peraza López, María Elena. 1986. Patrones alimentícios en Ichmul, Yucatán: Sus determinantes socioeconomicas, ecologicas y culturales. Tesis de licenciatura, Escuela de Ciencias Antropológicas, Universidad Autónoma de Yucatán.

Pfeiler, Barbara. 1997. El Maya: una cuestión de persistencia o pérdida cultural. In *Persistencia cultural entre los Mayas frente al cambio y la modernidad*, edited by Ramón Arzápalo Marín and Ruth Gubler. Mérida, Mexico: Universidad Autónoma de Yucatan, 55–71.

Press, Irwin. 1975. *Tradition and adaptation: Life in a modern Yucatan Maya village*. Westport, CT: Greenwood.

Prieto, Leonel. 2000. Land use on the Yucatan Peninsula: System and model description and land-use scenarios. In *Population, development and environment in the Yucatan Peninsula: From ancient Maya to 2030*, edited by Wolfgang Lutz, Leonel Prieto, and Warren Sanderson. Laxenburg, Austria: International Institute for Applied Systems Analysis, 228–255.

Primack, Richard, David Bray, Hugo Galletti, and Ismael Ponciano, eds. 1998. *Timber, tourists, and temples*. Washington, DC: Island Press.

Re Cruz, Alicia. 1996. *The two milpas of Chan Kom*. Austin: University of Texas Press.

Redfield, Robert. 1941. *The folk culture of Yucatan*. Chicago: University of Chicago Press.

———. 1950. *A village that chose progress: Chan Kom revisited*. Chicago: University of Chicago Press.

———, and Alfonso Villa Rojas. 1934. *Chan Kom, a Maya village*. Publication 448. Washington, DC: Carnegie Institute of Washington.

Redman, Charles. 1999. *Human impact on ancient environments*. Tucson: University of Arizona Press.

Restall, Matthew. 1996. *The Maya world*. Stanford, CA: Stanford University Press.

———. 1998. *Maya conquistador*. Boston: Beacon Press.

Rice, Prudence M. 1987. *Macanché Island, El Petén, Guatemala: Excavations, pottery, and artifacts*. Gainesville: University of Florida Press.

Rico-Gray, Victor, and José G. García-Franco. 1991. The Maya and the vegetation of the Yucatan Peninsula. *Journal of Ethnobiology* 11 (1): 135–142.

Rosales González, Margarita. 1988. *Oxcutzcab, Yucatán, 1900–1960: Campesinos, cambio agrícola y mercado*. Mexico City, Mexico: Instituto Nacional de Antropología e Historia.

Roys, Ralph. 1957. *The political geography of the Yucatan Maya*. Publication 613. Washington, DC: Carnegie Institute of Washington.

Rugeley, Terry. 2001. *Of wonders and wise men: Religion and popular culture in Southeast Mexico, 1800–1876*. Austin: University of Texas Press.

Sandoval-Palacios, J. M. 1982. Development of capitalism in Mexican agriculture. Its impact on the humid tropics: The case of the Yohaltún Project in the southeastern state of Campeche. Ph.D. dissertation, University of California at Los Angeles.

Santley, R. S. 1990. Demographic archeology in the Maya lowlands. In *Pre-Columbian population history in the Maya lowlands*, edited by T. P. Culbert and D. S. Rice. Albuquerque: University of New Mexico Press, 325–344.

Scarborough, Vernon L. 1996. Reservoirs and watersheds in the central Maya lowlands. In *The managed mosaic: Ancient Maya agriculture and resource use*, edited by S. L. Fedick. Salt Lake City: University of Utah Press, 304–314.

———, and Gary G. Gallopin. 1994. A water storage adaptation in the Maya lowlands. *Science* 251 (4): 658–662.

Scholes, France V., and Ralph Roys. 1968. *The Maya Chontal Indians of Acalan-Tixchel: A contribution to the history and ethnography of the Yucatan Peninsula*. Norman: University of Oklahoma Press.

SEMARNAT (Secretaría de Medio Ambiente y Recursos Naturales). 2002. Web pages. for the country of Mexico. Available at: http://sepultura.semarnat.gob.mx/ucamp/data. Accessed October 24, 2002.

Sharer, Robert J. 1994. *The ancient Maya*. 5th ed. Stanford, CA: Stanford University Press.

Siemens, A. H., and D. E. Puleston. 1972. Ridged fields and associated features in southern Campeche: New perspectives on the lowland Maya. *American Antiquity* 37: 228–239.

Sponsel, Leslie, Thomas Headland, and Robert Bailey. 1996. *Tropical deforestation*. New York: Columbia University Press.

Stephens, John Lloyd. 1841. *Incidents of travel in Central America, Chiapas and Yucantan*. 2 vols. (with drawings by Frederick Catherwood) New York: Harper and Brothers.

———. 1843. *Incidents of travel in Yucatan*. 2 vols. (with drawings by Frederick Catherwood) New York: Harper and Brothers.

Sullivan, Paul. 1989. *Unfinished conversations: Mayas and foreigners between two wars.* New York: Alfred P. Knopf.

Terán, Silvia, and Christian Rasmussen. 1994. *La milpa de los Maya.* Mérida, Mexico: Universidad Autónoma de Yucatan.

———. 1995. Genetic diversity and agricultural strategy in 16th century and present-day Yuctecan milpa agriculture. *Biodiversity and Conservation* 4: 363–381.

Tozzer, Alfred M. 1907. *A comparative study of the Maya and the Lacandones.* New York: Macmillan.

Turner, B. L., II. 1974. Prehistoric intensive agriculture in the Maya lowlands. *Science* 185 (4146): 118–124.

———. 1976. Population density in the Classic Maya lowlands: New evidence for old approaches. *Geographical Review* 66 (1): 73–82.

———. 1979. Prehispanic terracing in the central Maya lowlands. In *Mesoamerican archeology and ethnography*, edited by N. Hammond and G. R. Willey. Austin: University of Texas Press, 103–113.

Turner, John Kenneth. 1911. *Barbarous Mexico.* New York: Cassell.

Vargas, G. 1983. Origen y flora de las sabanas de la Provincia de Guanacaste, Costa Rica: Un análisis de fitogeografía histórica. *Revista Geográfica Américal Central* 17: 57–67.

Várguez Pasos, Luis A. 1981. *La milpa entre los mayas de Yucatán.* Mérida: Universidad Autónoma de Yucatán.

Villagutierre (Soto Mayor), Juan. 1933 (original 1701). *Historia de la conquista de la Provincia de el Itzá ... la de el Lacandón.* Guatemala City, Guatemala: Biblioteca Guathemala, Sociedad de Geografía e Historia.

Villanueva, Eric. 1985. *Crisis henequenera y movimientos campesinos en Yucatán 1966–1983.* Mexico City, Mexico: Instituto Nacional de Antropología e Historia.

Villa Rojas, Alfonso. 1945. *The Maya of east central Quintana Roo.* Washington, DC: Carnegie Institute of Washington.

Wagner, Phillip. 1964. Natural vegetation of Middle America. In *Handbook of Middle American Indians*, Vol. 1, edited by R. Wauchope. Austin: University of Texas Press.

Weaver, Muriel Porter. 1993. *The Aztecs, Maya, and their predecessors: Archeology of Mesoamerica.* 3rd ed. San Diego, CA: Academic Press.

Wells, Allen. 1985. *Yucatan's gilded age.* Albuquerque: University of New Mexico Press.

White, Christine, ed. 1999. *Reconstructing ancient Maya diet.* Salt Lake City: University of Utah Press.

2

Working with Peoples and Places Through Action: A History of the Participatory Research of Mary Lindsay Elmendorf

Alicia Re Cruz

I feel honored to present Mary Elmendorf's groundbreaking work on Maya studies in particular, and on anthropology in general. Elmendorf's outstanding contributions to the anthropological field certainly mark a shift in ways to do anthropology and ways to work with people. Free of theoretical pretensions, her work conveys a humanistic message in such a way that the ethnographic actors are presented through their own voices and work. Intertwined with this humanistic message is her advocacy for a methodology in favor of dialogue and participatory research to enhance people's awareness and confidence in order to empower their actions. Mary Elmendorf's prolific professional life is devoted to a particular population group: women, as she explains, "the over-used and misused human resources."

Let me briefly mention some crucial professional steps of Elmendorf's career to document the sophisticated humanist aspect of her work, which has brought her to the forefront of participatory research focused on women—for her, the main agents of change. Elmendorf, who grew up in the Carolinas, graduated from the University of North Carolina at Chapel Hill at the age of twenty with a B.A. in psychology and graduate credits in public administration and social work. After her graduation, she spent a decade as a social worker and administrator in various public and private agencies focusing on rural pockets of poverty in the South and in the urban slums of New England. She also worked for two years as a volunteer with the American Friends Service Committee (AFSC) during and after World War II—first in the prisons, and later as director of the Spanish refugee programs. These cross-cultural learning experiences made her realize the ethnocentricity of her training and led her to graduate studies in anthropology. In 1952, she opened the CARE (Cooperative for American Relief Everywhere) mission to Mexico. CARE, a nongovernmental organization (NGO), originally started as a temporary cooperative of U.S. agencies during World War II to bring direct relief to war-torn Europe. Elmendorf, as director of CARE, designed a pilot program for self-help and community development instead of continuing the usual relief and reconstruction programs designed for Europe. The CARE regional director for Latin America—and many others—expressed the opinion that a woman could not run an overseas office, certainly not one in Latin America, and that self-help and technical assistance were concepts that could not function in Latin America. Elmendorf decidedly proved him wrong; she became the first woman to negotiate, design, and direct an overseas office for CARE. In addition, she graciously showed how well women can communicate and work with Mexican men and women, and that self-help and technical assistance were far superior to relief and charity. These were the first steps of a long and fruitful career in serving people through cultural understanding. Elmendorf's blossoming professional life in Mexico, devoted to "putting people first," started with a project to help the villagers in Santa Maria, Mexico, dig a well for safe water in the village square. President John Kennedy used a case study of this project, which was developed with Quaker volunteers in the AFSC office in Mexico, to help win approval and support for establishing the Peace Corps.[1] Eight years as director of CARE in Mexico during the 1950s provided Mary Elmendorf with experience that allowed her to work very closely with Gonzalo Aguirre Beltran and Alfonso Villa Rojas in the Instituto Nacional Indigenista (National Indigenous Institute, or INI)[2] and also with the ministers of health, education, and agriculture. This experience gave her a unique opportunity to witness the multifaceted roles of women, particularly the power of rural women as agents of change. Elmendorf's ability to engage different powerful sectors in community development projects is a significant factor to take into account when evaluating the extent of her successes. From community members—both men and women—to teachers, planners, government officials in the Ministry of Agriculture, and so on, she was able to establish collaborative ties to design seminars and training courses, mainly among women. These pilot projects aimed at constructing a better future for Mexican communities through women's involvement. Now, at age eighty-five, she is still professionally active.

Certainly, her experience in Mexico ignited a passion for documenting the world of women, their dreams as much as the beauty and hardships of their everyday lives, a dimension of community life and culture still hidden from anthropologists, social workers, and social scientists in general. After coming back from Mexico to the United States in the early 1960s, Mary Elmendorf developed several proposals on the role of women as agents of change. In them, she expressed her continuing interest in the role of women as overused human resources in socioeconomic development, a research concern arising in the anthropological debate during the 1960s. Lessons learned from her different cross-cultural experiences moved her to focus on a specific group of women, the Maya women in Chan Kom, who were experiencing a transition from traditional to modern; she was—and still is—intrigued by their feelings and thoughts about technological and social changes in their lives and community.

Elmendorf completed her anthropology doctorate in 1972, at the Union Graduate School at Antioch, under the direction of Alfonso Villa Rojas, Dorothy Lee, Fernando Cámara-Barbachano, Eric Fromm, and Michael Maccoby. Mary Elmendorf's dissertation was published in 1973 in Spanish by the SEP-Setenta, a Mexican publication series directed by Gonzalo Aguirre Beltran for distribution in public schools across the nation.[3] Her doctoral work was first published in English by Ivan Illich in 1972, as a case study for a CIDOC seminar on the quality of life (Elmendorf 1972).[4] This doctoral study led to various consultant assignments for the Ford Foundation, the Mexican government, the United Nations Development Fund, the World Bank, USAID, and other agencies involved in analyzing the sociocultural aspects of development projects, with emphasis on the role of women and community participation in meeting basic needs. These basic needs fell into three main categories: population, energy, and water. In population studies, Elmendorf's efforts were concentrated on access to information about fertility control, presented in an understandable way for local groups. In her research on energy expenditure, Elmendorf focused on household energy, with a primary concern for conserving women's time and physical strength so that they would have more time and energy for themselves, their families, and their income-generating activities. Elmendorf's water projects focused on women and were mostly concerned with environmental sanitation and the need for sustained, effective community participation. The World Bank published *Sociocultural Aspects of Excreta Disposal* (Elmendorf 1977), which encapsulated case studies of successes and failures of intervention in sanitation problems throughout Latin America. Although she has been called "the four-letter-word lady," her projects brought a growing awareness of the importance of understanding taboos and habits in order to bring about changes in water use and excreta disposal. With this awareness, women as primary users and socializers became a priority in the agenda of community development projects. In sum, the quality of her work has convinced many agencies to ask for anthropological input and particularly for women's perspectives. She was one of the first to get the World Bank to consider women when designing and evaluating projects, convincing agency officials that many programs had failed because they did not include women as decision makers in their communities. She opened many doors for anthropologists as planners, and certainly she has demonstrated to economists and engineers, among many others, the need to integrate the sociocultural dimension into their development projects. Since 1976, when President Carter appointed her as advisor to the official delegation of the United States to the United Nations Conference on Water Resources in Mar del Plata, she has worked with both governmental and nongovernmental organizations within the UN system, in Asia, Africa, and Latin America. The techniques and tools developed in her work on water have been mainstreamed within the World Bank and have been adapted for use by most agencies specialized in this field, whether they are associated with the government or NGOs associated with the UN.

Let me now take Elmendorf's work in Chan Kom as a key to decode the cultural dimensions her research unveils in Maya culture by focusing on women's voices, thoughts, and souls. I would like to approach this analysis of her discoveries by explaining how I discovered her and how writing about her, reflecting on her work, and presenting her path in anthropology helps me discover yet more facets of her anthropological legacy.

The first time I saw Mayas in the flesh, after conceptualizing and imagining their lives and world through thousands of pages written about their past and present culture, was during the summer of 1986 while undertaking my first exploratory fieldwork in Chan Kom. I promise not to get into the intricate factors that drove me to Chan Kom, although I would like to stress how much I regretted ending up in Chan Kom at first, because of the villagers' constant references to other anthropologists who had studied them. First on the list was don Alfonso Villa Rojas, a quasi-mythical hero, "el maestro," to whom Mayas refer as the intellectual patriarch, one of the founders of the village. He was the first schoolteacher in Chan Kom, starting in the 1930s. Second on the list was "doña Mary," to whom women referred with great respect and affection, and whom men remembered as the woman who filmed a *ch'a chaak*,[5] the rainmaking ceremony performed by men in the forest.

Any student in Mesoamerican social science is well acquainted with the giants of peasant studies, Robert Redfield and Eric Wolf. I was also familiar with the work of Alfonso Villa Rojas, who collaborated with Robert Redfield in collecting ethnographic documentation to construct one of the most comprehensive, holistic models for rural communities, *Chan Kom, a Maya Village* (Redfield and Villa Rojas 1934). However, this and other ethnographic work previous to Mary Elmendorf's on Chan Kom (Redfield 1941, 1950; Goldkind 1965, 1966) was focused on the male peasant world, the production of corn. Men are the decision makers for milpa matters: how much energy to invest in milpa, the right time to burn the fields, the scheduling of agricultural rituals, types of seeds to use, and so forth.

Who was doña Mary? On my return to the United States, a more thorough investigation of the ethnographic encyclopedia on Chan Kom exposed doña Mary not only as the author of *Nine Mayan Women: A Village Faces Change* (Elmendorf 1976),[6] but as the architect of a groundbreaking chapter in that encyclopedia, the chapter devoted to Maya women. While for Redfield Maya peasant life is centered on milpa production, which embodies an intriguing symbiosis of natural resources and peasant ethics, Elmendorf's study in the

1970s brought to light women's economic performance as the sine qua non to develop and maintain that symbiosis. Certainly, Elmendorf's revolutionary shift in research focus evidenced how much milpa work had bewitched Redfield's analytic efforts by concentrating on male economic performance. Indeed, milpa represents an intriguing embroidery of worldviews, lifestyles, customs, sacred rituals, religious code, and social order; it encapsulates such a holistic cultural system that it has been considered the most relevant factor in the definition of Maya identity. However, talking about milpa production traditionally meant talking predominantly about male peasants. In filming the *ch'a chaak*, Elmendorf commuted between the world of men in the forest and the world of women in the home around the hearth, showing them performing roles they felt were equally important. She demonstrated the crucial contribution of female peasants to the performance and products of *milpa*. Focusing on the Maya female world, she unveiled ideological agendas hidden from former scientific procedures, as epistemological principles, as research foci, or as methodological frameworks. Elmendorf transformed the Redfieldian focus on change into one on women's awareness of their own economic power, which makes them promoters of change.

In fact, the early 1970s represented a crucial step in the transformation of Chan Kom: the opening of the road that connected the community with Mérida and Cancún. The opening of the road meant progress, modernization, and change, mainly for Maya women who started imagining the possibilities of initiating commercial enterprises (particularly, selling hammocks and embroidered *huipiles*—traditional female Maya dress—in major tourist centers) as a way to complement the customary milpa economy. The traditional connection between peasant men and milpa production, coupled with environmental degradation, set the scene for new economic alternatives for peasant survival; according to Mary Elmendorf, women's microeconomic creativity and innovations, pushed by the opening of the road, were the economic answer to future alternatives for survival for the Maya. Hence, she emphasized the fundamental role of women as participants in community struggles to meet basic needs.

Elmendorf's major contributions in Maya studies emerge from the 1960s' scholarly and political ferment in the United States. The Civil Rights and antiwar movements had grown and given birth to theory and activism concerned with environmental issues and American foreign policy, as well as the rights of gay, black, Latino, Asian, Native American, and other ethnic groups—and feminism. All these movements were influenced by, and, in turn, inspired intellectual shifts, a condition in which Mary Elmendorf's social paradigm provides concrete answers: the need to listen to people, particularly to women's voices. Her vision also embraces Latin American intellectual and social movements of the 1960s, favoring the creation of scientific paradigms in agreement with, and committed to, resolving social issues. (These contrast with Anglo American, machinelike, positivist models of social science.) Her participatory research owes much to the work and inspiration of Paulo Freire,[7] to his book *Pedagogy of the Oppressed* (1968), and to the practice and experience of sensitization to the problems of the marginalized in Latin America. The Freirian theme—poor and exploited people can, and should be, empowered to conduct their own analysis of their own reality—has been widely influential. Mary Elmendorf carried this theme of empowerment to the world of Maya women, and with this she adds a new chapter in the encyclopedic global knowledge of women's economic and social roles. In sum, in a 1960s milieu of shifting social and intellectual paradigms very similar to our postmodernism, Elmendorf proposed an epistemological twist that could provide answers to questions raised by multiple social realities. This twist has to do with several factors:

1. The need to learn the languages of other disciplines
2. The presentation of our findings in ways that local people can understand
3. Listening to people's voices as the key for understanding their everyday lives and the intricate composition of their ideologies and worldviews
4. Remembering that people's voices express both their communality and their differences throughout the world

NOTES

1. The documentary film *World Our Hands Can Make* EXPL1is based on this CARE/AFSC project. This documentary won honorary mention at the 1959 Venice Film Festival and is now in the National Film Archives at the Smithsonian.

2. Villa Rojas had published in 1985 a review of Maya ethnographies in Spanish, in addition to his original coauthored work on Chan Kom with Robert Redfield.

3. A second edition of *La mujer Maya y el cambio* was published for donation to the Organizacion de Estados Americanos (Organization of American States, OAS) at a special seminar on Indian women.

4. CIDOC is the Centro Intercultural de Documentación (Center for Intercultural Documentation).

5. Mary Elmendorf filmed different ritual and everyday activities of community life in Chan Kom during the 1970s. She donated these films to the Smithsonian Institution and in the 1990s converted some of them into a video format. I had the honor to be asked to return to Chan Kom, with Mary Elmendorf, to show her video to Chan Kom's people twenty years after the films were made. I am currently working on a video production that encapsulates an ethnographic view of Chan Kom using archival and current images of the village.

6. A revised, enlarged edition was published by Schenkman Press/John Wiley in 1976 (subsequent to the two publications in Mexico). The U.S. publication was reprinted by Schenkman Books in 1985 and 2000.

7. Freire was working at CIDOC when Mary Elmendorf's study was published.

REFERENCES

Elmendorf, Mary. 1972. *Mayan women and change*. Cuaderno No. 81. Mexico City: Center for Intercultural Documentation (CIDOC).

———. 1973. *La mujer maya y el cambio*. SEP-70. Mexico D. F., Mexico City: Secretaría de Educación Pública.

———. 1976. *Nine Mayan women: A village faces change*. New York: Schenkman.

———. 1977. *Sociocultural aspects of excreta disposal*. Washington, DC: World Bank.

Freire, Paulo. 1968. *Pedagogy of the oppressed*. New York: Seabury.

Goldkind, Victor. 1965. Social stratification in the peasant community: Redfield's Chan Kom reinterpreted. *American Anthropologist* 67: 863–884.

———. 1966. Class, conflict and cacique in Chan Kom. *Southwestern Journal of Anthropology* 22: 325–345.

Redfield, Robert. 1941. *The folk culture of Yucatan*. Chicago: University of Chicago Press.

———. 1950. *A village that chose progress: Chan Kom revisited*. Chicago: University of Chicago Press.

———, and Alfonso Villa Rojas. 1934. *Chan Kom, a Maya village*. Washington, DC: Carnegie Institute.

Villa Rojas, Alfonso. 1985. *Estudios etnologicos: Los Mayas*. Mexico, D. F., Mexico: Universidad Autónoma de México.

3

The Many Worlds of Mayan Women

Mary Lindsay Elmendorf

INTRODUCTION TO THE AUTHOR'S WORK WITH WOMEN, COMMUNITY DEVELOPMENT, AND ENVIRONMENTAL ISSUES

When we think of women today, we are much more conscious of the many worlds in which they play significant roles—their homes, their communities, the workplace, the nation, and the world—than we were fifty years ago. One of the chief themes throughout sixty of my eighty-five-plus years of life as wife, mother, social worker, teacher, anthropologist, and activist has been the importance of women within their homes and communities, and now globally, as agents of change. As gatekeepers of values and resources, as educators, managers, and decision makers in daily life, women are critically needed as full partners in development.

During the 1950s, I spent eight busy years in Mexico as the first woman director of an overseas program of the Cooperative for American Relief Everywhere (CARE, originally Cooperative for American Remittances to Europe). In that position, I worked with government officials and community leaders at all levels across the country and was continually impressed with what women could accomplish, especially the peasant women (Elmendorf 1953).

Already in the 1950s one could see the underestimated political power of Yucatec women. The wife of the governor of Yucatan broke a four-year deadlock against the duty-free importation of CARE milk into Mexico, which had been requested by the Ministry of Health in 1952 for school breakfast programs. After the governor's wife and I had delivered CARE packages of children's books to several schools in Mérida, where Mayan mothers were preparing breakfast for their children in makeshift kitchens with crude utensils, she asked whether surplus milk was still available from CARE. When I said, "Yes, but only if it can be imported duty-free," she responded, "We in Yucatan don't have to get permission from the federal government; we have our own port of Progreso. We'll talk about it with my husband tomorrow at breakfast. I'll have a report for you on the number of schools, children, and volunteers. I'll have my driver pick you up at 5:30." She did it all. She broke the embargo. Soon afterward, the wives of governors in other states, in order to get CARE milk for their programs, persuaded the Ministry of Hacienda to grant duty-free entry as they had done for years for all other CARE donations (Elmendorf 1997).

In 1959 a documentary film, *World Our Hands Can Make* based on a CARE/AFSC (American Friends Service Committee) project, showed the key roles women played in helping bring potable water to a Mexican village (Elmendorf 1959).[1] This project not only highlighted women's roles but demonstrated what foreign volunteers could do. AFSC volunteers lived in the village and worked with the people close cooperation with representatives of local, state, and national agencies, operating a well-drilling rig donated by CARE. At President Kennedy's suggestion, this project was written up in 1959 as a case study to help get approval of and funding for the Peace Corps by showing what volunteers and communities could do together (Elmendorf 1966, 1978). Twenty years later, the World Bank asked me to prepare another case study on this project to demonstrate sustainability. The elements of the project, which were a village savings account, community participation, and women's involvement, were adopted not only by CARE but also by the World Bank and many other aid agencies in later projects. In fact, these models are still used today (Elmendorf 1982a; Elmendorf and Buckles 1980; Witherell et al. 1995).

In 1960, we returned from Mexico to Providence, Rhode Island, where my husband was vice president of Brown University. I continued to work with Peace Corps, Overseas Education Fund of the League of Women Voters, and Brown University (and after 1965 with New College in Sarasota, Florida) on issues related to self-help, community development, off-campus studies, and women's roles. One of the most interesting assignments was for the State Department in 1967 when Balaguer was elected president of the Dominican Republic and replaced all the governors with women. I was invited, with a State Department leadership grant, by the League of Women Voters to visit the country. My task was to congratulate the women and see whether they would be interested in having the Overseas Education Fund design training programs to help them be more effective in their new positions (Elmendorf 1967). In fact, they performed so well that a resolution was introduced—but not passed—to eliminate the posts of governor! Earlier, I had been hired by the University of Florida as consultant on self-help and community development for the first training program for Peace Corps volunteers going to Latin America (Elmendorf

1966). First at Brown and then at New College, I designed internships and off-campus study programs for students to be undertaken in many Latin American countries. In 1952, CARE, which had been created during World War II to channel emergency donations from twenty-seven nongovernmental organizations (NGOs) to war-torn Europe, decided to negotiate an official agreement with the government of Mexico to initiate a technical assistance program channeling equipment through existing organizations both governmental and private. This pilot self-help program in Mexico was one of two tests of the feasibility of CARE continuing as a development agency. Richard Reuter, later director of CARE, who had worked at the AFSC in Philadelphia as chief of personnel during World War II, heard that John Elmendorf's wife had recommended two men for the new post in Mexico. He said, "If that is the same Mary Elmendorf who worked for the Quakers in Europe, get her! Forget the men." And so it was that in spite of warnings from the Latin American regional representative that a woman certainly could not work with Mexican men, I said, "Try me for six months—and then we can renegotiate." I became the first woman to be appointed as country director (chief of mission, then) of an overseas CARE office and held the post for eight exciting years (Elmendorf 1953, 1960).

Soon after I was appointed director, I suggested that we consider the Instituto Nacional Indigenista (INI, National Indian Institute) as a cooperating agency. This was just after Arnold Toynbee, the famous British cultural historian, following a visit to INI's first Regional Center in San Cristóbal de las Casas, had said publicly, "This is the best example of applied anthropology in the world." Mexico was far ahead of the United States in support and promotion of applied anthropology and use of anthropologists as top administrators and planners. Administrators in Mexico also were justly proud of their concern for the problems of indigenous people in their country and of their many projects and programs, both private and governmental. Manuel Gamio was director of the Inter-American Institute and Alfonso Caso was director of INI with his outstanding staff, which included Gonzalo Aguirre Beltrán, Julio de la Fuente, Gilberto Balam, Alfonso Villa Rojas, and Ricardo Pozas. Pozas, the first director, had just published his *Juan Pérez Jolote* (Pozas 1952), later translated as *Juan the Chamula* (Pozas 1962), a fictionalized autobiography based on careful research on the life of a Mayan Tzotzil-speaking Indian caught in the Mexican Revolution. Now Mayas, such as Rigoberto Menchu, can speak—and write—for themselves (Menchu 1983, 1992).

One of the first donations of CARE equipment was made through INI. Alfonso Villa Rojas, Mexico's great Mayanist (see Villa Rojas 1945, 1969, 1978, 1990a, 1990b), was director of the Regional Center in Chiapas at that time (Duby 1992; Nash 1992; Elmendorf 1994b; Glusker 1994a). Much to my surprise, the CARE equipment—sewing machines donated by a women's club in Chicago—was not put in the new community center with its clinic and classrooms but in a small adobe building nearby that had been set aside for the women, who were not supposed to use the clinic. (In fact, some men, according to rumors I heard at the time, went to the clinic to have injections if their wives were ill; then they would go home and have intercourse so their wives would get well!) The *presidente municipal*, the mayor, demonstrated to the women how to operate the Singer pedal machine, accompanied by rapid Maya interspersed with the words *máquina* and *CARE* (Elmendorf 1953).

When I returned to Chiapas the following year on CARE business, I enjoyed seeing how quickly the women had mastered the new tools and how delighted they were to be using machines. They had learned to make shirts and school uniforms with the sewing machines, and with the gardening tools and seeds they had started their first vegetable plot—with carrots! Women were publicly responding to the excellent puppet shows put on in local costumes and in the local languages by El Teatro Petul. They were even playing basketball! They were also going into the clinic, which they had been too shy to enter earlier. Young women were taking training courses along with the Maya men to become bilingual *promotores*—teachers, health workers, and the like. I observed, though, that to continue a professional life after marriage was another new role and difficult hurdle, for them and me. The first few Mayan women who did continue working were married to male *promotores*—fellow classmates at INI.

Just as these Mayan women were doing, I also faced criticism for working—from both the Mexican and American communities. Everything our children did wrong was because I was working. If they did right, it was in spite of my work. Some asked my husband whether he could not support us. We supported each other—morally and sometimes financially—while sharing our jobs and family and continued to do so until his death.

In later decades I continued research with these women, interspersing it with other work related to community development, women's changing roles, and environmental health. The lives and stories of Mayan women have helped me understand the important roles played by other women in mitigating the negative impacts of development and change, both in their lives and in the environment in which they live (Elmendorf 1972, 1973a, 1973b, 1975, 1976a, 1976b, 1977a, 1977b, 1979b, 1980, 1993, 2001a, 2001b, 2002). Later, I was invited by the Ford Foundation to participate in a research project, with Janet Giele of the Radcliffe Institute, on the changing roles of Latin American women (Elmendorf 1977b). Twenty years later the Ford Foundation sponsored my presentation of original research on Mexican women, including some from Mayan areas, to a new interdisciplinary women's studies program at El Colegio de Mexico (Elmendorf 1993). Many of the papers that I included in this presentation were not considered important in the 1970s, had never been published, or were out of print. They included work by both Mexican and American women, many of whom have continued their research on gender issues and are recognized scholars today (Chiñas 1971; Nash 1958, 1960; Pozas 1959, 1972; Piho 1975; Scrimshaw 1973; Schmink 1974; Rivera 1976; Vogt 1969). In fact, sharing the stories of the Mayan women at seminars and conferences also helped open departments of women's studies and gender issues in colleges and universities in the United States and overseas (Elmendorf 1979b, 1982d, 1984a, 1984c, 1984e, 2000b).

THE ENVIRONMENT OF THE YUCATEC MAYA

Today, the Mayan communities of Mexico's Yucatan Peninsula are linked by La Ruta Maya to linguistically related groups in Chiapas, Guatemala, Belize, El Salvador, and Honduras through a 1,500-mile network of existing roads, supplemented by, as often as possible, travel alternatives such as monorail or riverboat. These roads have been designed to provide visitors with access to new biosphere reserves and other protected areas, archaeological sites, and Maya communities, but they also facilitate travel and exchange between three million persons speaking various Mayan languages who live in these five countries, opening up exciting possibilities for sharing and cooperation. One hopes that this new convenience can be maintained without endangering the environment and that the increased tourism will benefit the living Maya (Faust 1991; Daltabuit-Godas 1992; Garrett 1989, 424).

Joann Andrews, founder and former president of the board of Pronatura, Península de Yucatan, has reported on a training course, held in cooperation with the Nature Conservancy, for men and women living within the Calakmul Reserve and its environs to manage and conserve the local resources (Andrews 1997). One of the strategies of the workshops was to combine the knowledge of the Maya with existing data on the flora and fauna of the region. At the same time, efforts for sustainable use of the land, promotion of ecotourism, and sale of forest products are underway in communities neighboring the Calakmul Reserve (many of whom are not Mayan, as migrants have moved in; see Ericson, chapter 5, this volume). Similar activities have started in Celestún, Río Lagartos, Punta Laguna, and Sian Ka'an, the other biosphere reserves that are so much a part of the Mayan world (Andrews 1993).

AN INTRODUCTION TO MAYAN WOMEN

My interest in Mayan women began in 1950, soon after the Pan-American highway had opened, when I first visited Chiapas in 1950. Pausing for a picnic on the outskirts of San Cristóbal de las Casas, I saw my first Mayan woman. Her clothing was drab black and gray. She trotted along with a heavy load of wood on her back hanging by a tumpline from her forehead. The handsome man beside her was dressed just like the colorful men I had seen along the highway earlier. As he trotted down the path with her, he had an identical load of wood hanging on a tumpline from his forehead. They stopped to rest across the road but seemed unaware of us. They were alone in their private world as they rested together. As I watched, they stood up, and very carefully he added his load of wood above hers on her forehead as she stooped to carry the heavy burden. I could not believe my eyes. He passed in front of her, and they continued into town with him several paces ahead, weaving a strip of straw parts for a new hat. I wondered then, as I have for the fifty years since, why public and private roles of men and women seem so different (Elmendorf and Isley 1982, 1983; Elmendorf 1990, 1995). Why had the man changed so suddenly from a partner, a coworker, to something else as he approached town?

During our trip, I never saw the sharing again within the markets and streets of the city. In my short visits to the villages outside, the men and women were always in separate spaces. Most of the men looked young, handsome, and fun-loving, or old and wise. The women looked tired and drab, surrounded by girls dressed like them and boys dressed as the men were (Lombardo-Otero 1944; Pozas 1959, 1972).

During that first trip to Chiapas, I had another unforgettable day with the Maya when I accompanied Frans Blom and Gertrude Duby, his Swiss photographer wife, on horseback deep into the tropical forest to visit a group of Lacandóns with whom they were doing research and mapping. There were no roads, just paths through the lush hardwood forests, until suddenly we were in a small settlement with a group of people—men and women—all dressed alike in long handwoven white shifts. Except for trimmed bangs on a few, their hair looked uncut. It was hard to tell which were women and which were men. When people started talking, their voices helped identify sex, and nursing mothers gave clear clues. But there was no gender-stereotyped clothing to which we have become accustomed, and there were no beards, mustaches, or even stubble. Trudy Duby's exhibition the next year of her collection of photographs of the Lacandóns at the Mexican-American Cultural Institute received rave reviews. She continued working as an environmentalist and advocate for the Lacandóns in their struggle to keep their forests until her death in 1996, just before her 100th birthday (see Duby 1992).

WATER AND SANITATION CONCERNS

One day as we were returning from a field trip in Chiapas, we passed Yalcuc, one of the most interesting villages. Yalcuc was no different from its neighboring communities except that every thatched hut had a tile-covered latrine nearby with a well-worn path to it. I wondered why. Later, Alfonso Villa Rojas (1969) told me that Evon Vogt, director of the Harvard Chiapas Project, had just assigned one of his first graduate students, Frank Miller, to live there and find out. He found that instead of rejecting pit toilets as did most of the neighboring villages, 88 percent of the Yalcuc households installed latrines and 65 percent used them. The acceptance rate for latrine use was the highest in Chiapas and perhaps in all of rural Mexico at the time (Miller 1965).

Twenty years later, in 1977, when I had accepted an assignment for the World Bank Research Project on Appropriate Technologies for Water Supply and Sanitation to design and supervise the preparation of eight case studies in

Latin America, I recommended that Frank Miller and his wife Cynthia Cone return to Yalcuc to do one of the case studies (Miller and Cone 1979, 1980). I was sure they would do an excellent job of demonstrating the need for longitudinal studies by anthropologists to help us begin to understand acceptance and effective use of new technologies and the behavioral changes involved. Miller documented the fact that the original initiative had come from the community leader, who became a believer in modern medicine when he was successfully cured of dysentery in the local clinic and was shown a large preserved tapeworm, which he never forgot. It was this leader who persuaded the men in the community to sign commitments to the collective project to install latrines. However, the behavioral changes in excreta disposal in this isolated Maya community, even with a strong leader, could never have been made so dramatically without the cooperation of the women, who worldwide are the trainers of the children and the daily managers of human waste (Chiñas 1971; Elmendorf 1981a, 1981b, 1981c, 1981d, 1982a, 1982b, 1982c, 1983a, 1983b, 1983c, 1984a, 1984b, 1984c, 1984d, 1990, 1992a; Elmendorf and Isely 1982, 1983a, 1983b).

When the latrine pits were filled with excreta, the tile roofs were lovingly and carefully removed, tile by tile, to cover latrines rebuilt over new pits. Fresh fruit from trees donated by INI and planted in the old composted pits provided added incentives to rebuild the latrines. The Maya belief in excreta as a valuable resource was proven by the luscious fruit (Elmendorf 1972, 1973a, 1973b, 1975, 1982a, 1982b, 1982c)! Latrine use became an expected daily habit, the local accepted custom. Twenty years later, when new families moved into Yucatan, they were criticized if they did not adapt. Some 75 percent of the households were still proudly using the latrines, even though they had had to be relocated three or four times during the nearly four decades (Miller and Cone 1979, 1980).

WOMEN'S ROLES

What relationship is there between the worlds of Mayan women in Guatemala and in Mexico? Between those of Chiapas and Yucatan? Between their old worlds and their worlds today? Even before the 1970s, historians and anthropologists were disagreeing about the roles of women in pre-Columbian society. For instance, Gamio (1916) and Leon-Portilla (1958) felt that women had high prestige and played vital cultural roles. However, Bandelier concluded that among the Aztecs, "the most degrading epithet that could be applied to any Mexican, aside from calling him a dog, was that of woman" (as cited in Gruening 1928, 623).

A Swedish anthropologist, Anna-Britta Hellbom, disagreed in 1967, just as Traci Ardren does in her book *Ancient Maya Women* (2002), as the codices are rapidly being decoded and new techniques for skeletal analysis are developed. Hellbom noted in her excellent pioneer study using material from the Aztec codices that the roles of women were reported in their relationships as wives and mothers. In many of the pictures that richly illustrate the book, a woman is portrayed in her home, with husband and children, with beauty and dignity. "There is no explicit evidence of disdain of any of the Aztec woman's activities" (Hellbom 1967, 229). In fact, she went on to point out that women are given great importance in the codices, even though few historians or analysts have noted this. While the father was educating his sons, the mother was educating her daughters to perform their special roles, which were in most cases confined to the home and were sex-defined but not, perforce, considered of less importance—just as I found in Chan Kom in the 1970s so vividly during the filming of the *ch'a chaak*, when I commuted between the two spheres—the forest with the men and the hearth and home with the women (Elmendorf 1973b). The *hetzmek*, a Maya ceremony, is performed for boys at four months of age, to signify the four corners of the milpa, the cornfield. For girls the ceremony is at three months, because of the three stones of the hearth—the center of the home, where women turn corn dough into tortillas, both basic and ceremonial food. Even though defined by gender, the roles seem equally important to the Maya. Both are related to corn—the staff of life, with related Mayan gods and beliefs (Elmendorf 1972, 1973a, 1973b, 1976a, 1976b; Merrill-Sands 1984; Teran, Rasmussen, and Cauich 1998; Vail 1989). Even today this traditional ceremony is still performed—with additional symbols for girls and boys—in the urban areas, including the capital city of Mérida and even Cancún, as described by Cervera in her presentation for the panel, Yucatec Maya: Political and Cultural Ecology at the 1997 annual meeting of the American Anthropological Association (Cervera 1997).

The interesting volume *Pre-Hispanic Maya Agriculture* (Harrison and Turner 1978), which focuses on the question of how the ancient Maya fed themselves, has no references to women. This book pointed out evidence that the early Mayan civilizations were not built solely on swidden (slash-and-burn) agriculture but included sophisticated drainage and storage systems. At the American Anthropological Association's centennial panel in honor of Margaret Mead's life, I recalled how her work influenced my own in Chan Kom (Elmendorf 2001a). Mead drew our attention not only to female and male roles, but also to the devastating results of ignoring the roles played by women in population control and food production, preservation, and preparation (Mead 1976, 10).

Esther Boserup, the famous Danish economist, had already pointed out women's various contributions around the world in "male and female farming systems in subsistence economy," past and present (Boserup 1970; 16–17). Over time, various scholars added specific details to gender issues, including Isabel Pozas (1959, 1972), Nancy Modiano (1968), Beverly Chiñas (1972), June Nash (1970, 1975, 1992), Cornelia Flora (1973), Marianne Schmink (1974), Carmen Diana Deere (1975), Ruth Dixon (1983), Deborah Merrill-Sands (1984, 1986), Vina Mazumdar (1989), Betty Faust (1988, 1998), Ellen Kintz (1990a, 1990b), and others. Much of the research was related to food production and agriculture. At the same time, United Nations specialized agencies such as the Food and Agriculture Organization (FAO) and International Labor Organization (ILO) came to understand that women in developing countries were contributing more than half the world input of food products. UNIFEM (United Nations Development Fund for Women, originally the United Nations Voluntary Fund for Women) announced that "50% to 80% of production, processing, and marketing of food are in the hands of women" (UNIFEM/USA 2001, 2). Boserup noted in a chapter, "Industry: From the Hut to the Factory," that Manning Nash had observed:

> Among the Mayan Indians in the Guatemala Highlands factory work was also considered "not quite fitting job" for women, but the prejudice is not an absolute one; when Latin America's largest textile mill was set up in the region, the Indian women accepted employment in large numbers, since they could thereby add handsomely to family income, and recruited small girls as domestic servants to take care of their children during factory hours. (Boserup 1960, 116–117)

MAYA WOMEN AND ENVIRONMENTAL SANITATION

In the field of environmental sanitation, I refer briefly to several published case studies where Yucatec Mayan women have made great differences in project outcomes. Michael McGarry, an engineer working for the Canadian International Development Research Centre (IDRC), joined me to carry out a brief study in Chan Kom. The results were published by the World Bank in two books (Elmendorf and McGarry 1980, 1982). Even though both men and women in many families in the village expressed interest in having an alternative to the customary use of the bush or patio for excreta disposal, their choices were sex stereotyped. We designed a questionnaire with Mayan-speaking interviewers administered separately to selected groups of men and women in Chan Kom. Much to everyone's surprise, except mine, the women's choice was for a colorful water-flush toilet inside or adjacent to their thatched huts, while the men preferred a composting latrine outside. Each expressed her or his own needs: the women wanted the toilet where they bathed. "Then we could use the dirty water to flush with, and the waste would go out into the garden through a drain underground, so our plants will grow."[2] Both men and women thought of garbage and excreta as valuable resources to be worked into the kitchen garden or to be composted and recycled. Men, women, and children bathe each night at sunset with a bucket, saving the water for watering plants or even animals. By 1998, a few homes had separate cement rooms, some with tile floors, for bathing, but only one, Eusebia Cime's, had a porcelain toilet.

The lead article in the December 1998 issue of *WATERfront* (UNICEF), picked up on the themes which I and others summarized in 1996 at the twenty-second WEDC (Water, Engineering and Development Centre) conference in Delhi, Reaching the Unreached (Elmendorf 1996), and later during the seminar

Water Is Life at the new International Women's University in Hannover, Germany (Elmendorf 2000b), highlighting the key roles women play in selecting and managing improved environmental sanitation in their homes and communities. Women are an integral part of a global effort to improve maternal and child health and decrease diarrheal disease, the leading cause of infant deaths, claiming an estimated 6,000 children every day.

In addition to the case study by McGarry and me, three others were carried out in Maya communities for the World Bank series Appropriate Technology for Water Supply and Sanitation, one in Yalcuc, Chiapas, by Frank Miller and Cynthia Cone (1979, 1980), mentioned earlier, and two in Guatemala—one in Sololá, San Pedro de la Laguna, by Bertha Salinas and Roberto Caceres (1980), and the others in El Quiché, by Patricia Buckles (1976, 1979, 1980). Buckles also helped me edit the 1980 executive summary of the case studies, which was published in Spanish, French, and Portuguese (Elmendorf and Buckles 1980; see also Black 1998).

Another case study discusses a low-cost housing project on the southern outskirts of Mérida, inhabited mostly by Maya who had migrated from the abandoned henequen fields for employment in the state capital (Schmink 1984, 1989). A Yucatec woman architect, Josefina Mena, proposed her invention, the Sistema Integral de Reciclamiento de Desechos Organicos (SIRDO; the name means "integrated system for recycling organic waste"; see Mena 1992, 1993) for use in a public housing project. This technique, which uses both human and vegetable waste, not only improves sanitary conditions but also offers possibilities for income generation through reuse or sale of the resulting compost. For several years against many odds, the Mayan women, through hard work and innovative strategies, made it a successful cooperative, which they called Mochuc-Baex, a Mayan term meaning "let's get together" (Schmink 1984, 5).[3] The resulting compost was even called "beautiful dirt," or *tierra bonita* in Spanish. In 1989, in an analysis of women in development projects worldwide, Vina Mazumdar of India selected the SIRDO project as the most interesting (Elmendorf 1990), explaining her reasons as follows:

> [T]he SIRDO project from Mexico mobilized women to control their environment through an age-old tradition devised by women: the conservation and recycling of waste. The instinct is old but in this case it addresses a problem that is a creation of our present day civilization. Extension of the debate about women and development from a narrow definition of economics to encompass the wide range of challenges facing humanity today such as environmental problems will be . . . the great leap forward. (Mazumdar 1989, 217)

MAYA WOMEN IN COMMUNITY DEVELOPMENT

Yucatec Maya women today are proud, as were their ancestors in pre-Classic times, of their domestic production (Nash 1992). Hammocks are an essential part of Maya life, not just as the coolest, most comfortable way to sleep but also as a sales item (Littlefield 1976). Even Hammacher-Schlemmer carries Mayan hammocks, which are produced by several microenterprises in the Yucatan. In fact, in 1978, when Brigitte Jordan and I participated in an INI training course for Mayan midwives, we strung our hammocks in the same area where we all slept—in fact, in the same space we used to show Jordan's films of birth in four cultures (Jordan 1978; Elmendorf 2002).

In 1975, Virve Piho quoted Sahagún describing activities during the Spanish Conquest: "The commoner woman was especially active in the market as retail vendor but also, in times, functioning as wholesale merchant. And of course women excelled in the productive process as in the fields of weaving and feather-work" (p. 200).

Today the native dress, often crafted by women, is still a symbol of cultural identity. It provides a viable microenterprise for women (Nash 1992). Carmen L. Pettersen (1976, 95), in her beautifully illustrated *Maya of Guatemala: Life and Dress*, says of the women of Chichicastenango, "The embroidery and brocading of the clothes record their origin, the creation of the race, their history and how they settled in the highlands."

As for marriage, the husbands of both Mayan and Aztec women were chosen for them for life by the gods. Adultery was punishable by death for men as well as women. Women were strangled; men were stoned (Wolf 1958, 34–39). Are the punishments different but equal? Both result in death. On the other hand, Anna Macias (1971, 1) said, "In general, Mexicanists in sympathy with the emancipation of women usually argue that in Aztec and Maya society the position of women was inferior. A feminist in Yucatan asserted in 1916 that Maya women had always been 'beasts of burden and the slaves of men.'"

This was the feeling I had as I watched the Maya women in the marketplace in San Cristóbal de las Casas in 1950, but it was tempered by my first observation of the sharing by that Maya man and woman in private as they walked alone from their village with their loads of wood. In fact, this vivid picture of private versus public roles greatly influenced my research designs in fertility determinants and excreta disposal, both taboo subjects among many cultures (Elmendorf and Buckles 1980; Elmendorf and Isely 1983; Elmendorf and Ma 1984, 1985). To understand present behavior and prepare information on options for appropriate technological change, one has to be able to communicate in that private world by learning the special vocabularies and cultures as well as gender issues.

In 1973, many felt that so little was left of the pre-Columbian cultures that we must depend on reconstructing what must have been from archaeological remains and a few findings such as the Bonampak murals, figurines from Jaina, and the codices, which themselves had rarely been interpreted from the woman's perspective. An interesting exception was a study of the Tzeltal women, including pregnancy, birth, rituals, and women's roles within the home and community in Oxchuc, Chiapas, in the 1940s (Otero-Lombardo 1944). Other sources were isolated Indian groups in Mexico, many of whom had kept their language, culture, and customs amazingly intact (Redfield and Villa Rojas 1934; Villa Rojas 1945, 1969; Redfield 1950; Cámara-Barbachano 1966; Nader 1964). Efforts in the late 1960s and early 1970s such as those by Isabel Pozas (1959, 1972), Laura Nader (1969), June Nash (1970), Pozas and Pozas (1971), Rosario Castellanos (1973), Lourdes Arizpe (1975), Marie-Odele Rivera (1976), and Gertrude Duby (1992), along with my studies of the Yucatec Maya, shed light on what the pre-Columbian societies might have been like through present-day ethnographic studies. I have reviewed some of this history for Instituto Nacional de Antropología e Historia (INAH), particularly the contributions of Cámara-Barbachano (Elmendorf 2000a.)

Perhaps there were also differences among the Maya due to the personalities or administrative dictates of sixteenth-century Catholic bishops such as Diego de Landa in Yucatan (Tozzer 1941) and Bartolomé de las Casas in Chiapas. And of course there were differences in social stratification among the various Maya groups even within regions. In Chichicastenango, according to Pettersen, the designs of the embroidery on the women's blouses identify six different classes with clearly defined rights and responsibilities. For men, "The embroidery on the trouser panels shows the wearer's standing in his community from childhood to youth to virility to old age" (Pettersen 1976, 91–96).

During the last twenty years, many scholars who have continued this reinterpretation of the archaeological and historical materials have confirmed and extended earlier findings. For example, the section on Latin America and the Caribbean in *Gender and Anthropology* (Morgen 1989, 14) discusses "the gender differential impact of colonialism and slavery on indigenous peoples and the ways gender intersected with race and class in the ruling ideologies." In fact, many Mexicans and others from the 1950s into the 1970s felt that portrayals of "machismo" and the "culture of poverty" (Lewis 1949, 1956, 1962) were only a part of the picture (Lenero-Otero 1968; Acosta 1973; Fromm and Maccoby 1970; Stevens 1973a, 1973b) of Mexico's mosaic of family life and structure. Even though many of the Indian groups experienced great poverty, there was often sharing and respect and a different quality of life (Urquidi 1963), as had also been noted earlier by Leon-Portilla (1958). For example, Indian women's coping strategies as street vendors in Mexico City, after they had migrated from their villages to the city, were innovative. In traditional clothing, usually with small children helping, they sold nuts and candies on street corners while attempting to make a home with children and friends wherever they could find space (Arizpe 1975). In 1975, as Mexico prepared to host the first United Nations Conference on Women, the vendors were banned from the streets even though no other place was offered for them to sell their wares. Similarly, in 1990 the Mayan women vendors were banned from selling in the main plaza of San Cristóbal de las Casas, the cultural and commercial center of Chiapas, and headquarters since 1950 of INI's regional center. Even though they were not unionized, they organized a protest with members of their social and kin networks and marched in the *zocalo* (town center). They eventually regained their vending territory, but were constantly harassed and sometimes robbed and raped. During the summer of 1991 in Chiapas, Kathleen Sullivan, a student of June Nash, designed and obtained funding for an unusual program. She trained a group of the street vendors, mostly young women, in the martial arts. They were excited to learn and have since shared their new knowledge with their younger siblings and other vendors. Kathleen Sullivan (1995, 2) advises, "Since there is often an urgency in the lives of the people with whom we work, demanding a humane response, we should follow Tylor's advice 'to leave the world a better place than we found it.' Assume the challenge and intervene." Perhaps providing an example of participatory action research (see Gibson 2002), both Gary Gossen (1994) and June Nash (1994, 1997, 2001) have written on the resistance of present-day Mayas, as has Fabri (1990), continuing Victoria Bricker's previous history of Maya resistance movements and rebellions (1981) and reports by Mayas in Allan Burns's (1983) collection of the oral literature of the Yucatec Maya.

My own research among the Yucatec Maya of Chan Kom was reported in several publications, including *Nine Mayan Women: A Village Faces Change* (Elmendorf 1976a), with a Spanish edition (Elmendorf 1973a), and various articles in addition to many project reports and presentations of papers (see references). I am delighted that so many younger scholars such as Alicia Re Cruz (1989, 1991, 1996), Betty Faust (1988, 1991, 1998a, 1998b), Ellen Kintz (1990a, 1990b), Grace Bascopé, and Susannah Glusker (1994a, 1994b), who have chapters in this volume, as well as Deborah Merrill-Sands (1984, 1986), Christine Eber (1990), Antonella Fabri (1990), Susan Morfit (1994), and Marcia Good-Maust (1994, 2000; Good-Maust et al. 2002) have found my material useful as they enrich the field of women's studies and understanding of Mayan women. Equally important are my male colleagues, such as E. N. Anderson, David Forrest, and John Frazier, who also have chapters in this volume.

My original research, *Mayan Women and Change* (Elmendorf 1972), was published by Ivan Illich at CIDOC (Center for Intercultural Documentation) as a case study to use in the seminar on quality of life in which Paulo Freire was participating.[4] Paulo Freire's now classic writings, especially *Pedagogy of the Oppressed* (Freire 1968), were essential to the discussions. My study demonstrated how women made personal sacrifices for their children, cooperated with each other, and planned strategies to survive better and maintain traditions they valued. Many of the Mayan women felt that development did not always improve the quality of their lives. For instance, in 1978 I invited four of the Chan Kom women, all of whom were making embroidery for sale, to visit a new sewing cooperative supported by a World Bank loan in Sotuta, a nearby community just north and west of Chichén Itzá. With great excitement and effort, they got permission to leave town. Dressed in their whitest and most elegantly embroidered *huipiles*, they piled into my little Volkswagen Beetle and we were off. After a long visit with good questions from them and lots of interest in the expensive electric sewing machines, they decided on the way home that they were better off selling hand-embroidered products, using the traditional designs they knew and loved, than going into debt to buy fancy machines and changing their designs, a decision many others have made as they get encouragement and support from their peers. Others have chosen to buy treadle sewing machines that are less expensive to purchase and maintain than the electric ones, while still speeding the work. As we neared Chichén Itzá on our way to Chan Kom, I asked how many had visited the ruins. Only one had. "Would you like a soda and a tour of the ruin?" They were delighted. It was wonderful to be with them as they scampered up the pyramid with obvious pleasure and pride.

In 1976, the World Bank supported a field study by Deborah Merrill-Sands and me to analyze the impact of various development projects initiated in Chan Kom between 1971 and 1976 on the socioeconomic and behavioral patterns of men and women (see Elmendorf and Merrill-Sands 1976). Merrill-Sands concentrated on the roles of children and supplements to subsistence agriculture, including beekeeping and cattle (Merrill-Sands 1984, 1986). Comparisons were made with baseline data obtained earlier by both researchers and from other studies. The Mayan women and men felt that the opening of the road was the single most important change. With it, of course, came many of the other changes, both for better and for worse (Elmendorf and Merrill-Sands 1976, 1977, 1978). As the women had told me during my fieldwork, "Now a doctor can come, but strangers will bring new diseases."

POPULATION GROWTH AND ENVIRONMENTAL IMPACT

Alfonso Villa Rojas and I carried out applied research between 1977 and 1979 in Chan Kom and elsewhere in Yucatan, on issues of fertility determinants among the Maya and on socioeconomic variables affecting their decision making. The project, Fertility Determinants Among the Maya, was sponsored by the Research Institute for the Study of Man (RISM) and funded by the Office of Population, Bureau for Development Support, United States Agency for International Development (USAID). The research was designed to augment the basic knowledge and behavior of Mayan women in Yucatan in relation to demographic and technological changes, and the impact of these changes on fertility. Field work in Yucatan consisted of studying the changing roles and status of women; household economy and the contributions of men, women, and children; spatial mobility, including migration; communication; and decision making. The variables included the value of children, the level of education, contact with the outside world, and acceptance of new technologies, including family planning. The purpose of the study was to understand how Mayan women view family planning and how they handle the cultural constraints and supports as well as to provide basic information for an improved module for maternal and child health and family planning. This study was greatly enriched by the participation of Deborah Merrill-Sands, Brigitte Jordan, and several Mexican colleagues including several Maya, Hilaria Mass Colli (Mexico's first Mayan woman anthropologist), Elsie Cime Hu (a primary health care worker), and numerous traditional midwives, especially the ones in Chan Kom (Elmendorf 2002).

Regular meetings were held with staff of the Office of Health in Mérida, the capital, with Dr. Gilberto Balaam, then director of the regional center of INI in Valladolid, and with Dr. José Estrella,[5] then director of maternal child health at the regional hospital in Valladolid, both to share relevant findings and to discuss mutual concerns. Sociologist María del Carmen Elu Lenero, then national director of maternal/child health in the Mexican Ministry of Health in Mexico City, visited us in Yucatan and invited us to several of the national and regional conferences relevant to the topic. Elu Lenero (1969, 1971, 1972, 1977) used her excellent national studies of the Mexican family and those of her husband, Luis Lenero-Otero (1968), which had had ongoing support from the Ford Foundation, in combination with culturally specific findings such as ours to redesign national and regional strategies and programs. Elu Lenero was particularly interested in ways of incorporating training of health volunteers and midwives in rural communities.[6] As I broadened my network of midwives, I fully realized that the most knowledgeable, the wife of the *h-men*, was considered by some of my other friends to be a witch (for a similar case see Faust 1988).

On June 13, 1999, more than twenty years after our research, the lead editorial in the Sunday *New York Times*, "Latin America's Birth Surprise," noted the reduced birthrate in Latin America, detailing the change in Mexico. "In the mid-1960s, Latin American women averaged more than 6 children. Today the average is 2.9." A family-planning campaign initiated by the government in 1974 established a network of clinics and a campaign of billboards touting the benefits of small families. I wondered as I read that article about whether the success was due in part to the outstanding efforts made by Elu and others during those formative years, including information from and about Mayan women. Earlier Graciela Duce had told *The News* (Mexico City, September 1, 1993, "Mexico to Continue Reducing Population Growth") about Mexico's amazing strides in sex education and its accomplishments in "reducing its population growth from 3.4% to 1.9%." Much of this success is related to the determined research, training, and publishing in gender issues and sex education by Esther Corona Vargas (1973). I first met Corona in 1952 when she was a student at the National University of Mexico majoring in gender issues—for which she was ridiculed! Later she was invited to write the first sections on sex education for the Ministry of Education's official textbooks, which were not used in the next administration. During the same period, the family planning clinics so bravely started by Dr. Edris Rice-Wray were often padlocked and their staff threatened. With a grant from Syntex and private donations, she tested birth control pills synthesized from the Mexican yam. Friends of mine, both Americans from the U.S. foreign colony and Mexicans from the city and villages, and I were willing guinea pigs. Neither we from the United States nor the peasant women from Mexico, some facing death from another pregnancy, felt that we were being misused. I well recall my distress at hearing some feminists, at a lecture in Sarasota, Florida, in the 1970s, accuse Dr. Rice-Wray of endangering the lives of Third World women. To the contrary, we all felt confident about the careful medical attention all of us were receiving as participants in an important—very important—experiment.

Another factor was the RISM/USAID Yucatan project, Fertility Determinants Among the Maya (Elmendorf 1979a, 1979c; Elmendorf and Villa Rojas 1979). Accumulated data from the longitudinal study of thirteen families in Chan Kom showed that one of the most significant results of the opening of a feeder road in 1971 was the postponement of marriage by young women as they became aware of options other than motherhood and increased their decision-making power with cash incomes (Elmendorf 1980); once married, they chose to have smaller families and to have them later in life (Elu Lenero 1969, 1971, 1972, 1977; Elmendorf 1979a, 1979b, 1979c).

In 1971, there was only one unmarried woman over seventeen in Chan Kom. Marriage was still the state expected of all adults a few years after puberty, just as it had been in 1930 (Redfield and Villa Rojas 1934, 95). Marriages were still arranged by families, with little or no participation of the young people, and often without their knowledge (Elmendorf 1973a, 1973b, 1975, 1976a, 1976b). By 1978, the pressure to marry had decreased dramatically: of all the women seventeen and over, twenty-five were unmarried in 1978, compared to only one in 1971. Even the three who were over twenty-two did not yet seem worried about not having married, nor did their parents; the age of becoming an "old maid" had quickly changed from seventeen years to twenty-two-plus! Most significantly, young women were being given, or had assumed, responsibility for choosing their mates, a role traditionally held by their parents (Elmendorf 1979a, 1979b, 1979c, 1980).

One concrete result of research done by RISM in collaboration with USAID was a revised information sheet on contraceptive choices for Yucatec Maya, which I prepared in cooperation with the Programme for the Introduction and Adaptation of Contraceptive Technology (PIACT; Elmendorf 1976c) and pretested with women in Chan Kom (Elmendorf 1979b). Another concrete result was a request from Dr. Gilberto Balam, director of the regional center for INI, that I participate in their training courses for empirical midwives. I was honored to be invited but amused, since I had never seen a live birth except those of my son and daughter in the mirror above my bed as cesarean sections were performed. However, I had learned a lot from the Mayan midwives in Chan Kom as we discussed problems in the village related to maternal and child health.

At my suggestion, Dr. Balam agreed to invite Brigitte Jordan, a medical anthropologist whose research for *Birth in Four Cultures* (Jordan 1978) included valuable insights, as well as an excellent ethnographic film of a Yucatec midwife in action (Jordan 1975). During the evening activities at our training course, we watched films showing births in Holland, Sweden, and the United States, as well as Yucatan, which the midwives thoroughly enjoyed and critiqued. Many, if not most of them, had never before seen a film. When Jordan stopped the projector to discuss a point just after the birth of a baby, there was great tension in the room. Someone said loudly, "But the placenta must come too!" There was laughter as the movie continued. The film stimulated an open, frank discussion of problems they had encountered and solutions they had employed in their own work.

This training course was one of the first opportunities to show the importance of involving traditional midwives in interactive research and nondidactic teaching (Jordan 1989). INI acted on our recommendations to incorporate Jordan and her films into their next training course. We also recommended starting a Mayan school of midwifery. There was no financial support for this but a great deal of interest. Decades earlier, Isabel Kelly had published seminal papers suggesting anthropological perspectives in the training of midwives (Kelly 1955, 1956), but there was very little follow-up then. CARE donated technical equipment in the 1950s to graduates of some early training courses organized by the Mexican Ministry of Health (Elmendorf 2002).

In 1977, when Jordan and I presented our preliminary findings to the USAID office in Washington, one of the reviewers scathingly asked, "Why are we talking about midwife training? What in the world does conception have to do with contraception?" (Elmendorf and Jordan 1977). There is certainly a relationship between conception and contraception within the minds and lives of the people, especially the women (Cosminsky 1983). The role of midwives as "cultural brokers" in both conception and contraception has emerged more clearly in later years with integrated approaches to holistic health care (Good-Maust 1994). In fact, in June 2002, CASA Professional Midwifery School hosted the first International Conference on Professional Midwifery and Self-Regulation on their campus in San Miguel de Allende with support from USAID (Elmendorf 2002). Among the keynote speakers was Robbie Davis-Floyd (2000, 2001), who is completing an international study of midwifery and maternal child health, focusing in part on the frightening increase of cesarean sections worldwide. Marcia Good-Maust, Miguel Güemez Piñeda, and Robbie Davis-Floyd have written *Midwives in Mexico: Continuity, Controversy and Change* (2002; see also Good-Maust 2000).

At the request of CEDPA (Center for Education, Development, and Population Activities), I drafted a working paper, "Environmental Strategies: A Partnership Approach in Mexico" for expanding their family planning programs to include environmental projects. Later this paper became a background document at the UN preparatory conference in Mexico preceding the 1994 United Nations International Conference on Population and Development (ICPD) in Cairo.

At a workshop in San Miguel de Allende, representatives from the CEDPA network in Mexico enthusiastically discussed their environmental efforts. Esther Corona's group at Asociación Mexicana de Educación Sexual (AMES,

Mexican Association of Sex Education) was intrigued to add environmental projects, particularly alternative waste and garbage disposal, which could improve family living conditions in the fringe areas of Mexico City. Squatter families were being threatened with evacuation on the basis of accusations that they were polluting the watershed. Wealthy families were building weekend homes on these same restricted hillsides, but they could get (buy) permits. In fact, three of the communities of squatters, who had built homes on the volcanic slopes of Ajusco, had tried SIRDO, a solution for sewage and wastewater disposal, which was designed by a Yucatecan and tested first in Mérida, as noted earlier. Local enthusiasm here, just as in Mérida, was not enough (Elmendorf 1992a, 1992b; Schmink 1974, 1984). Lack of any continuing support from city or state government, coupled with the fact that the squatter families had no ownership of the land, made it impossible to keep the system operating in spite of spurts of community activity and dynamic leadership. SIRDO was working perfectly in the wealthier suburbs of Mexico City, where cooperative housing projects of young professionals were delighted with the technology (Elmendorf 1992a, 1992b). Simpler home-based techniques with environmental sanitation were being tested by AMES with the hope that the squatter families would not be evacuated.

For instance, at a workshop in Oaxaca in 1992, Discovering the Linkages Between Environment and Population (sponsored by InterAction and the Pathfinder Fund), the importance of women's roles in this as well as other household and community issues relating to population and the environment became more and more obvious (Elmendorf 1992a, 1992b). Just as the Maya women in Chan Kom were the focal points for delivery of basic needs, so were women in many cultures (Elmendorf 1984b, 2001a).

By 1999, CEDPA partners from Mexico and around the world were even more active at the ICPD+5 PrepCom (preparatory conference to the five-year follow-up of the International Conference on Population and Development) than they had been at the 1992 UNCED (United Nations Conference on Environment and Development) in Rio de Janeiro. Women were challenging the United Nations to help maintain and advance the implementation of the Cairo program of action adopted by 179 nations at the 1994 ICPD:

> Governments, international agencies, non-governmental organizations and other concerned parties should undertake timely and periodic reviews of their development strategies with the aim of assessing progress towards integrating population into development and environment programs that take into account patterns of production and consumption and seek to bring about population trends consistent with the achievement of sustainable development and the improvement of the quality of life. (Preamble, Program of Action, ICPD)

Earlier, in 1986, with a team of four Guatemalan consultants, I prepared a midterm evaluation of a $2.5 million USAID grant to two NGOs, CARE and

Agua de Pueblo (ADP; see Buckles 1976), for designing projects to bring water and sanitation to Guatemala's rural populations, 75 percent of whom are without adequate coverage. Approximately 100 Maya communities were included in the 172 villages, with beneficiaries totaling 116,000 people. The team's midterm evaluation recommended more community involvement in planning and implementing services in order to have sustainability and effective use by the communities. Research clearly demonstrated that failures in existing systems upset users, especially women, just as the Mayan women in Chan Kom had complained to the regional authorities when the mayor had not repaired the pump in their village. Women disliked having to go back to traditional sources such as wells with muddy water, polluted rivers, and crowded public standpipes. The Mayan women in the craft communities were well aware of the time lost in collecting water and of the difficulty in maintaining environmental and personal sanitation in the home (Elmendorf et al. 1986). The funding was stopped temporarily until the so-called software was added by both agencies (Elmendorf 1984a).

Community participation was a very basic concept in this program, as it had been in the 1958 CARE/AFSC pilot project in Santa María, Mexico. I will never forget my feelings when, twenty-six years later, I heard an elderly Mayan man stand in the town meeting in the Chimaltenango area of Guatemala and say in broken Spanish, "We, the men in San Felipe, want to say to Agua de Pueblo that not only will we dig trenches to lay the pipes to connect the wonderful clean water we are getting from the mountain spring to our homes, but we promise to dig the trenches to the homes of all the widows here who have lost their husbands in this dreadful conflict." This event, in that Guatemalan village, returned as vividly in my memory as a video when I read the following:

> In the Chimaltenango area of Guatemala, 450 widows of the thirty-six-year civil war that ended in 1996 now support their families. They weave traditional local patterns into tablecloths, placemats, and napkins, using subtly hued, colorfast yarns selected to appeal to a sophisticated consumer market. Mayans believe the Goddess Ixchel created weaving; illustrated native manuscripts show her using the same backstrap loom the weavers use today. (Gianturco and Tuttle 2000, 240)

In 1997, UNIFEM sponsored the first Congress of Mayan Women in Mérida. At this meeting, representatives of bilateral and international agencies and governmental and nongovernmental agencies had an opportunity to listen to Mayan women expressing their own concerns and suggestions about projects they would like to develop to improve the quality of their lives and their children's lives. The bilingual report of the conference, *Na Molay* (Espinosa 1997), gives a detailed and colorful presentation of the Mayan women's views, remarkably similar to those expressed to me in my research for *Nine Mayan Women* (Elmendorf 1976a).

UNIFEM's work with Mayan women has involved collaboration with the United Nations Development Program (UNDP) through its Small Subsidies Programme (PPS) on the Yucatan Peninsula. It also receives support from ROSDES, A.C., a network of nongovernmental organizations with wide-ranging experience in the region, which in conjunction with the PPS has lent support both to the women's congress and to UNIFEM's subsequent activities in the region (Espinosa 1997, 5).

In November 1999, the Gulf Coast Florida chapter of UNIFEM formed its first travel committee. According to the president, Florence A. Jackson, "The short-term goal of this committee is to visit UNIFEM projects to gain first-hand experience with the women these projects support. The long-range goals include gaining the ability to speak knowledgeably about UNIFEM sites in an effort to solicit more funds for these and other projects." Not only did we meet the director of the UNIFEM Embroidery Project (*Bordado Maya*, Maya Embroidery), we met with the women in two village workshops and a butterfly studio in Izamal. Many of the beautiful images these women create with their embroidery are of local flowers, birds, and butterflies. Some even include iguanas, fish, crocodiles, and turtles. Views of village life with surrounding forest are also done. The subsequent journey, in November 2001, was fully booked within a few months. It began with organizational meetings in June (Elmendorf 2001b). During these activities, I gave two talks and several workshops on the history and future of work with Maya women (Elmendorf 2001a, 2001c). More visits to Mayan women for chapters in other countries could turn out to be an effective way to build a broader base of support for the microenterprises of Mayan women as well as better understandings of their hopes and dreams.

As Margaret Snyder, UNIFEM's founding director, noted in reviewing our strategy that "We would have to devise the types of development co-operation that would deal with women as agents of change rather than as victims of circumstances.... Recognizing women as economic agents, the Fund does not merely provide for material needs but insists that people participate in its activities in order to increase their self-reliance and gain self-confidence" (Snyder 1995, xvi).

It was the Yucatec Mayan women in Chan Kom who originally put me in touch with the larger world of both Mayan women and Mexican women—and later women around the world. The Spanish edition of my basic microstudy was translated and published in 1973 by the Mexican Ministry of Education as *La Mujer Maya y el Cambio*. Several thousand copies were distributed to all public schools and universities in Mexico, including rural schools in villages like Chan Kom. In 1973, while I was doing a research project on roles and status of Mexican women for the Ford Foundation (Elmendorf 1975), I was invited by the Inter-American Indian Institute to be an observer at the Organization of American States (OAS) seminar on the problems of Indian women. Each country in the Americas was represented by two Indian women leaders. The

Mexican government had donated copies of *La Mujer Maya y el Cambio* for the seminar and asked me to make the presentation. I felt honored, but I was very aware of the possibility of being suspect or mistrusted as an anthropologist. The two official delegates from the United States were an Apache and the assistant to Manuel Lujan, the representative of Indian Affairs in the U.S. Congress. The two indigenous leaders from Mexico were a Purépecha from Michoacán and a Yucatec Maya, Olga Marie Yah Uc, a teacher from Felipe Carrillo Puerto in Quintana Roo.

The scene of this seminar is vivid to me still. The Inter-American Commission on Women (Comisión Interamericana de Mujeres, CIM) of OAS had arranged the meetings for the Indian women leaders in a very formal setting in the auditorium of the Bank of Guatemala in Guatemala City. But this did not intimidate the Indian women who, at the end of the formal openings and distribution of the agenda, requested permission to meet together at lunch and to have the first half hour of the afternoon session for reporting. An Otovalena from Ecuador chaired the afternoon session. She opened it by saying, "We want to thank you very much for bringing us together here, but since you invited us as Indian women leaders to discuss our problems, we feel we should prepare our own agenda." As she dramatically tore up the CIM agenda distributed earlier, she requested time to call on all the delegates around the table so that they could decide together on problems they wished to discuss during the seminar. The CIM organizers were visibly shaken, but I was quietly impressed by the performance, so like the 1968 student rebellions I had experienced vividly as the wife of a college president at the Democratic National Convention in Chicago. The new agenda prepared by the delegates covered most of the same issues—basic needs, literacy/education, environmental and family health—but added legal protection and political participation. The Ecuadorean said proudly, "I plan to be the first woman president of Ecuador."[7]

The Mayan women in Guatemala had boycotted the seminar at the beginning because there had not been a preparatory national seminar in Guatemala as there had in most of the other countries. When they were assigned seven official representatives instead of two, five from Quetzaltenango and two from Quiché, they joined the group. Gradually eighteen observers from Guatemala were allowed in. Some were indigenous women leaders. Others were nuns, teachers, social workers, psychologists, linguists, and journalists. The concerns *of* and *for* the highland Maya in Guatemala were evident at the seminar. On May 20, 1993, large headlines on the front page of the local newspaper, *La Tarde*, read, "In America, there exists racial, religious and sex discrimination against Indian groups" with a subtitle "Of great concern to the UN and the Catholic Church." Included was a quotation from Dr. Gonzalo Rubio Orbe, director of the Inter-American Indian Institute, "When the Indigenous people become organized, you are going to see a very different population and I hope that it will not degenerate into violence."

VIOLENCE AND RESPONSE

Violence toward the Mayan population in Guatemala accelerated in the late 1970s and into the 1980s, with repeated struggles to find solutions throughout the 1990s. Entire villages disappeared, and estimates of civilian casualties have ranged from 50,000 to 100,000 or more. A Maya refugee from this violence, Rigoberta Menchu, was awarded the 1992 Nobel Peace Prize for her efforts. Her presentations drew large audiences, particularly in 1993, the UN Year of Indigenous Peoples, and at the 1995 UN Women's Conference in Beijing. In 1997, she was one of the negotiators concerned with the Maya uprising in Chiapas. Even though critiques rage in the academic and political world about her two books (Menchu 1983, 1992), *Me llamo Rigoberto Menchu y así me nacia la concencia* and *I, Rigoberto Menchu: An Indian Woman*, there is no doubt that her memoirs raised the level of world consciousness about the atrocities against the Maya in Guatemala, as noted in an article, "Exchange," published in the June 14, 1999, issue of *The Nation*. A woman filmmaker sensitive to the voices of Maya women among the refugee settlements has made an excellent film, *Mayan Voices: American Lives* (Carrescia 1992), while Allan Burns (1993) has documented the lives of Guatemalan Maya women and men living in exile.

Soon after the final peace accord had been signed between the Guatemalan government and the guerilla movement, ending thirty-five years of civil war, the Human Rights Watch reported that courageous workers were celebrating their achievement in bringing to the bargaining table Phillips-Van Heusen's foreign managers of local *maquiladoras* (foreign-owned factories). As reported in an article in *AWID News* (May 1997, vol. 2, no. 1), Liz O'Connor, of the STITCH delegation to Guatemala,[8] was quoted as saying that after six long years of organizing, "these women surprised their bosses by sustaining a long, difficult campaign. They've sent a message of hope to workers throughout Guatemala and the U.S." It is estimated that about 70,000 workers, many of them Mayan women, are employed in the decades-old clothing assembly industry in Guatemala, which exports to brand-name U.S. companies.

Another very moving moment was a meeting with representatives of three Mayan villages during my 1985 fieldwork in El Salvador, supported by USAID. One elderly man stood up proudly to say how hard their communities had worked together to bring potable water down from a mountain source, water that they were now sharing. "To have safe water near our homes is important to us and our wives and children, but the most important thing that we have learned, by cooperating on the water project, is to know and trust our neighbors again" (Elmendorf 1995). Having known the sometimes violent struggles in both El Salvador and Peru, Paul Doughty recommended "A return to the basics of community development is in order if societies are to become self-reliant, enduring and rewarding places to live" (Doughty 1987, 156; see also Doughty 2002). I would add that community includes women. The importance of women's role in making technological changes and making sanitation sustainable and effectively used cannot be overemphasized (Elmendorf 1959, 1978, 1981a, 1981b, 1981c, 1981d, 1982a, 1982b, 1982c, 1983a, 1983b, 1983c, 1984a, 1984b, 1984d, 1984e, 1984f, 1984g, 1985, 1990, 1992a, 1992b, 1994a, 1995, 1996, 2000a; Elmendorf and Isely 1982, 1983a, 1983b).

The violence in Guatemala continued in spite of the peace accord, and new violence erupted in Chiapas on January 1, 1994. More than four decades earlier, in 1952, the government of Mexico, through INI, as mentioned earlier, had undertaken an ambitious set of programs to address the needs of the Maya population of Chiapas (Acosta 1996a, 1996b). The effort sought to make available the benefits of development without imposing the aggression and dominance so common in dealings between more- and less-powerful groups. It drew on the finest and most creative minds and won international recognition for innovation and creativity. Where did the vision go? Somehow, between the optimistic view of the 1950s and the dark assessment of 1994 following the beginning of the Zapatista revolution (of the Ejército Zapatista de Liberación Nacional, EZLN, or Zapatista National Liberation Army), the vision of a mutually satisfactory relationship between modern Mexico and indigenous Mexico disappeared. June Nash (2001, 1) has presented an analysis of the EZLN military strike on January 1, 1994: "Faced with the possible loss of their collective resource base as marginalized cultivators in the global society, they said, 'Basta!' (Enough!)." During the years between the formation of INI and the rebellion, peaceful demonstrations of both workers and peasants had been met with military action and arrests. As she stated in an earlier presentation:

> As they assert their rights to land, political participation, and their distinctive cultures, Mayas of Chiapas are redefining the modernist ideals of justice, liberty, and democracy for a postmodern age. Accustomed to cultural diversity, they have learned to live without attempting to eradicate or dominate the others in their midst. Their vision of progress still contains the communal values found in mythopoetic traditions from the preconquest period. But far from being primordial remnants of the past, these values have been enacted continually in everyday life since the conquest and may offer a model for pluriethnic and pluripolitical institutions as we enter the third millennium. (Nash 1994, 1)

There are more than 600 million indigenous persons belonging to more than 5,000 different ethnic groups, estimated at between 10 and 15 percent of the world population, while their traditional homelands cover 20 to 30 percent of the land surface of the earth (Clay 1993, 65–68). June Nash (2001, 29) finds that, "the internationalization of indigenous political movements combined with the convergence of indigenous women's political participation is revolutionizing the way global trends are being defined and resisted." In spite of attempts to reach a peaceful settlement, the conflict continues in Chiapas but has increasing numbers of witnesses, both members of Mexican organizations and outsiders (Acosta 1996a, 1996b; Nash 1994, 1997; see also chapter 11, this volume).

CONCLUSION

Now, in the new millennium, while violence against the Maya continues to occur sporadically in parts of Chiapas and Guatemala, it is not visible in the Yucatan. The threats here are more subtle: population growth, environmental degradation, loss of access to resources, poor sanitation, and water pollution. These are global problems, compounded in the Maya case by the conflict between tradition and modernity and by lack of political power to participate in the planning of programs affecting Maya communities. Even when the Maya are included, their informed consent is often compromised by (1) lack of access to information concerning environmental impact and (2) the pressure of immediate threats to their livelihood. Communal resources were somewhat protected under Mexico's agrarian reform laws, but 1992 changes to the national constitution now allow privatization, including sale of land to foreign agribusiness firms and tourism enterprises.

When in the mid-1960s the Mexican government, from air-conditioned buildings in Mexico City with the latest computer techniques, designed their newest tourist attraction, Cancún, they did not really consider the impact on either the environment or the Mayas, many of whom were unemployed because of the collapse of the henequen market. People from all over Mexico flooded into the peninsula, taking jobs, both menial and professional, that Yucatecans who needed employment could have done well. Now tourists from all over the world flood the area, bringing some of the worst aspects of our global consumer society and seriously threatening the environment they come to enjoy.

Many Mayas have new sources of income but, like the young people from Chan Kom who headed into Cancún as day laborers, dishwashers, and maids, only a few of them have been able to become their own bosses. Through their entrepreneurship, some own small stores, others have trucks and construction businesses, and all have much larger cash incomes than their relatives back in the village of Chan Kom, which they consider home. Some of the Cancún residents have become active in the still powerful Partido Revolucionario Institucional (PRI, Institutional Revolutionary Party), while most of the families of the *milperos*, those still planting their own cornfields, are allied with the center-right Partido para Acción Nacional (PAN, National Action Party). The resulting political conflict between these two groups became violent several years ago. The mayor of Chan Kom was threatened and his office barricaded. Alicia Re Cruz (1996, 37) identified "two phenomena that are seen [locally] as factors exacerbating their social crisis: *la Politica* and *el Gilberto. La Politica* refers to the crisis of authority and power in the village initiated by the increasing interference of external government forces in community affairs. The environmental devastation caused by *el Gilberto* (Hurricane Gilbert) perpetuated the existing perception of crisis." More recently, a gradually dawning awareness of shortages in common property resources has created new concerns among some people, while others focus on securing more income for their own families in order to purchase the new "necessities."

The new necessities are part of a process of increasing interaction with a wider world. Unfortunately, this process has carried with it a feeling of inferiority, of shame at being Maya. This was pointed out to me by the visiting regional supervisor of schools when I was in Chan Kom briefly in 1998. I had stopped by the mayor's office to leave a video copy of the ethnographic film I had taken in the early 1970s so it could be kept in the village archives. In fact, the mayor had put a large television and chairs in the area outside his office so that Alicia Re Cruz and I could show the video and tape people's reaction during the November fiesta of San José. Several of the village families with relatives in Cancún who had their own videotapes had asked whether they could make copies for their families and, more important, the schoolteachers had asked whether they could use them in the school. The supervisor said, "You have no idea how important it is to have material about our traditions for the schools. We need more information about the Maya to help our young people know about their past, to have pride in their ancestors, to regain their dignity. I will make copies for other schools." Perhaps some of the excellent films about the Maya from the National Geographic Society and other sources could have Spanish soundtracks for use on television and in schools in regions where there are Maya living, or perhaps I should say where there are living Maya.

The Yucatec Maya women are increasingly gaining experience in politics. At a 1994 seminar at the University of Florida, a Yucatec Maya, Arcelly Cob Cumi, the *diputada* from Maxcanú, discussed her experience as a Maya woman politician. Lucilda Cime, a defeated PAN candidate for mayor of Chan Kom, joined us at the 1994 meeting of the Society for Applied Anthropology in Cancún to relate her experiences and in 2001 ran for mayor again but lost. After all, Yucatan gave women the right to vote in 1922 (thus being the first state in Mexico to do so). Four women had been elected deputies by 1923 (Macias 1971). In 1993, both the governor of the state of Yucatan and the mayor of Mérida, its capital, were women, even though from different parties—one from the PRI, the other from PAN! This too was a first for Mexico—female heads of government in both a state and its capital city. In 1997, with the help of a large financial donation from the Japanese, UNIFEM sponsored a conference for 150 Yucatec Maya women leaders in Mérida to design projects and programs for empowering them.

Will the Mayan women in Yucatan, in Chiapas, in Guatemala, and in the other countries continue to mediate between tradition, development, and protection of their environment? Can they, with their mothers and grandmothers, pass on the values they treasure, the skills and knowledge they have, to their children and grandchildren? Can Mayan women, with their families, help all of us learn what sustainable development can be, as together we respect our shared environment?

I feel just as I did in the 1970s, in the seminar preceding the 1975 United Nations Conference on Women in Mexico City, that in spite of continuing violence and increasing degradation of the environment, Mayan women in general, like the women of Chan Kom,

> are more fortunate than their sisters in other rural areas; they have a rich cultural heritage which provides a sense of dignity, identity, and stability, and a marketable skill which provides a means of livelihood without having to leave the village. For others, unfortunately, the only alternative seems to be migration to the slums, where they have little choice other than to be come domestics, low-paid factory workers, prostitutes, or street vendors.... Thus far the women of the village have maintained their self-esteem and cultural identity while at the same time accepting—and even initiating—change that they feel will be beneficial for them and their children. (Elmendorf 1976b, 201)

Recently, the Mayan women of Chiapas have been making news, nationally and internationally, first through their military involvement in the EZLN in Chiapas and later through their active participation in the construction of a pluralistic civil society. They also participate in the House of the Writer in San Cristóbal de las Casas (Laughlin 1997). An associated theatre group developed *Todos para Todos* (*All for All*), an outstanding dramatization of the present conflict between Mayas and Ladinos (outsiders or non–Maya speakers) and between development and the environment as epitomized by the animals and spirits of the forest. It has been well received in tours of the United States and Mexico. In the drama, the women, although outnumbered by men three to one, participate actively in the decision making about how to solve their domestic and community problems. A women's group has also prepared their own theatrical productions with Mimi Miriam (Myers 1997).

Since the 1950s, Mayan women have been in the forefront of concern for the environment, working to improve sanitation and delivery of clean water, while improving infant and child health and reducing population growth through family planning. As other aspects of their environment increasingly suffer from inappropriate development, they can be expected to participate in decision making concerning appropriate technologies to conserve resources for future generation, as well as to protect their communities from pollution.

ACKNOWLEDGMENTS

I would like to thank again the women of "the village that chose anthropologists," (Serrie 2001, 10) for sharing their hopes, their dreams, and their fears with me so freely. I thank them not only for myself, but for many others—researchers, engineers, planners, and especially women in developing countries. From their original openness in the 1930s to questions from Robert Redfield and Alfonso Villa Rojas through Redfield's 1950 restudy, *A Village That Chose*

Progress: Chan Kom Revisited, on to those of myself and others into the present, they have been wise, critical evaluators of the appropriateness of new technologies—from contraceptives to latrines, from loans for fancy pigs to intricate sewing machines or irrigation systems—over a period of seventy years. As they struggle to continue living creative, proud, peaceful lives within their community, I hope they will be able to pass on to their children and grandchildren some of the wisdom they inherited from their ancestors as they face political, social, and economic pressures on their culture and environment.

NOTES

1. The film, which was directed by Benito Alazraki, with Walter Reuther as photographer, was released by R. K. Tompkins in time to win honorable mention at the 1959 Cannes film festival, thanks to Peter Rathvon, the producer. The original 16 mm film, complete with full documentation, is now in the National Human Film Archives at the Smithsonian Institution.

2. People in Chan Kom are still waiting, and are even more willing to cooperate than they were twenty years ago, but there has been no local agency to give technical assistance, even though INI offered to help back in 1978.

3. Even though the successful SIRDO system in Yucatan ultimately was abandoned, there are four extremely successful cooperatively run SIRDOs in suburban Mexico City. After several years of research, Mena improved on the design and implementation of this innovative, environmentally friendly system, which she has presented to the World Bank and to the Pan-American Health Organization.

4. Later, Illich published *Retooling Society* (1972) following his earlier study, *Outwitting Development* (1968), as part of questioning the development models. There is also a more recent book addressing this issue by Godotti (1994), *Reading Paulo Freire: His Life and Work*.

5. Dr. Estrella played a very important role in this research. He was the first Mayan to be trained as a physician in the United States after graduating from the Presbyterian boarding school in Xochenpich. (The guest house in Xochenpich was my "launching pad" for my research in Chan Kom. I stayed there the night before I left for Chan Kom to start my research.) This was the same Dr. Estrella to whom the first CARE delivery of medical equipment was made in 1952. And it was to his small clinic in Pisté that I had taken Mayan midwives with their patients (and the husbands of their patients) in my little Volkswagen from Chan Kom during difficult deliveries.

6. A similar study on highland Maya was done both before and after the Guatemalan earthquake (Glittenberg 1976, 1994). Both Paul Doughty and I visited some of the communities for the evaluation. An extremely sensitive study of violence in Guatemala during this period was done by Victor Montejo (1987).

7. Maybe she will be president later. For many years I have tried to keep in touch with various participants in this seminar. I would like to recommend that someone sponsor another seminar with as many of these women as possible to discuss the same issues now, nearly thirty years later.

8. A national network of North American women supporting Guatemalan women organizing to improve their work lives can be reached at 4933 S. Dorchester, Chicago, IL 60615, stitch@igc.apc.org.

REFERENCES

Much of the unpublished material cited below may be found in the Mary Elmendorf Collection, Archives of the University of Florida, Gainesville, and National Anthropological Archives of the Smithsonian Institution (see Finding Guide for Mary Elmendorf Papers at the University of Florida, on the Internet at http://special.uflib.ufl.edu/guides/Elmendorf.htm). The collection includes manuscripts, lists of cassette tapes of interviews with individual women (including midwives), government documents, photographs, and so on. Related material on video can be obtained from the Human Studies Film Archives in the Smithsonian Institution.

Acosta, Mariclaire. 1973. Los estereotipos de la mujer mexicana en la cultura de masas: El caso de las fotonovelas. Presentation for panel, Man in the Americas, American Association for the Advancement of Science meeting, Mexico City.

———. 1996a. Los rostros del racismo. Mexico: Comisión Mexicana de Defensa y Promocion de los Derechos Humanos, A. C. July.

———. 1996b. Human Rights in Mexico: An Overview. Mexico: Comisión Mexicana de Defensa y Promoción de los Derechos Humanos, A. C. July.

Andrews, Joann. 1993. *Kambul Biodiversidad* (Newsletter of Pronatura Península de Yucatan, Mexico) 3: 131.

———. 1997. Act for the future. *Kambul Biodiversidad* 7: 2–3.

Ardren, Traci. 2002. Death became her: Images of female power from Yaxuna burials. In *Ancient Maya women*, edited by Traci Ardren. Walnut Creek, CA: AltaMira Press, 68–87.

Arizpe, Lourdes. 1975. *Indigenas en la Ciudad de México: El caso de las Marías* [Indian Women in México City: The Case of the Marías]. SEP/Setentas, Mexico: Secretaria de Educación Pública.

Black, Maggie. 1998. *Learning what works: A twenty-year retrospective on international water and sanitation cooperation*. Washington, DC: UNDP and World Bank, Water and Sanitation Program.

Boserup, Esther. 1960. Industry: From the hut to the factory. In *Labor, commitment, and social change in developing areas*, edited by W. E. Moore and A. S. Feldman. New York: St. Martin's Press, 106–117.

———. 1970. *Women's role in economic development*. London: Allen and Unwin.

Bricker, Victoria. 1981. *The Indian Christ, The Indian King*. Austin: University of Texas Press.

Buckles, Patricia. 1976. Documentation of the CARE/Ministry of Health Potable Water and Latrine Project: Chijtinimit. Mimeograph. CARE.

———. 1979. The introduction of potable water and latrines: A study of two rural communities in Guatemala. In *Seven case studies of rural and urban fringe areas of Latin America*, edited by Mary Elmendorf. World Bank Research in Appropriate Technology in Water Supply and Sanitation. Washington, DC: World Bank, 35–94.

———. 1980. Chijtinimit and Chontala, El Guiché. In *Socio-Cultural Aspects of Water Supply and Excreta Disposal*, edited by Mary Elmendorf and Patricia Buckles. Appropriate Technology in Water Supply and Sanitation, Vol. 5. Washington, DC: World Bank, 13–17. (Mimeo, 1978, translated into Spanish, French, and Portuguese.)

Burns, Allan. 1982. *An epoch of miracles: Oral literature of the Yucatec Maya*. Austin: University of Texas Press.

———. 1993. *Maya in exile*. Philadelphia: Temple University Press.
Cámara-Barbachano, Fernando B. 1966. *Persistencia y cambio cultural entre Tzeltales de los Altos de Chiapas*. Mexico: INAH.
Carrescia, Olivia. 1992. *Mayan voices: American lives*. Ethnographic video. LASA Award, N.Y.
Castellanos, Rosario. 1973. *Mujer que sabe Latín*. Mexico: SEP/Setentas.
Cervera, María Delores. 1997. Becoming a Yucateca Maya: The first three years of life. Paper presented at the American Anthropological Association Annual Meeting, Washington, DC.
Chiñas, Beverly L. 1971. Women as ethnographic subjects. In *Women in cross-cultural perspective*, edited by L. Jacobs. Urbana: University of Illinois Press, 21–31.
———. 1972. *The Isthmus Zapotecs*. New York: Holt, Rinehart and Winston.
Clay, Jason W. 1993. Looking back to go forward: Predicting and preventing human rights violations. In *State of the peoples: A global human rights report on societies in danger*, edited by Marc S. Miller. Boston: Beacon Press, 64–71.
Corona Vargas, Esther. 1973. Sex education. Presentation for panel, Man in the Americas, American Association for the Advancement of Science (AAAS), conference, Mexico City.
Cosminsky, Sheila. 1983. Traditional midwifery and contraception. In *Traditional medicine and healthcare coverage*, edited by Robert Bannerman, John Burton, and Ch'en Wen-Chieh. Geneva: WHO, 143–162.
Daltabuit-Godas, Magali. 1992. *Mujeres Mayas*. Mexico: Instituto de Investigaciones Anthropologicas.
Dauber, Roslyn, and Melinda L. Cain. 1980. *Women and technological change in developing countries*. Boulder, CO: Westview Press.
Davis-Floyd, Robbie. 2000. Global issues in midwifery: Mutual accommodation or biomedical hegemony? *Midwifery Today* (March): 12–16, 68–69.
———. 2001. La partera profesional: Articulating identity and cultural space for a new kind of midwife in Mexico. *Medical Anthropology* 20: 185–243.
Deere, Carmen Diana. 1975. The division of labor by sex in agriculture: Peasant women's subsistence production on the Minifundios. Ph.D. dissertation, University of California, Berkeley.
Dixon, Ruth. 1983. Land, labor, and sex composition of the agricultural labor force. *Development and Change* 14: 347–372.
Doughty, Paul. 1987. Against the odds: Collaboration and development of Vicos. In *Collaborative research and social change: Applied anthropology in action*, edited by Donald D. Stull and Jean J. Schensul. Boulder, CO: Westview Press, 129–157.
———. 2002. Ending serfdom in Peru: The struggle for land and freedom in Vicos. In *Contemporary cultures and society of Latin America*, edited by Dwight Heath. Prospect Heights, IL: Waveland Press, 225–243.
Duby, Gertrude. 1992. Los Lacandónes: Su pasado y su presente. In *Homenaje a Alfonso Villa Rojas*. Mexico: Instituto Chiapaneco de Cultura (Cultural Institute of Chiapas), 407–417.
Eber, Christine. 1990. Before God's flowering face: Juana, a Tzotzil Maya shaman. Paper presented at Annual Meeting of American Anthropological Association, New Orleans.
Elmendorf, Mary. 1953. CARE, Que es CARE? Mimeograph. New York: CARE.

———. 1959. *World Our Hands Can Make*. Technical notes for files. Washington, DC: Smithsonian Institution.

———. 1960. *Final report to CARE board of trustees on CARE de Mexico pilot program 1952–1960*. New York: CARE.

———. 1966. Community participation: Volunteers as facilitators. Presentation, Peace Corps Training, Arecibo, Puerto Rico, for University of Florida.

———. 1967. The women governors of the Dominican Republic: Tools and techniques for new solutions. Report to the Overseas Education Fund of the League of Women Voters for the U.S. State Department.

———. 1972. *Mayan Women and Change*. Cuaderno no. 81. Cuernavaca, Mexico: CIDOC (Centro de Intercultural Documentacion).

———. 1973a. *La mujer maya y el cambio*. Mexico: Secretaría de Educación Pública, SEP/70s.

———. 1973b. Mayan women, their daily lives and hopes. Paper presented at panel, Man in the Americas, American Association for the Advancement of Science, June.

———. 1975. The Mayan woman and change. In *Women cross-culturally: Change and challenge*, edited by Ruby Leavitt. The Hague, Netherlands: Mouton, 111–128.

———. 1976a. *Nine Mayan women: A village faces change*. Rochester, VT: Schenkman.

———. 1976b. The dilemma of peasant women: A view from a village. In *Women and development*, edited by Irene Tinker and M. Bo Bransen. Washington, DC: Overseas Development Council, 184–201.

———. 1976c. Suggestions for a revised bulletin on family planning for the Yucatec Maya. Presentation at Programme for the Introduction and Adaptation of Contraceptive Technology (PIACT).

———. 1977a. The changing roles of Mayan women in the home and in the market place. Paper presented in the First Mexican/Central American Symposium on Research About Women, El Colegio de México, Mexico, November.

———. 1977b. The many worlds of women: Mexico. In *Women: Roles and status in eight countries*, edited by Janet Giele and Audrey Smock. New York: John Wiley, 129–172.

———. 1978. Public participation and acceptance. In *Environmental impacts of international civil engineering projects and practices*; edited by Charles G. Gunnerson and John M. Kalbermatten. New York: American Society of Civil Engineers, 68–70.

———. 1979a. Changing roles and the status of Maya mothers and daughters in relation to marriage and the family in the Yucatan. Paper presented at symposium, Changing Attitudes and Behavior Patterns of Yucatec Maya Regarding Marriage and Fertility. Annual Meeting of Society for Applied Anthropology, Philadelphia, PA, March.

———. 1979b. Changing family structures and development. In *Impact on the child: Population growth and urbanization in Latin America: The rural-urban interface*. Mimeographed conference report, no pagination. Lansing: Michigan State University.

———. 1979c. Anita: A Maya peasant woman copes. In *Learning about peasant women*, edited by S. Zeidenstein. *Studies in Family Planning* 10 (November): 11–12.

———. 1980. Changing roles of Maya mothers and daughters. In *Women and technological change in developing countries*, edited by Roslyn Dauber and Melinda L. Cain. Boulder, CO: Westview Press, 149–178.

———. 1981a. Women, water and waste: Beyond access. In *Health needs of world's poor women*, edited by Patricia W. Blair. Washington, DC: Equity Policy League,

92–95. (Previously presented at the 1980 UN Conference on Women in Copenhagen at the Workshop on Women and Health).
———. 1981b. *Women, water and the decade.* WASH Technical Report 6. Arlington, VA: Water and Sanitation for Health.
———. 1981c. Women, water and waste: Beyond access. In *Proceedings of 7th WEDC conference: Water, people and waste in developing countries.* Loughborough, England: Water, Engineering, and Development Centre (unpaginated offprint in Archives).
———. 1981d. Beyond access to improved water supply and sanitation: Impacts on and of women and children. In *The impact of interventions in water supply and sanitation in developing countries*, edited by James D. Lindstrom. Washington, DC: U.S. Agency for International Development, 93–103.
———, ed. 1982a. *Seven case studies of rural and urban fringe areas of Latin America.* World Bank Research in Appropriate Technology in Water Supply and Sanitation, Vol. 8. Washington, DC: World Bank.
———. 1982b. Water and sanitation-related health constraints to the economic development of communities. In *Proceedings of conference on women, health and international development.* Lansing: Michigan State University.
———. 1982c. "Women's Roles in Irrigation Projects." Presentation, Workshop in Irrigation and Human Welfare, Rutgers University, New Brunswick, NJ.
———. 1982d. Women: The underused human resource. *Educación Médico y Salud* 16 (4). Washington, DC: PAHO/WHO.
———. 1983a. Women as managers of human waste: Training for new roles, training for old roles. In *Seminar on human waste management for low-income settlements*, Vol. 1. Bangkok: Environmental Sanitation Information Center, Asian Institute of Technology, 114–124.
———. 1983b. The role of women as participants and beneficiaries in water supply and sanitation programs. WASH Technical Report No. 11. Arlington, VA: USAID.
———. 1983c. Women's roles in irrigation projects. Presentation at Rutgers University Workshop in Irrigation and Human Welfare, New Brunswick, New Jersey.
———. 1984a. International development, women and water. Seminar, Virginia Polytechnical Institute and State University, Blacksburg, Virginia, February.
———. 1984b. Generalizations and linkages: Pertinent research in the literature on women and sanitation. Paper presented at the Institute for Training and Research on Women Seminar, Cairo, March 13–16.
———. 1984c. Pertinent research: Generalization and linkages drawn from a preliminary review of the literature on women in water and sanitation. Paper presented at INSTRAW Interregional Seminar, Cairo, March.
———. 1984d. Women's roles and changing society: The case of water. Paper presented at Research Alternatives and Priorities Panel, Cornell University conference, Women and Development: Redefining the Issues. Ithaca, New York, April.
———. 1984e. The softer side of software. In *Proceedings of 10th WEDC conference, water and sanitation in Asia and the Pacific* (Singapore, August). Loughborough: WEDC (unpaginated, in Archives).
———. 1984f. Women as change agents: A look at action research. In *Proceedings of IDRC seminar, women's issues in water and sanitation* (Manila, September). Ottawa: IDRC.

———. 1984g. Women, water, and waste: The hidden decision. Paper presented at the Interfama Conference, Singapore.

———. 1985. Water: The basic resource. Paper presented at panel, Resources and Development, at the Association of Women in Development (AWID) Conference, Washington, DC, November.

———. 1990. *The international drinking water and sanitation decade (1980–1990) and women's involvement*. Geneva: World Health Organization.

———. 1992a. Report to IDRC on socio-cultural obstacles to acceptance of the SIRDO. Mimeo.

———. 1992b. Trip report to IDRC: A failed success, the SIRDO in Yucatan. Mimeo.

———. 1992c. Trip report–participation in (1) global forum at UNCED in Rio de Janeiro, and (2) World Urban Forum in Curitiba, Brazil, for CEDPA, Washington, D.C.

———. 1993. Una mirada retrospective: los diversos mundos de la mujer en México: 1973–1993. Presentation, El Programa Interdisciplinario de la Mujer, El Colegio de Mexico, September.

———. 1994a. Environmental strategies: A partnership approach for Mexico. CEDPA Working Paper. Washington, DC: CEDPA.

———. 1994b. Chiapas, the highland Maya, 1950–1994. Paper presented at Panel in Honor of Alfonso Villa Rojas. Society for Applied Anthropology, Cancún.

———. 1995. Lessons learned. One-page handout, incorporated in report *Evaluación del proyecto: Mejoramiento de los servicios públicos de El Salvador*, edited by R. Witherell, L. B. Taylor, J. L. Roush, and M. L. Elmendorf. Cambridge: Cambridge Consulting Corporation for USAID/El Salvador, annex.

———. 1996. Priorities, challenges, and strategies: A feminine perspective, 1975–1995. In *Proceedings of the 22nd WEDC Conference, Reaching the Unreached: Challenges for the 21st Century*, New Delhi (unpaginated paper).

———. 1997. Review of CARE Programs 1945–1965. Collection of historical documents of CARE, prepared for their archives, includes reference to the film *World Our Hands Can Make*, based on 1958 CARE/AFSC Well-Drilling Project.

———. 2000a. The United Nations and me: Ten specialized agencies viewed through the eyes of a consulting anthropologist. Presentation, United Nations Day Celebration, Sarasota, FL, October 23.

———. 2000b. Water is life: A view through the eyes of women. Review paper in the proceedings of the seminar Water Is Life at the new International Women's University, Hannover, Germany, August. See http://www.Int-Frauenuni.de.

———. 2001a. Nine Mayan women and technological change in developing countries. Paper presented in the Mead Centennial Session: Margaret Mead's Legacy: Children, Families and Contemporary Human Issues, Annual meeting of the American Anthropological Association, Washington DC, November.

———. 2001b. Background information on the Yucatec Mayan women and the world they live in. Presentation, Gulf Coast Chapter of UNIFEM. (See Espinosa 1997. Papers available in Elmendorf Collection, University of Florida, Gainesville.)

———. 2001c. Women, water, and UNIFEM. Paper presented at panel, The Power of One, Annual Meeting of the United States Committee for UNIFEM, Sarasota, FL, June.

———. 2002. Memories of midwife training in Mexico, 1952–2002: Perspectives of an 85-year-old anthropologist. Prepared for the First International Conference on

Professional Midwifery and Self-Regulation. San Miguel de Allende, Guanajuato, Mexico, June 19–22.
———, and Patricia Buckles, eds. 1980. *Socio-cultural aspects of water supply and excreta disposal.* Series: Appropriate Technology in Water Supply and Sanitation, Vol. 5. Washington, DC: World Bank.
———, and Raymond Isely. 1982. The roles of women in water supply and sanitation. *World Health Forum* 3: 229–230.
———, and Raymond Isely. 1983a. Public and private roles of women in water supply and sanitation. *Human Organization* 42: 195–204.
———, and Raymond Isely. 1983b. Water and Sanitation-Related Health Constraints on Women's Contributions to the Economic Development of Communities. WASH Technical Report 17. Arlington, VA: USAID.
———, and Brigitte Jordan. 1977. Preliminary presentation of RISM research project, Fertility Determinants Among the Maya, to USAID.
———, and Ma Yansheng. 1984. Strategy paper II: How to involve women? Women's Task Force of the IDWSSD Steering Committee. New York: UNDP.
———, and Ma Yansheng. 1985. *Lessons from the field: How to involve women.* New York: UNICEF.
———, and Michael McGarry. 1980. What is appropriate technology? A Maya village asks. In *Socio-cultural aspects of excreta disposal*, edited by Mary Elmendorf and Patricia Buckles. Appropriate Technology for Water Supply and Sanitation, Vol. 5. Washington, DC: World Bank, 34–36.
———, and Michael McGarry. 1982. A behavioral case study: Chan Kom, Mexico. In *Seven case studies of rural and urban fringe areas in Latin America*, edited by Mary Elmendorf. Appropriate Technology for Water and Sanitation, Vol. 8. Washington, DC: World Bank.
———, and Deborah Merrill-Sands. 1976. Socio-economic impact of development projects on Chan Kom 1971–1976. Executive summary and full report to Department of Development Economics, The World Bank, Washington, DC.
———, and Deborah Merrill-Sands. 1978. Socio-economic impact of development in Chan Kom, 1971 to 1976: A shortened version. Presented at Symposium on Regional Development, Yucatan, at annual meeting of Society for Applied Anthropology, Mérida, Mexico.
———, C. Solares, R. Alvarado, E. Hurtado, and L. Vielman. 1986. Mid-term evaluation of USAID-financed projects to Aqua del Pueblo and CARE Guatemala. International Science and Technology Institute (ISTI), Spanish and English reports. Washington, DC: USAID.
———, and Alfonso Villa Rojas. 1979. Changing patterns of fertility: The impacts of contraceptive technology on a Maya village. Presentation, USAID panel, Washington, DC.
Elu Lenero, Maria del Carmen. 1969. *Hacia donde va la mujer mexicana?* Mexico City: Instituto Mexicano de Estudios Sociales (IMES).
———. 1971. *Mujeres que hablan.* Mexico City: Instituto Mexicano de Estudios Sociales, IMES.
———. 1972. La mujer en México. *Punto Crítico* 8 (August): 26–32.
———. 1977. Socio-familial aspects of the working woman. Man in the Americas, American Association for the Advancement of Science Encuentro de Mujeres (Grupo 7).

Espinosa, Guadelupe. 1997. *Na molay, primer congreso de mujeres Maya, Na molay, first of Mayan women.* Mexico: UNIFEM, Oficina Regional de Mexico, Centroamericano, Cuba y Republica Dominica.

Fabri, Antonella. 1990. Tactics of resistance of Maya women in Guatemala. Paper presented at 89th annual meeting of the American Anthropological Association, New Orleans.

Faust, Betty B. 1988. When is a midwife a witch? A case study from a modernizing Maya village. In *Women and health: A cross-cultural perspective*, edited by P. Whelehan. Granby, MA: Bergin and Garvey, 21–39.

———. 1991. Maya culture and Maya participation in ecotourism. In *Ecotourism and resource conservation*, Vol. 1, edited by John Cuslar. Berne, NY: Association of Wetlands Managers.

———. 1998a. Cacao beans and chili peppers: Gender socialization in the cosmology of a Yucatec Maya curing ceremony. *Sex Roles* 39: 603–642.

———. 1998b. *Mexican rural development and the plumed serpent: Technology and Maya cosmology in the tropical forest of Campeche, Mexico.* Westport, CT: Greenwood.

Flora, Cornelia Butler. 1973. The passive female and social change: A cross-cultural comparison. In *Male and female in Latin America*, edited by Ann Pescatello. Pittsburgh: University of Pittsburgh Press, 103–109.

Freire, Paulo. 1968. *Pedagogy of the oppressed.* New York: Seabury.

Fromm, Erich, and Michael Maccoby. 1970. *Social character in a Mexican village.* New York: Prentice Hall.

Gamio, Manuel. 1916. *Forjando patria* [Forging a nation]. Mexico: Pomia.

Garrett, Wilbur E. 1989. La Ruta Maya. *National Geographic*, October, 424–477.

Gianturco, Paula, and Toby Tuttle. 2000. *In her hands: Craftswomen changing the world.* New York: Monacelli Press.

Gibson, Jane W. 2002. Where do you hang your hat? A case for participatory action anthropology. *Practicing Anthropology* 24: 39–45.

Giele, Janet, and Audrey Smock, eds. 1997. *Women: Roles and status in eight countries.* New York: John Wiley.

Glittenberg, JoAnn. 1976. A comparison of fertility in two highland Guatemalan towns. Ph.D. dissertation, University of Colorado, Boulder.

———. 1994. *To the mountain and back: The mysteries of Guatemalan highland family life.* Prospect Heights, IL: Waveland Press.

Glusker, Susannah. 1994a. From Chan Kom to Chicago, Carnegie to CONACYT. Paper presented at Panel in Honor of Alfonso Villa Rojas. Society for Applied Anthropology, Cancun, Mexico.

———. 1994b. Rights among the Maya: Background and issues. Paper presented at American Anthropological Association annual meeting, Washington, DC.

Good-Maust, Marcia. 1994. The urban Maya midwife as culture broker. Paper presented at National Symposium on Indigenous Knowledge and Contemporary Social Issues, University of South Florida, Tampa, FL.

———. 2000. Making bodies: Cesarean narratives in Mérida, Yucatan. Ph.D. dissertation, University of Florida at Gainesville.

Good-Maust, Marcia, Miguel Güemez Piñeda, and Robbie Davis-Floyd. 2002. *Midwives in Mexico: Continuity, controversy, and change.* Austin: University of Texas Press.

Gossen, Gary H. 1994. From Olmecs to Zapatistas: A once and future history of souls. *American Anthropologist* 96: 553–570.

Gruening, Ernest. 1928. *Mexico and its heritage*. New York: Century.

Harrison, Peter D., and B. L. Turner II, eds. 1978. *Pre-Hispanic Maya agriculture*. Albuquerque: University of New Mexico Press.

Hellbom, Anna-Britta. 1967. *La participacion cultural de las mujeres*. Stockholm: Ethnological Museum.

Illich, Ivan. 1968. *Outwitting the "developed" countries*. Cuernavaca, Mexico: CIDOC.

———. 1972. *Retooling society*. Cuaderno no. 80. Cuernavaca, Mexico: CIDOC.

Jackson, Florence. 2000. "President's Corner." UNIFEM/USA Gulf Coast Chapter, Florida, Vol. 5, issue 4, April.

Jordan, Brigitte. 1975. Commentary on ethnographic film, *Birth in four cultures*. Michigan State Film Archives.

———. 1978. *Birth in four cultures: A Cross-cultural investigation of childbirth in Yucatan, Holland, Sweden, and the United States*. Prospect Heights, IL: Waveland Press.

———. 1989. Cosmopolitical obstetrics: Some insights from the training of traditional midwives. *Social Science and Medicine* 28(9): 925–944.

Kelly, Isabel. 1955. El adiestramiento de parteras en Mexico, desde el punto de vista anthropologico. *America Indigena* 15: 109–117.

———. 1956. An anthropological approach to midwifery training in Mexico. *Journal of Tropical Pediatrics* 1: 200–205.

Kintz, Ellen. 1990a. Planting by the moon: The myths and reality of Yucatec Maya women. Paper presented at the 89th annual meeting of American Anthropological Association, New Orleans.

———. 1990b. *Life under the tropical canopy: Tradition and change among the Yucatec Maya*. Fort Worth, TX: Holt, Rinehart and Winston.

Lenero-Otero, Luis. 1968. *Investigación de la familia en México*. Mexico: Galve.

Leon-Portilla, Miguel. 1958. La mujer en la cultura indígena. *Nicaragua Indígena* 21 (July-August): 6–8.

Lewis, Oscar. 1949. Husbands and wives in a Mexican village: A study of role conflict. *American Anthropologist* 51 (October–December): 602–610.

———. 1956. A day in the life of a Mexican peasant family. *Marriage and Family Living* 18 (February): 3–13.

———. 1962. *The children of Sanchez*. New York: Random House.

Littlefield, Alice. 1976. *La industria de las hamacas en Yucatan, México*. Mexico: Estudio de Anthropología Económica, INI.

Lombardo-Otero, Rosa María. 1944. La mujer Tzeltal [The Tzeltal Woman]. Prologue by Alfonso Villa Rojas. Mexico: Privately published.

Macias, Anna. 1971. Mexican women in the social revolution. Paper presented at the American Historical Association Conference, New York.

Mazumdar, Vina. 1989. Political commentary. In *SEEDS: Supporting women's work in the third world*, edited by Ann Leonard. New York: Feminist Press, 213–217.

Mead, Margaret. 1976. A comment on the role of women in agriculture. In *Women and world development*, edited by I. Finley and M. B. Branson. New York: Praeger, 9–11.

Mena Abraham, Josefina. 1992. Transforming waste into flowers. In *Proceedings of Global Assembly of Women and the Environment, UNEP*. Washington, DC: World WIDE (unpaginated presentation).

———. 1993. Reporte final: IDRC/GTA SIRDO processo, 1989–92. Mimeo. Mexico: Grupo de Technologia Alternativa.

Menchu, Rigoberto. 1983. *Me llamo Rigoberto Menchu y así me nacía la concienca.* Barcelona: Argus Vergara.

———. 1992. *I, Rigoberto Menchu: An Indian woman in Guatemala*, edited and with an introduction by Elzabeth Bargos-Detray. London: Verso.

Merrill-Sands, Deborah. 1984. The mixed subsistence-commercial production system in the peasant economy of Yucatan, Mexico: An anthropological study in commercial beekeeping. Ph.D. dissertation, Cornell University, Ithaca, New York.

———. 1986. Farming systems research: Clarification of terms and concepts. *Explorations in Agriculture* 22: 87–104.

Miller, Frank C. 1965. Cultural change and decision-making: A Tzotzil example. *Ethnology* 4: 53–65.

Miller, Frank, and C. A. Cone. 1979. Latrines in Yalcuc: A twenty-year perspective. In *Seven case studies of rural and urban fringe areas in Latin America*, edited by Mary L. Elmendorf. Appropriate Technology for Water Supply and Sanitation, Vol. 8. Mimeo. Washington, DC: World Bank, 95–117.

———. 1980. Yalcuc, Chiapas. In *Socio-cultural aspects of water supply and excreta disposal*, edited by Mary Elmendorf and Patricia Buckles. Appropriate Technology in Water Supply and Sanitation, Vol. 5. Washington, DC: World Bank, 18–20.

Modiano, Nancy. 1968. A Chamula life. *Natural History* 77: 58–63.

Montejo,Victor. 1987. *Testimony: Death of a Guatemalan village.* Willimantic, CT: Curbstone Press.

Morfit, Susan. 1994. Portrait of a birth. Paper presented at National Symposium on Indigenous Knowledge and Contemporary Social Issues, University of South Florida, Tampa.

Morgen, Sandra. 1989. *Gender and anthropology: Critical reviews for teaching and research.* Washington, DC: American Anthropological Association.

Myers, Robert. 1997. Mayan women find their place is on the stage. *The New York Times*, September 28, p. AR 4.

Nader, Laura. 1964. An analysis of Zapotec law cases. *Ethnology* 3: 404–419.

———. 1969. The Trique of Oaxaca. In *Handbook of Middle American Indians*, Vol. 7, edited by Robert Wauchope. Austin: University of Texas Press, 400–416.

Nash, June. 1970. *In the eyes of the ancestors: Beliefs and behavior in a Maya community.* New Haven, CT: Yale University Press.

———. 1975. Resistance as protest. In *Women cross-culturally, change and challenge*, edited by Ruby Leavitt. The Hague: Mouton.

———. 1992. Producción domestica en el mercado mundial. In *Antropología Mesoamericana: Homenaje a Alfonso Villa Rojas*, edited by Victor Manuel Esponda Jimeno, Sophia Pincemín Deliberos, and Mauricio Rosas Kifuri. Tuxtla Gutierrez, Chiapas: Instituto Chiapaneco de Cultura, 155–187.

———. 1994. The revindication of indigenous identity: Mayan responses to state intervention in Mexico and Guatemala. Paper presented at the Society for Applied Anthropology.

———. 1997. The fiesta of the word: The Zapatista uprising and radical democracy in Mexico. *American Anthropologist* 99: 261–274.

———. 2001. *Mayan visions: The quest for autonomy in a world of globalization.* London: Routledge.

Otero-Lombardo, Rosa Maria. 1944. *La mujer Tzeltal.* Mexico: author.

Pettersen, Carmen L. 1976. *Maya of Guatemala: Life and dress.* Guatemala: Ixchel Textile Museum.

Piho, Virve. 1975. Life and labor of the female textile worker in Mexico City. In *Women cross-culturally: change and challenge*, edited by Ruby Rohrlich-Leavitt. The Hague: Mouton, 247–260.

Pozas, Isabel H. de. 1959. La posición de la mujer dentro de la estructura social Tzotzil. *Ciencas políticas y sociales* 5: 565–575.

———. 1972. Motherhood as a social institution. Presentation for panel, Man in the Americas, American Association for the Advancement of Science meeting, Mexico City.

Pozas, Ricardo. 1952. *Juan Pérez Jolote.* Mexico: Universidad Nacional Autónoma de México.

———. 1962. *Juan the Chamula.* Translated by Lysander Kemp. Berkeley: University of California Press.

Pozas, Ricardo, and Isabel H. de Pozas. 1971. *Los Indios en las clases sociales de México.* Mexico: Siglo XXI.

Re Cruz, Alicia. 1989. Conflicto social y tradición oral: Los modernos uayes y los aluxes tradicionales. In *Proceedings of primer congreso internacional de Mayistas.* San Cristóbal de las Casas, Chiapas, Mexico.

———. 1991. Modern times and the winds of transformation: The sacred and profane of Maya identity. Paper presented at the International Congress of Americanists, New Orleans, July 7–11.

———. 1996. *The two milpas of Chan Kom: A study of socioeconomic and political transformations in a Maya community.* Albany, NY: State University of New York Press.

Redfield, Robert. 1950. *A village that chose progress: Chan Kom revisited.* Chicago: University of Chicago Press.

Redfield, Robert, and Alphonso Villa Rojas. 1934. *Chan Kom, a Maya village.* Washington, DC: Carnegie Institute.

Rivera, Marie-Odele. 1976. *Una communidad Maya en Yucatan.* Mexico: Secretaría de Educación Pública, SEP/Setentas No. 261.

Salinas, Bertha, and Roberto Caceres. 1980. San Pedro de la Laguna. In *Socio-cultural aspects of excreta disposal*, edited by Mary Elmendorf and Patricia Buckles. Appropriate Technology for Water Supply and Sanitation, Vol. 5. Washington, DC: World Bank, 7–12.

Schmink, Marianne. 1974. Dependent development and the division of labor by sex. Paper presented at Latin American Studies Association fifth national meeting, San Francisco.

———. 1984. *Community management of waste recycling in Mexico: The SIRDO.* New York: Population Council.

———. 1989. Community management of waste recycling in Mexico: The SIRDO. In *SEEDS no. 8 Supporting women's work in the Third World*, edited by A. Leonard. New York: Feminist Press.

Scrimshaw, Susan. 1973. *Lo de nosotros: Pudor and family planning clinics.* New York: International Institute for the Study of Human Reproduction, Columbia University.

Serrie, Hendrick. 2001. Chan Kom: A village that chose anthropology. *Newsletter of the Society for Applied Anthropology* 12: 10.

Snyder, Margaret. 1995. *Transforming development: Women, poverty and politics*. London: Intermediate Technology Publications.

Stevens, Evelyn P. 1973a. Machismo and marianismo. *Society* 10: 57–63.

———. 1973b. The prospects of a women's liberation movement in Latin America. *Journal of Marriage and the Family* 35: 313–321.

Sullivan, Kathleen. 1995. Helping people help themselves: Teaching martial arts to women street vendors. *Practicing Anthropology* 17: 1–2.

Teran, Silvia, Christine H. Rasmussen, and Olivio May Cauich. 1998. *Las plantas de la milpa entre los Mayas*. Mérida: Fundación Tun Ben Kin.

Tozzer, Alfred M., ed. 1941. *Landa's relación de las cosas de Yucatan*. Papers of the Peabody Museum, Vol. 18. Cambridge: Harvard University.

Urquidi, Victor. 1963. The children of Jones. *America Indígena* 23: 359–363.

Vail, Gabrielle. 1989. Ritual continuity among the lowland Maya: An examination of rain-making, harvest, and Uayeb ceremonies. *Human Mosaic* 23: 21–44.

Villa Rojas, Alfonso. 1945. *The Maya of east central Quintana Roo*. Pub. no. 559. Washington, DC: Carnegie Institute.

———. 1969. The Maya of Yucatan. In *Handbook of Middle American Indians*, Vol. 7, edited by Robert Wauchope. Austin: University of Texas, 244–275.

———. 1978. *Los elegidos de Dios: Etnografía de los Mayas de Quintana Roo*. Serie Antropología Social, coleccion SEP/INI No. 56. Mexico City: Instituto Nacional Indigenista.

———. 1990a. *Notas sobre la etnografia de algunas comunidades tzeltales de Chiapas, 1939*. Translated by Sophia Pincemín. Tuxtla Gutierrez, Mexico: Instituto Chiapaneco de Cultura.

———. 1990b. *Etnografia tzeltal de Chiapas, modalidades de una cosmovision prehispanica*. Mexico City: Gobierno del Estado de Chiapas, Consejo Estatal de Fomento a la Investigación Difusión de la Cultura and Miguel Angel Porrua.

Vogt, Evon Z. 1969. *Zinacantan: A Maya community in the highlands of Chiapas*. Cambridge, MA: Belknap Press.

Witherell, Ronald A., L. B. Taylor, Jr., James L. Roush, and Mary Lindsay Elmendorf. 1995. Evaluación del proyecto: Mejoramiento de los servicios públicos de El Salvador. Report of Cambridge Consulting Corporation for USAID/El Salvador, Washington, DC, March.

Wolf, Eric R. 1958. The Virgin of Guadalupe: A Mexican national symbol. *Journal of American Folklore* 71: 34–39.

Related Audio-Visual Materials Available in the Elmendorf Collection at the University of South Florida

Some of these reference materials are still in the Mary Elmendorf Collection in Sarasota, Florida, destined for the University of Florida Archives.

Carrescia, Olivia (producer and director). 1992. *Mayan voices: American lives*. Film. New York: LASA Award.

Elmendorf, Mary Lindsay. 1993. *Daily life and ritual in a Mayan village: Chan Kom, 1971–76*. Washington, DC: Elmendorf Human Film Archives, Smithsonian Institution.

———, and Robert Champlain. 1959. CARE Mexican Film Story Outline: *World Our Hands Can Make*.

Teatro Petul. 1955. Puppet show, as described in *América Indigenista* no. 24, Mexico (Instituto Nacional Indigenista).

Todos para todos: All for all. 1994. Play. San Cristóbal de las Casas, Chiapas, Mexico: The House of the Writer.

World our hands can make. 1959. 29-minute film. Washington, DC: Smithsonian Institution.

4

Shifts of Strategies and Focus of the Conservation Efforts of Pronatura on the Yucatan Peninsula: A Personal History

Joann M. Andrews

Fostering the conservation of natural resources by directly involving civic associations is a concept that evolved in Mexico only during the late 1980s and 1990s. This chapter recounts the establishment of one of these nonprofit civic organizations, Pronatura Península de Yucatán (PPY). As the founder of PPY, I will describe the evolution of this nongovernmental organization (NGO) over a period of twelve years. The activities of the organization have been affected by policy shifts within the Mexican government as well as changes of focus by international conservation organizations. Responding to these shifts is a challenge. Care must be taken to protect resources needed for urgent local problems and to continue the assistance needed to stabilize new and fragile conservation programs in rural communities. This chapter relates the challenges faced by PPY in its attempt to combine community development and natural resource conservation in a context of shifting government policy and changing criteria for funding from international donor organizations.

At a recent workshop, members of the board of directors and the staff of PPY were asked their opinions on the overall performance of the organization. I candidly replied, "It still functions." That may appear to be an uninspiring response, especially coming from its founder and president at that time. Anyone, however, who has attempted to establish a private, nonprofit organization in Mexico during the last two decades would understand such a remark. The fact that PPY, the pioneering association devoted to the preservation of the Yucatan Peninsula's natural resources, not only survived its first two decades but has even expanded its activities and its reputation, is in itself an accomplishment.[1]

Indeed, in retrospect, establishing a private conservation organization with a structure strong enough to ensure permanency was the principal challenge, and the reason is evident to any keen observer. During the past two decades, Mexico has been brutally buffeted by major economic crises, which have severely limited the private sector's capacity to provide financial support for conservation efforts.

Simultaneously, the government has been unable to allocate sufficient funding for the adequate management of the areas under its protection. Suspicious of private groups' direct intervention, the public sector did not encourage, and sometimes even impeded, NGO initiatives. Fortunately, wary government officials gradually changed their attitude toward conservation NGOs, as successful working relationships developed.

The account that I present here is based on my impressions of an era (the 1980s and 1990s) in which interest in Mexican conservation was awakened and, little by little, private groups exerted greater influence over the destiny of Mexico's natural resources. Often ignored in such personal reminiscences are fortuitous factors that can be recognized only in retrospect. In my case, archeology and orchids were the two fields of study that triggered my concern for the peninsula's natural resources.

The saga of Pronatura's establishment on the Yucatan Peninsula begins, quite aptly, in the very heart of the peninsula—deep within the forests of southern Campeche, near the border with Guatemala. In 1968, my husband, the late Dr. E. Wyllys Andrews IV, launched a major archaeological expedition there, centered on the sites of Becán and Chicanná. Logistically, the venture was a real challenge, as the transpeninsular road from Escárcega to Chetumal was a veritable dust bowl in the dry season and miles of often-impassable red mud in the rainy season. We built our campsite near the town of X'puhil, which possessed several houses, a roadside restaurant, an unpredictable truck for hire, and an abbreviated airstrip. Our base consisted of five thatched-roof bungalows and a long, screened laboratory next to a murky *aguada*.[2] Lack of social services, electricity, water supply, telephone lines, and adequate transportation, not to mention fear of malaria, leishmaniasis, and dysentery, discouraged human settlement in the area and challenged our presence there daily. During the first couple of years, therefore, my husband and I shuttled back and forth from Mérida to X'puhil in his single-engine plane, loaded down with everything from machetes and trowels to food and water supplies. We had no flight instruments to guide us, nor did we have any natural landmarks to reassure us that we were directed toward X'puhil. Below our tiny craft lay only a vast expanse of forest, occasionally punctuated by an *aguada*.

Indeed, because of the many obstacles to human settlement in this area, its tropical vegetation was not marked by the many patches of abandoned cornfields seen elsewhere on the peninsula. The few lumber roads, if not cleared in the dry season, were rapidly covered within one rainy season by fallen trees and undergrowth and within a few years were undetectable. During our frequent passes over this wilderness, I was much impressed by this unmarred landscape. Several years later I would be equally struck by the rapid settlement of families along the transpeninsular road, once it had an all-weather surface.

My relationship with the forests deepened with my growing enthusiasm for studying the remarkable variety of orchids. Even before participating in the archaeological expedition in Campeche, I had already begun making a tentative list of Yucatan orchids, having identified about half of the twenty-one species of orchids cited by Paul Standley (1930) in his *Flora of Yucatan*. Examining orchids in this unexplored area of Mexico was like living in an earlier century, knowing that I could contribute new, basic information to the scientific world and at the same time discovering how much I could learn from the local people. Imagine my surprise and delight when I was able in just one field season at X'puhil to triple the number of orchid species identified and recorded for the peninsula. As the ruins at Becán and Chicanná were cleared for excavation and restoration, the workers saved the orchids for my study. Soon they took delight in bringing me new ones. Once again I was impressed, though this time by the powers of observation of these men, many of whom during the rainy season returned to the forest as *chicleros*, collectors of chicle, the basic ingredient in chewing gum.

To appreciate the role played by this community of forest dwellers, one must understand a bit about the history of chicle extraction on the Yucatan Peninsula. Although the extraction of latex from the chicosapote tree (*Manilkara zapota*) was an ancient Maya activity, the tons of processed latex exported to the United States to make chewing gum were the product of an extractive industry that boomed on the Yucatan Peninsula in the early part of the twentieth century until artificial substitutes were invented during World War II. Although the United States is no longer a major importer of chicle, other markets such as Japan and Italy still purchase the processed latex, but in reduced amounts.

The chicosapote tree is found in abundance principally in the southern part of the Yucatan Peninsula, northern Belize, and the department (similar to a state or province) of Petén, Guatemala. The men who harvest the latex generally live in communities near the forest and tend their cornfields when not extracting chicle. Because chicle can only be collected during the rainy season and the trees can be only be harvested every five to ten years, knowledge of the location of these trees and the ability to survive under the harshest conditions are the prerequisites for being a successful *chiclero*. The industry still produces men who know the trees better than any other user or student of these forests, as well as the names and medicinal uses of the plants, the activities of the forest animals night and day, and the location of Maya ruins, some still unrecorded.

Needless to say, without the involvement of the *chicleros* in the excavations at Becán and Chicanná, we would never have been able to accomplish our objectives. At the beginning of the second year of excavations, when my husband was desperately concerned with finding workers in this seemingly remote niche around X'puhil, Juan Briceño de la Cruz, a former *chiclero*, suddenly appeared at the camp and asked us if we would need workers that season. My husband's reply was affirmative. "How many?" queried Juan. "To start with, twenty, and then later up to ninety," responded my husband. "When do you want them?" "Next Monday." And next Monday the workers appeared as if out of nowhere. Juan always found them through the network of *chicleros*. We never really knew how his system of communication worked, without modern telephones and transportation, but it was so effective—as was his handling of the workers—that he became the chief of the labor corps and later the official guardian of the ruins of Becán. His knowledge of the sites in the area, as is true of most of the *chicleros*, was astonishing. My husband always said that archaeologists do not discover Maya ruins in that area but rather record them, because the *chicleros* know already where they are. For them, the archaeological sites are a beacon for survival, since they invariably were built near precious sources of sustenance: an *aguada* for water and ramón trees (*Brosimum alicastrum*), which provided food for the ancient Maya and the best forage for the *chicleros*' mules and horses.

As my quest for orchids continued, with an increasing commitment by the *chicleros* to find new species, I relied heavily for scientific advice on Eric Hagsater. Founder and former president of the Asociación Mexicana de Orquidología (AMO, Mexican Association of Orchids), editor of the magazine *Orquidea*, and creator of the AMO herbarium, Eric Hagsater is the person responsible for involving me and many others in the conservation and investigation of orchids in Mexico. One of the major ways in which Hagsater promoted biological conservation was to help found and later become president of Pronatura, a private, nonprofit organization dedicated to the conservation of biological diversity in Mexico. In 1981, Hagsater joined several other Mexicans, many of them ornithologists and bird lovers, to form Pronatura A.C. with Arq. Hector Ceballos Lascurrain as its first president, Dr. Jesús Estudillo López as vice president, and Dr. Leopoldo Paasch Martínez as treasurer. Many of Mexico's foremost biologists, including Dr. Arturo Gómez Pompa, Dr. Gonzalo Halffter Salas, and Dr. Mario Ramos Almos, were on the original board. Some of the members of the original board of directors, like Eric Hagsater Gartenberg and Andrés Sada Garza, later became presidents. Brianda Domecq, who succeeded to the board on the death of her father, became its president from 1983 to 1985.

It was at this time that Hagsater asked me to become Pronatura's Yucatan representative. I was hesitant to accept, because of many other commitments: raising two small children as a widow, sustaining the archaeological activities and research library of my late husband, and documenting the preliminary list of the peninsula's orchids. My concern for the barely touched stretch of forest in the south and for Yucatan's fragile coastal complex of wetlands, estuaries, and thin strips of beach responded to Eric Hagsater's persuasive manner. Little did I know that my love for orchids and sea turtles would put me square in the middle of budding Mexican conservation efforts. Welcomed by its then president, Brianda Domecq, and armed with stationery with Pronatura's letterhead, I was given a free hand to organize Pronatura's activities in Yucatan as I wished. It was a challenging task. In the early 1980s, practically the only successful private, nonprofit organizations in Yucatan were Catholic charities and the Red Cross. Saving Mexico's natural resources was simply not considered a matter of concern for the private sector. Mutual distrust, which marked the relationship between government agencies and business and professional communities under the administrations of Luis Echeverría and José López Portillo, was not easy to dissipate, especially in Yucatan, known traditionally for its rebellious attitude toward federal control. Investigators at universities and research institutions, like Centro de Investigaciones y Estudios Avanzados (CINVESTAV, Center of Investigation and Advanced Studies), Universidad Autónoma de Yucatán (UADY, Autonomous University of Yucatan), the Instituto Tecnológico de Mérida (ITM, Technological Institute of Mérida), and the Centro de Investigaciones Historicas y Sociales de la Universidad Autónoma de Campeche (UAC, Center of Historical and Social Research, Autonomous University of Campeche), worked in isolation, frustrated that their research would have no impact on government or private management of natural resources.

Even though the challenges to conservation were indeed mighty at the time, Mexico has, in fact, a long-standing tradition of regulating the use of its forest for timber exploitation, going back to pre-Columbian times. Lane Simonian (1995), in his book *Defending the Land of the Jaguar: A History of Conservation in Mexico*, cites a conservation law imposed by Nezahualcoyotl (1418–1472) to deter his people's abuse of woodlands in the Valley of Mexico. There are other examples of pre-Columbian respect for forest resources, including the Maya ceremony still performed today that requests permission of the gods to alter the forest by clearing it for cornfields. However, there is also evidence of soil erosion from overuse at the end of the ninth century, just prior to the Classic Collapse. Postclassic populations were considerably less dense than those of the Classic period. More recently, population has grown again and, after a long period of heavy exploitation of natural resources during the last century, the pendulum has begun to swing back toward a deeper regard for the natural world (Faust 2001).

The concept of conserving modern Mexico's biodiversity was just beginning to be popularly recognized in the 1980s, as worldwide studies of flora and fauna demonstrated that Mexico's species richness in certain taxonomic groups was among the world's greatest. A brief overview of presidential policies easily demonstrates the upward trend of legal protection for Mexico's natural resources. To begin, although President Luis Echeverría (1970–1976) enacted new regulations on pollution, no areas were declared protected during his administration. Under President José López Portillo (1976–1982), however, a subtle change took place, with 863,783 hectares of land placed in federal protected areas. This included two coastal sanctuaries in Yucatan, now biosphere reserves: Celestún and Ría Lagartos. The federal administration under President Miguel de la Madrid (1982–1988) improved upon that of his predecessor, more than tripling acreage in federal protected areas.

For the peninsula, his conservation strategy was of particular importance, since it included the creation of a major reserve, Sian Ka'an, with 528,147 hectares. Sian Ka'an Biosphere Reserve had multiple significance: it established the importance of conserving aquatic as well as terrestrial habitats in an integrated system, and it brought to the peninsula the UNESCO concept of man and the biosphere, which acknowledges the importance of integrating human activities in protected areas. Most important, however, the masterminds behind the reserve's establishment, especially the then governor of Quintana Roo, Pedro Joaquín Coldwell, encouraged the establishment of a private conservation group to secure the long-term conservation of the reserve by serving as a channel for private sector funding. The area was declared a biosphere reserve in 1986; the nonprofit private association Amigos de Sian Ka'an, headed by Barbara MacKinnon de Montes, was established the same year. This new approach to conservation efforts in Mexico permitted a government-sanctioned private association to tap foreign and private Mexican sources to help fund the protection of a federally declared reserve. It proved to be a model funding mechanism, adopted for other protected areas in Mexico, now known as Areas Naturales Protegidas (ANPs, Natural Protected Areas).

Since the 1970s, these ANPs have known five major reorganizations, being moved from one government department to another. The power struggles, confused lines of authority, and uncertain and delayed budget allocations all resulted in poor management of the ANPs. In 1983, the Secretaría de Desarrollo Urbano y Ecología (SEDUE) was created, transferring control of most natural resources from the Secretaría de Agricultura y Recursos Hidráulicos (SARH, Secretary of Agriculture and Hydraulic Resources). It was precisely at this time that Pronatura A.C. asked me to be its representative in Yucatan. With no precise mission dictated from the parent organization in Mexico and no funds at all, I decided to make contacts with the private sector, government agencies, and scientific institutions. To activate PPY, I called on my personal friends in the private sector, ones who I knew shared my love of nature.[3] In the federal government, I contacted the head of SEDUE in Yucatan, Arq. Manual Castillo Rendón, who gave us permission to work directly with members of his staff in charge of various projects. Scientists associated with CINVESTAV, involved in protecting the recently declared wildlife sanctuaries of Ría Lagartos and Celestún, volunteered their assistance in setting Pronatura's priorities and establishing cooperative agreements between Pronatura, SEDUE, and CINVESTAV.

In retrospect, it was an effective move. It established the basic philosophy behind PPY's activities on the Yucatan Peninsula: collaborative efforts involving the government agencies responsible for conservation, local scientific institutions, and members of the business and professional community represented through PPY's local board of directors. Like many other "do good" organizations without funding, we depended on the goodwill of friends, who happily threw themselves into the task of forming a private conservation force in Yucatan with a collective energy and enthusiasm that attracted local and international attention. We worked out of my library with its one telephone and one overused computer. Biologists, artists, and volunteers from the United States and Mexico City pitched in to help—preparing bird lists and Pronatura-Yucatán brochures, and giving talks to our informal group of founding members. Others offered their artistic talents designing T-shirts, which became our major funding source for environmental educational programs.

In those early, unsophisticated days, without any authorized mission directives, we felt that our main objective was simply to show the community that PPY could be effective and react rapidly to emergency conservation situations. SEDUE was struggling with a limited budget and sometimes with no authorized funding, so our initial support went to some of their projects, which included finding donations of corn for the deer at the SEDUE wildlife sanctuary in southern Yucatan, providing food baskets for the guards in the Celestún and RiaLagartos sanctuaries, and helping fund the SEDUE sea turtle project in El Cuyo.

From contacts with SEDUE biologist Isabel Aguirre, whose efforts Pronatura was supporting, I learned about the bureaucratic rivalries within the federal government. The El Cuyo beach was divided into two sections, one monitored by personnel of the Secretaría de Pesca (PESCA) and the other by those of SEDUE. The biologists were not allowed by their respective departments to share information. Each patrolled her part of the beach, often seeing one another at night, but not communicating. I found this absurd and immediately asked them to come to lunch at my home. Since that lunch, PPY has helped organize ten annual meetings of representatives of the peninsular sea turtle programs. In these meetings, researchers have shared their findings, thus building a common base of knowledge that is critically important for conservation efforts (e.g., Frazier 1993). These meetings for the presentation of research results to fellow researchers and the interested public not only have been locally useful but have also garnered serious commitments by U.S. Fish and Wildlife Service to support various sea turtle programs, including those run by PPY. Sea turtle protection was PPY's first important conservation program, and it remains one of our strongest today, with our organization active in the three states of the peninsula and funded in large part by the Fondo Mexicano para la Conservación de la Naturaleza (FMCN, Mexican Fund for the Conservation of Nature), the U.S. Fish and Wildlife Service and the Mexican construction company, CALICA.

Also during this early period, PPY built up strong relationships with governmental, scientific, and private institutions working on the coast and in the tropical forests of Campeche. The organization was formalized as a chapter of the national organization, Pronatura A.C., with an expanded board of directors. When in 1988 Hurricane Gilbert struck the Yucatan Peninsula with the greatest force ever recorded, PPY forged even closer ties to both national (Mexican) and international conservation organizations, and to NGOs from the United States that wished to help. With the support of the editor of *National Geographic* magazine, Wilbur Garrett, who offered the National Geographic Society auditorium in Washington, DC, PPY presented an afternoon of lectures, slides, and videos to show the destruction to the coastline. Over 200 people attended, including representatives from most of the prominent conservation NGOs in Washington, DC. I mention this meeting because it was a turning point in international recognition of our struggling association.

Ever since I had been introduced to southern Campeche in the late 1960s, I was convinced that here lay Mexico's best opportunity to save vast expanses of unpopulated forest, rich in biodiversity and refuge to many endangered animals and plants. The principal protagonist of the protection of much of central and southern Campeche was Dr. William Folan, director of the Centro de Investigaciones Históricas y Sociales of the Universidad Autónoma de Campeche. His unflagging dedication to this vision was finally realized when the Calakmul Biosphere Reserve, comprising over one and a half million acres, was declared in 1989.

Before the Calakmul Biosphere Reserve had been established, PPY received its first major grant in 1988 from the World Wildlife Fund (WWF) in the United States. Dr. Mario Ramos, WWF's Mexico desk officer at the time, gave us the go-ahead to form a team of specialists to collect the basic data on flora and fauna and human settlements for the proposed Calakmul Reserve. It was an act of faith on his part, as SARH was blocking SEDUE's efforts to protect this forested area in Campeche, where SARH officials handled timber rights. Moreover, PPY had no proven ability to prepare such extensive and diverse studies.

What Pronatura accomplished was quite remarkable; we assembled a team drawn from diverse Mexican institutions with experts in the fields that we wanted to study. Dr. Eckart Boege and Raul Murguía of CINVESTAV, in collaboration with the University of Campeche, undertook the socioeconomic studies. ECOSUR (El Colegio de la Frontera Sur) (then the Centro de Investigaciónes de Quintana Roo, CIQRO [Center for Research in Quintana Roo]) elaborated the management plan drafts, the geographic analysis, and studies of community subsistence hunting. The Instituto de Ecología (Ecological Institute) of Jalapa, Veracruz, did the jaguar and mammal studies, while the University of Yucatán researched the flora. PPY did the ornithological studies, and Biocenosis did the initial work on the organizational structure in collaboration with SEDUE-Campeche. PPY provided the directional, logistical, and administrative support. WWF's assistance was complemented by the Mexican division of the U.S. Agency for International Development (USAID).

Joe Quiroz of The Nature Conservancy (TNC) recognized that PPY, with a project of that magnitude as well as others it had established with CINVESTAV and SEDUE for coastal collaboration, needed administrative support. A generous grant from TNC allowed the PPY team to move from my library into its own office with telephone, fax, computer, and a director. It was a great day! Soon after, to facilitate its activities and administration, the Yucatan chapter of Pronatura officially became Pronatura Península de Yucatán A.C., maintaining the same general objectives as the parent organization, Pronatura A.C., but being responsible for its own accounting, fund-raising, and programs.

Most of our activities during the 1980s concentrated on specific protection of certain flagship species, like endangered sea turtles or flamingoes. A subtle change took place at the beginning of the 1990s, as the concept of the biosphere reserve drew attention to the necessity of balancing wildlife protection with sustainable economic activities for the communities in the protected areas. There has been no greater challenge for the conservation community, and specifically PPY, than this dilemma.

PPY's first community endeavor was focused on a small Maya community, Punta Laguna, located on the frontier between Yucatan and Quintana Roo. The story of Punta Laguna and our efforts there was emblematic of the challenges of community-based work we faced in the peninsula. The village is on a paved road leading to the archaeological site of Cobá, much visited by tourists from the Mexican Caribbean coast. In the tropical forest surrounding Punta Laguna and several other settlements, a sizable group of endangered spider monkeys make their home and can be viewed by adventurous tourists.

In this part of the peninsula, Maya Indians, some of whom are descendants of Caste War refugees (Reed 1964) eke out a subsistence living, based on traditional slash-and-burn agriculture, producing their own corn and beans, as is typical in most rural areas of the peninsula. Their small cash income is acquired from the sale of honey. Unfortunately, in the past decade droughts have reduced grain production, hurricanes have wiped out beehives, and the lack of educational facilities (no schools until very recently) has left many monolingual Maya-speaking children with limited opportunities for the future. Rural communities like these, during the various economic crises that Mexico has faced in the past twenty years, have been the most severely affected, and their very limited purchasing power has been reduced considerably through inflation and devaluation of the peso. In order to support their families, many men have left their villages to join the unskilled labor force in the exploding resort towns of the Caribbean coast, or in the Mérida pool of migrant workers.

In many cases, the breakdown of traditional subsistence agriculture has resulted, as the younger men who migrate no longer attend to the cornfields; their remittances provide cash income for their families, which is often poorly spent on junk food and soft drinks, television sets, and stereos in their villages. Because of population growth and an increasingly concentrated rural population,[4] the traditional system of a twenty-year fallow period has been reduced to seven, with subsequently reduced yields (Redfield and Villa Rojas 1934; Terán and Rasmussen 1994; Faust and Bilsborrow 2000). In addition, many small farmers no longer regard commercial corn farming as profitable. The price of corn has fallen as a result of decreases in government subsidies and the importation of less expensive corn from the United States. People still plant corn for domestic consumption and plant extra as a hedge against losses due to climate fluctuation. When there is a good harvest, they sell some; but they do not count on the sale of corn as an important part of the family's food security and are exploring other alternatives for earning money (Kintz 1990; Faust 1998; Bascopé, chapter 8, this volume; Ericson, chapter 5, this volume). Increasing desiccation due to large-scale forest destruction for cattle ranches and commercial, mechanized agriculture has also made corn farming less productive per hectare. The 1998 ENSO (El Niño-Southern Oscillation) resulted in such unusually dry conditions that the traditional burning of plots for subsistence sparked large-scale destructive forest fires that were difficult to extinguish. This initiated an increase in government efforts to control burning in and near protected areas. Pronatura joined various universities and research centers in exploring the introduction of green manure crops as an alternative to the burning involved in traditional shifting agriculture.

Nonetheless, the inhabitants of many rural communities in the peninsula, be they Mayas or landless immigrants from other states or countries, have demonstrated to PPY that they are willing to agree to the protection of the surrounding forest, if it offers them a better life. And herein lies the dilemma that PPY faces in its community development and conservation programs in protected areas. What can an NGO with limited funds and personnel offer poor rural or coastal communities that will help them protect their resources?

We have learned from experience that there is no single answer. Yet if conservation groups wish to preserve natural resources, they must provide ecologically compatible and economically sound projects for the rural and coastal communities with which they are collaborating. That is what PPY has attempted to do in the communities where it is working. We have introduced an array of small projects that are appropriate for the skills and interests of the rural communities and complementary to the natural resources of the area. In Punta Laguna, Calakmul, and the coastal reserves of Celestún and Ría Lagartos, we have trained local farmers and fishermen to be ecotourism guides and armed them with information about their valuable natural resources; we have promoted the identification of medicinal plants, many of them found in the forest, and encouraged women in the Calakmul Reserve to produce herbal medicines for sale. Reforestation programs, diversification of crops, home gardens for women, and intensive organic agriculture have been priorities for PPY and its work with the Calakmul populations; technical advice for improved beekeeping and the commercialization of local honey is a continuing activity in Punta Laguna and Calakmul. In the Calakmul Biosphere Reserve, improved beekeeping programs have been implemented in fourteen ejidos (communities with legally established communal property rights to agricultural and forest lands).

PPY has also been an ardent supporter of gender-based work, trying to involve women in its programs—no mean task where male leadership of the ejidos and family tradition have generally limited women's commercial activities (outside of family businesses) to selling foods and beverages from the home. An example of this support for women is a project in the Calakmul area, where PPY took the innovative and successful step of organizing three groups of women beekeepers, with ten members in each group. This was a reintroduction of techniques used by the ancient Maya with the native stingless bees, *Melipona* sp. They do not produce as much honey as the imported European bees that are becoming Africanized, but their honey has medicinal properties recognized in traditional Maya medicine.

This project provides a good example of the delicate task of introducing new activities in communities where economic resources are so scarce. In some cases, the money earned by the women during the first year caused family friction and required Pronatura personnel to play the role of marriage counselor.

In Celestún, Pronatura works specifically with women producing shell artwork and embroidery, and processing blue crabs. In the Punta Laguna region, PPY has started a project with women to produce embroidered items valued by tourists. From our experiences working with women in these communities on the peninsula, we can easily characterize them as very hardworking, creative, financially reliable, and motivated to earn a cash income (many for the first time). However, having little experience in leadership roles or in collaborative efforts with other women, and often mired in long-standing inter- and intrafamily village conflicts and jealousies, it is difficult to help them develop techniques for cooperation in collaborative work. This problem is not by any means peculiar to Maya women but rather is common to women attempting to organize for collective action, even in industrialized countries (see Harragan 1977; for problems of Mexican women see Townsend 1999).

Arroya Irigoyen, in her chapter "Yesterday and Today's Crafts" in the UNIFEM publication *Na Molay*, aptly describes the drastic economic changes in the past decade on the Yucatan Peninsula, which has forced many men to leave their villages and which has turned many women into heads of household and often key breadwinners by commercializing their traditional skills in handicrafts. The author emphasizes the many challenges in marketing these handicrafts and cites efficient organization as the principal one (Arroya Irigoyen 1999).

These are just some of the many lessons we have learned in our direct work with small, rural communities, the same ones frequently learned in other parts of the world where there are problems of poverty and isolation with histories of family tensions, often caused by conflicts over resources. Overcoming these obstacles requires great patience, awareness of the formal and informal leadership patterns in this environment, and sensitivity to the need not to disrupt the economic and political structure of the community. I can recall when one of

Pronatura's biologists reported that he had to rotate the laundering of his clothes among different households, so that each family in the small community could benefit from the twenty pesos he paid weekly. Despite limited funding and personnel, Pronatura has succeeded in creating innovative models for community development. In a search for alternatives to traditional slash-and-burn agricultural methods, our projects have been varied and unique: testing the feasibility of cornfields interspersed with heavy leafed legumes to build soil and produce nitrogen, successful experimentation with double harvesting, and the introduction of multiple crops (including the planting of citrus and lumber trees as windbreaks and for future financial benefit). When we first introduced the concept in the Calakmul region, we started with experimental plots, ridiculed as "postage stamp" plots. Yet this intensive organic agricultural program soon won converts and has expanded every year in the region; in 1999, eighteen local farmers, including four women, served as extensionists, and twenty-five ejidos are now introducing intensive organic agriculture with multiple crops as a method to conserve part of the forest and avoid extensive slash-and-burn agriculture. The Calakmul municipality has migrants from twenty-three different Mexican states living in the area, some of whom originally dreamed of owning ranches or extensive farms, which they could not possess in their own states. Nevertheless, many of them have shown amazing adaptability and even inventiveness in Pronatura's agricultural programs (Eastmond and Faust in press), experimenting with plants and trees they knew in their own part of Mexico and showing a genuine willingness to conserve the designated protected areas in the Calakmul Biosphere Reserve (see Ericson, chapter 5, this volume).

The expansion of PPY's scope of work into community development paralleled a remarkable period of conservation activities under the administration of President Carlos Salinas de Gortari (1988–1994). Among the most important were the decrees protecting over 6 million hectares (Figure 4-1), including El Vizcaino Reserve with 2,546,790 hectares, the Calakmul Biosphere Reserve with 723,185 hectares, and Laguna de Terminos Reserve with 705,018 hectares. In May 1990, President Salinas de Gortari declared a total and permanent prohibition on the killing of sea turtles and the use of their eggs and other products. This decree has given legal backing to private and government forces protecting these endangered animals. He also convened an international conference on Mexico's biodiversity, bringing in eminent scholars from all over the world who called for immediate action to protect the diverse flora and fauna that make Mexico one of the four leading regions in the world for biodiversity and endemism. This conference, which received worldwide press attention, stimulated international conservation NGOs to designate Mexico as a priority area because of its great diversity. As a result, major funding mechanisms were established to channel money for the protection of its ANPs. During the same period, the federal government manifested a radical change of attitude toward conservation NGOs. Instead of the suspicion and distrust that had often marked relations between private conservation groups and the federal government in Mexico

FIGURE 4-1. Hectares in federally protected natural areas by presidential administration, (6-year terms) clearly showing the effort made during Carlos Salinas de Gortari's administration to expand significantly Mexico's protected areas. *Sources:* (1) Mexican government (Sistema Nacional de Información Ambiental—Instituto Nacional de Ecología); (2) Carabias Lillo et al. 2000. Los tesoros de la Nacion, *Desarrolla Sustentable* 2(14): 28–34.

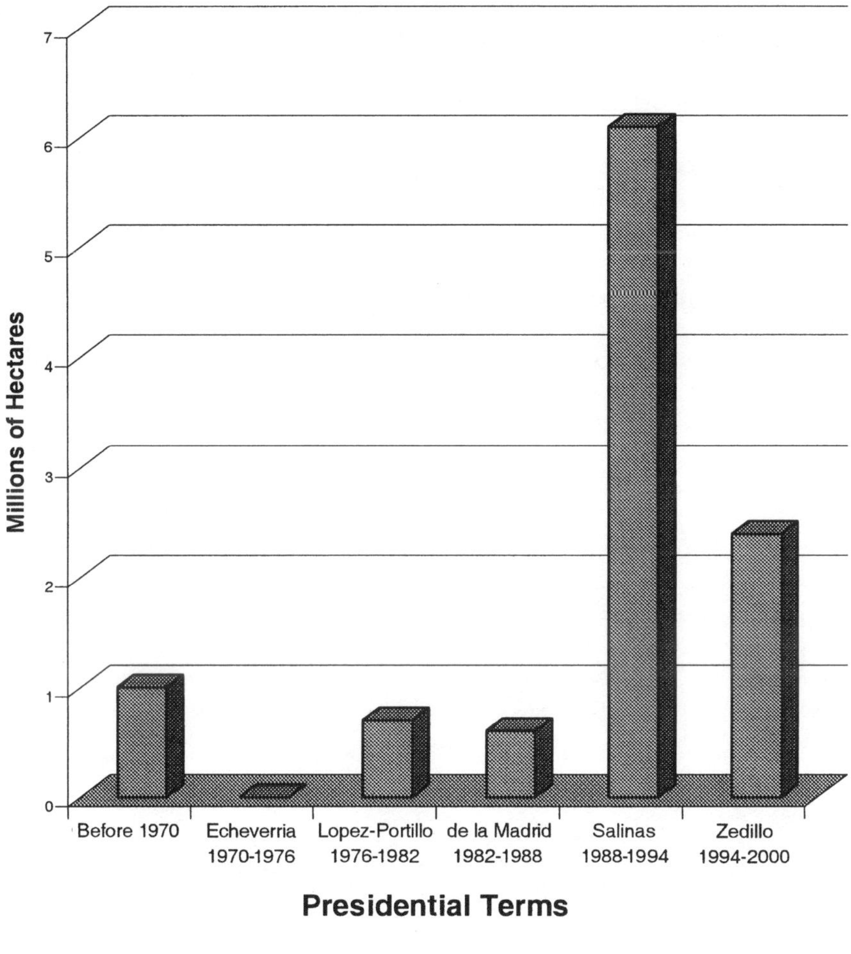

City, NGOs were embraced as important players in the protection of Mexico's natural resources. It was a refreshing change that allowed groups like PPY to expand their role, participating directly in the planning and management of the reserves.

The succeeding administration of President Ernesto Zedillo Ponce de León (1994–2000) followed Salinas's lead, reorganizing the Secretaria de Medio Ambiente, Recursos Naturales y Pesca (SEMARNAP, Secretary of Environment, Natural Resources and Fisheries), as well as giving financial security and organizational structure to the ANPs.[5] The respected and dynamic head of SEMARNAP, Julia Carabias Lillo, insisted on establishing permanent trained staff and a predictable and adequate budget; she succeeded in getting approved management plans for many of the highly diverse parks and reserves.

The experience that PPY gained through its participation in the TNC-USAID Parks in Peril Program in the coastal reserves of Celestún and Ría Lagartos and in the Calakmul Biosphere Reserve served as valuable experience for the collaborative programs initiated by SEMARNAP in the 1990s. International banking institutions like the World Bank, international conservation organizations like WWF and TNC, and national conservation organizations joined in the effort by SEMARNAP to convert reserves from parks that existed on paper to viable, sustainable ANPs.

As I look back on these two decades of PPY's establishment and growth on the peninsula, I want to emphasize that from its inception it has always tried and has frequently succeeded in bringing together private, academic, and government forces concerned with conservation. In that complicated mosaic of interests, certainly the everyday users are the most important elements. Yet their participation in decision making and long-term strategies in natural resource planning is often not given sufficient attention. In rural communities, where indigenous people often have a different perspective on resource conservation, it behooves groups like Pronatura to give priority attention to their ideas. That should include greater recognition of women's roles in natural resource preservation. Mary Elmendorf (1976), in her sensitive study of Maya women (*Nine Mayan Women: A Village Faces Change*), stressed the Maya women's readiness for change. As part of their overall conservation strategy, it is the responsibility of associations like Pronatura with long-standing relationships of trust with women on the peninsula to recognize their decisive role in natural resource preservation and offer them opportunities to improve their own lives and those of their family in a manner harmonious with nature.

NOTES

1. In 2001, Pronatura Península de Yucatán had a workforce of forty-five, of whom the large majority were employed in the field in different parts of the peninsula, either on PPY projects or on the staff of other organizations that were receiving funds from PPY. Its budget in 2001 amounted to 6,406,953 Mexican pesos (approximately U.S.$711,884). A breakdown of the sources for funding shows that approximately 37.8 percent comes from international foundations; 45.9 percent from Mexican foundations; 12.9 percent from Mexican businesses and private donors; 0.5 percent from private foreign donors; and 0.2 percent from other sources. The way the monetary resources are used reflects Pronatura's overall objectives: over 91.5 percent was directed to its projects, whereas 8.5 percent was committed to the operations of the organization. These figures were presented to Pronatura A.C. for use in its annual report, presented at the annual assembly in April 2002.

2. A silted-up pond frequently found in the area and essential as a water supply, and, in our case, conveniently located for the daily washing of the potsherds.

3. It is gratifying to note that the original group included various members who have remained supportive of PPY over the years: Carlos Abrahám Mafud, María Elena Achach de Chapur, Hernán Patrón Vales, and Fernando Ponce García.

4. Many hamlets have been abandoned as people flock to communities large enough that the government can afford to provide them with schools, electricity, and piped water systems.

5. By June 2000, in Mexico there were 119 ANPs including reserves, parks, refuges, and natural monuments, covering 15,848,016 hectares. The additional ANPs established during the Zedillo administration included twenty-seven new ANPs covering 2.4 million hectares (Carabias et al. 2000).

REFERENCES

Arroya Irgoyen, L. 1999. Yesterday and today's crafts. In *Na Molay, First Congress of Mayan Women*, edited by G. Espinosa. Mexico City: UNIFEM.

Carabias, J., J. de la Maza, D. Gutierrez, and O. Piguet. 2000. Los tesoros de la nación. *Desarrollo Sustenable* 2 (14): 28–34.

Eastmond, Amarella, and Betty B. Faust. In press. The challenge of introducing green manure and cover crop systems into traditional Maya agriculture. *Landscape and Urban Planning*, (Special Issue, editors R. Smardon and B. Faust).

Elmendorf, Mary L. 1976. *Nine Mayan women: A village faces change*. New York: Shenkman.

Faust, Betty B. 1998. *Mexican rural development and the plumed serpent: Technology and Maya cosmology in the tropical forest of Campeche, Mexico*. Westport, CT: Bergin and Garvey Press.

———. 2001. Maya environmental successes and failures in the Yucatan Peninsula. *Environmental Science and Policy* 4 (4/5): 153–169.

———, and Richard Bilsborrow. 2000. Maya culture, population and the environment in the Yucatan Peninsula. In *Population, development and environment in the Yucatan Peninsula: From Ancient Maya to 2030*, edited by Wolfgang Lutz, Leonel Prieto, and Warren Sanderson. Laxenburg, Austria: International Institute for Applied Systems Analysis, 77–107.

Frazier, J., ed. 1993. *Memorias del IV taller regional sobre programas de conservación de Tortugas Marinas en la Península de Yucatán, 11–13 Marzo, 1991, Mérida, Yucatán, México*. Mérida, Mexico: Universidad Autónoma de Yucatán.

Harragan, Betty Lehan. 1977. *Games Mother never taught you: Corporate gamesmanship for women*. New York: Rawson Associates.

Kintz, Ellen. 1990. *Life under the tropical canopy: Tradition and change among the Yucatec Maya*. New York: Holt, Rinehart, and Winston.

Redfield, Robert, and Alfonso Villa Rojas. 1934. *Chan Kom, a Maya village*. Publication 448. Washington, DC: Carnegie Institute.

Reed, Nelson. 1964. *The Caste War of Yucatan*. Stanford, CA: Stanford University.

Simonian, Lane. 1995. *Defending the land of the jaguar: A history of conservation in Mexico*. Austin: University of Texas Press.

Standley, Paul C. 1930. Orchidaceae. In *Flora of Yucatan*, edited by Paul C. Standley. Pub. 279, Botanical Series, Vol. III, No. 3. Chicago: Field Museum of Natural History, 237–240.

Terán, Silvia, and Cristian Rasmussen. 1994. *La milpa de los Maya.* Mérida, Mexico: Universidad Autónoma de Yucatán.

Townsend, Janet G. 1999. *Women and power: Fighting patriarchies and poverty.* London: Zed (St. Martin's Press).

5

Population and Land Use: An Essential Link for Conservation in the Calakmul Biosphere Reserve

Jenny Ericson

Integration of the life experiences of local people into the process of land use planning for effective biodiversity conservation is critically important. The following vignettes provide brief glimpses into the lives of three individuals who have settled on the borders of the Calakmul Biosphere Reserve in the state of Campeche, in the southwestern portion of the Yucatan Peninsula.

Vignette 1: Doña Adelina[1] arrived in the Calakmul region three years ago, fleeing the civil unrest of clashes between Zapatistas and the Mexican military in the neighboring state of Chiapas. Doña Adelina is a Tzeltal Indian. She has no formal education; in fact, she speaks only a few words of Spanish. She has given birth to fourteen children, eleven of whom have survived. She is thirty-five years old. She and her family live in an ejido called Nueva Vida,[2] which is located just beyond the border of the reserve. Doña Adelina's husband is a *poblador*, which means that he does not have land rights. He is dependent on the good graces of the ejidatarios in his ejido who do have access to land to lend him a hectare or two during the growing season.

Vignette 2: Don Diego is an ejidatario whose ejido, La Guadalupe, is located partially inside the reserve. He and his wife left their home in the densely populated state of Veracruz thirteen years ago in search of land and greater opportunity. Government officials directed them to the state of Campeche, where unoccupied land was still abundant. They have four children, three girls and one boy. When money gets tight, don Diego often has to travel great distances to look for work with government-sponsored assistance projects. On these occasions, his wife sometimes has to entertain paying male "guests" to make ends meet.

Vignette 3: Doña Lucha is of Yucatec Maya origin. She lives in the ejido 20 de Noviembre, located along the main road that cuts through the reserve. Like most Yucatec Maya people, doña Lucha speaks Maya with friends and family members, and Spanish with outsiders. She is fifty-five years old and was born in the state of Campeche a few hundred kilometers to the north of the Calakmul region in another ejido. She and her husband were part of a large group of Yucatec Maya families that migrated south three decades ago because the communities to the north were becoming overcrowded and land was getting scarce. Many of doña Lucha's children and grandchildren have grown up in 20 de Noviembre and are now ejidatarios. The family still maintains a house in the north.

Within a period of two to three decades, a growing population of colonists has carved out an agricultural frontier in the vast tropical forests that once covered the southern part of Campeche. Much of this area is now under protection, having been declared a biosphere reserve, but many of the ejido communities and small landowner settlements were established prior to the founding of the reserve. Religion, ethnicity, language, culture, and varying educational levels internally divide many of the communities, and a subsistence-level economy, based primarily on slash-and-burn agriculture and the extraction of forest products, keeps the region's settlers on the edge of survival. These people are pioneers who share the opportunity of starting new lives and creating something from nothing in a challenging ecological and social environment.

A mosaic of causes related to regional population dynamics and land-use strategies employed by colonists has raised concern regarding the future of the rich biological diversity contained within the remaining forested areas in and around the reserve. The rapidly changing frontier zone is characterized by fluctuating rates of in- and out-migration, overall high fertility levels, and lack of local-level land use planning. Diverse political, environmental, and economic interests indicate potential to nurture conflict and pose a challenge to regional planning. Human population levels in the county[3] of Calakmul, which includes the reserve and surrounding areas, increased by 400 percent between 1980 and 1995, from 6,000 to more than 24,000 (Ericson, Freudenberger, and Boege 2001).

Shortly after the reserve was established in 1989, a variety of national and international, governmental and nongovernmental assistance programs took notice of the region. During this time, the Mexican government encouraged nongovernmental organizations to participate in the design and implementation of management strategies for the reserve (Boege 1995). The Mexico Program of the World Wildlife Fund (WWF), with support from the U.S.-based program, in partnership with a local nongovernmental organization, Pronatura Peninsula de Yucatán A.C., established integrated conservation and development programs in a number of the ejidos surrounding the reserve. With the primary goal of conserving biodiversity by reducing deforestation and habitat fragmentation, these organizations promoted "ecodevelopment" projects designed to generate economic incentives for resource conservation and to foster effective stewardship of natural resources. Local government officials had long noted the importance of human population dynamics in the region, but in 1995 the director of the reserve brought his concern to the attention of WWF and Pronatura. As a result, these organizations decided to incorporate a population dynamics component into their integrated conservation and development approach in order to broaden understanding of this factor and its relation to land use. This chapter examines issues identified during a two-year participatory research program carried out in collaboration with WWF and Pronatura during the mid-1990s in the ejidos bordering the Calakmul Biosphere Reserve.

THE STUDY SITE

Covering 723,185 hectares, the Calakmul Biosphere Reserve is the largest tract of protected tropical forest in Mexico and an important site for biodiversity conservation. Located in the state of Campeche, it was established in 1989 by the Mexican federal government and accepted into the UNESCO network of biosphere reserves in 1993. Calakmul is an important element in a larger system of protected areas on the Yucatan Peninsula, forming an ecological corridor of over 2 million hectares (Figure 5-1).

Calakmul represents a patchwork of tropical forest fragments including mature, disturbed, and secondary growth vegetation less than twenty-five years old, as well as savannah-type floodplains. Four types of forest have been identified in the region: (1) high, twenty-five meters or more, (2) medium humid to dry forest, primarily semideciduous; (3) low forest, deciduous and semideciduous; and (4) seasonally flooded forest. These four broad types can be divided into at least sixteen vegetative associations based on floristic composition and structure (Martínez Salas and Galindo-Leal 2002). The most abundant tall trees include chicosapote (*Manilkara zapota*) and ramón (*Brosimum alicastrum*) (Miranda and Hernandez-Xolocotzi 1963).

Two centuries of human activity throughout the region, in conjunction with natural factors such as local variations in soils and rainfall patterns, have significantly influenced the basic composition of vegetative species found in these forests, beneath which lie the ruins of the Classic Maya civilization, which prospered with high population densities from approximately 200 to 800 C.E. (Dominguez and Folan 1995). The current state of the forest within both core and buffer zones, as well as in the ejido communities that surround the reserve, is the result of timber extraction and forest clearing for agriculture and cattle grazing. Prominent commercial tree species are mahogany (*Swietenia macrophylla*) and Spanish cedar (*Cedrela odorata*). Calakmul is also home to charismatic threatened and endangered species such as jaguars (*Felis onca*), howler monkeys (*Alouatta pigra*), spider monkeys (*Ateles geoffroyi*), and tapirs (*Tapirus bairdii*). Over 20 percent of the bird species recorded in the reserve

FIGURE 5-1. Map of Yucatán Peninsula showing ecological corridor and protected areas

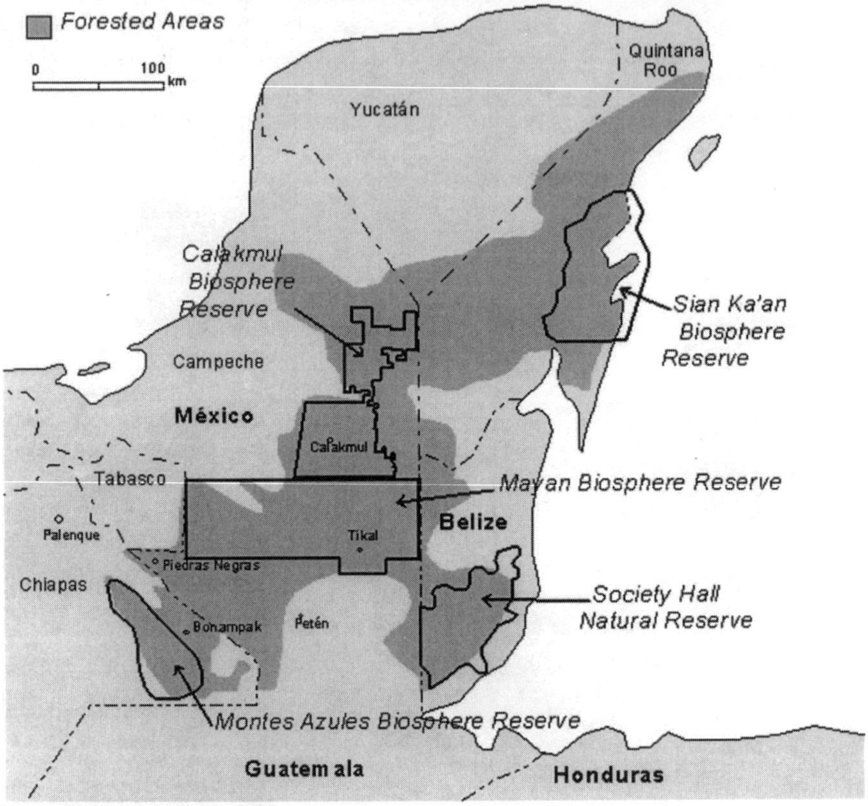

Source: Modified from Boege, 1995

breed in the United States and Canada and use these tropical forests as their wintering grounds (Berlanga and Wood 1991).

A PARTICIPATORY RESEARCH PROGRAM

Once it was recognized that links between population and environment issues were likely to play a critical role in the future of the Calakmul Biosphere Reserve, a participatory research program was initiated to generate dialogue among stakeholders about the impact of population growth and distribution on biodiversity conservation around the eserve, and in so doing lay the groundwork for a culturally and politically appropriate participatory land use planning process. To do this, it was necessary first to broaden existing knowledge of regional population dynamics, to gain a macrolevel view of social and environmental changes in and around the reserve, and then to assess local-level population dynamics by carrying out case studies in selected ejidos.

Selection of the case study ejidos was based on a matrix of criteria, including population growth rates, ethnic composition, and geographic relation to the reserve. All three of the chosen ejidos experienced rapid population growth from 1990 to 1995, and each one has a unique mix of other indigenous groups from southeastern, northern, and central Mexico.[4] One of these ejidos, Conhuas, is located along an interstate road and extends into the buffer zone of the reserve.[5] Another of these ejidos, 11 de Mayo, is located in the remote southeast near the border of Guatemala, and 10 percent of its territory is located within the core area of the reserve. The third ejido, Nueva Vida, is just north of the region's most urbanized area and does not border the reserve. Conhuas has had minimal exposure to conservation and development projects, while extensionists working with Pronatura frequently visit the other two.

Fieldwork for the case studies was carried out by a six-member team trained in participatory data collection and analysis methods by the Grupo DIP of the Universidad Autónoma de Yucatán (UADY, Autonomous University of Yucatan) and ranged between fourteen and twenty-one days of residency in each community. Team composition was diverse in terms of gender, age, ethnic origin, and educational background. Four members of the team live in the region and arrived there fifteen to twenty years ago as immigrants. Three of them are ejidatarios who play active roles in the social, economic, and political development of their respective communities. The fourth resident is a rural health assistant born in Chiapas but who has lived most of her life in the Calakmul region. All speak Spanish in addition to one of the indigenous languages of the Maya family, including Chol and Tzeltal. The team was managed by two other people: a field-based project assistant who was a social anthropologist of Yucatec Maya origins, and a field-based project coordinator, a U.S. citizen with a graduate degree in natural resource management. The results of the participatory exercises carried out in the ejidos were reviewed and discussed with community members at the completion of each case study for their approval prior to final documentation and dissemination to outside groups.

Stakeholder debate and collaboration can generate a foundation for formulation of appropriate policy responses to migration, and they can shape an approach to land use planning in the region at large, as well as the individual communities within and around the reserve. Thus, upon completion of individual case study documentation, the team compiled and analyzed the results, using them to initiate dialogue about the threats posed by migration and inappropriate land use practices to the long-term economic, social, and ecological viability of the reserve and its surrounding population.

REGIONAL DYNAMICS

Population densities in the state of Campeche increased from 2.1 persons per square kilometer (pp/km^2) in 1950 to 12.3 pp/km^2 in 1995. In-migration appears to have had a greater impact on Campeche's population growth during these years than fertility rates. In 1992, 25 percent of the population was reportedly born outside of the state, while overall natural increase is recorded to have fallen from 3.8 live births per woman in 1970 to 2.2 live births per woman in 1995. In the late 1980s, Campeche's immigration rate was between 3.5 and 10 percent, which at that time was greater than that of Mexico City and less than that of the neighboring state of Quintana Roo (Partida Bush 1995).

Government campaigns during the 1960s, 1970s, and 1980s encouraged people living in more densely populated areas of the republic to establish ejido communities on Campeche's "underutilized" forest lands, mainly in the southern portion of the state. The effort was also driven by the constant tension of the border conflict between Campeche and Quintana Roo. The border marker was lost to the jungle many years ago, creating ambiguity about territorial lines and prompting each state to intensify colonization in the region (Ericson 1997).

Numbers of inhabitants in the communities within and around the reserve are still relatively low, averaging about 200 people per community. Typical of a frontier zone, population growth rates vary dramatically between communities, and populations within individual communities can fluctuate sharply over relatively short periods of time. Calculations based on the five-year period between 1990 and 1995 indicate crude growth rates as high as 16 and 23 percent for Nueva Vida and 11 de Mayo, respectively, while the growth rates of other communities indicate population declines. The geographic size of these communities also varies widely, ranging between 1,500 and 60,000 hectares; thus, community-level population densities vary from less than 1 pp/km^2 to more than 13 pp/km^2. Older ejidos tend to have larger land holdings with extensive forest resources, whereas younger ones were created to accommodate a large landless population and are more limited in land size and richness of forest resources (Stedman-Edwards 2000).

Since the early 1900s, colonists have been drawn to the region by such factors as the abundance of forest resources, the availability of land, and the tranquillity of a sparsely populated frontier. They have migrated from their places of origin because of insufficient economic opportunities, scarce land, and social unrest, and in some cases they have been displaced by large-scale agricultural projects and ecological disasters. Most of the region's inhabitants have lived in a variety of different places prior to settling in the Calakmul area (Ericson and Maas 1998; Haenn 2001).

Since the beginning of the twentieth century, five distinct waves of immigrants have arrived in the region. *Chicleros* settled in seasonal camps around the beginning of the twentieth century to harvest latex from the chicosapote trees. Loggers and sawmill workers began arriving in the 1940s to fell, process, and export hardwoods from the region. These two early waves of immigrants were primarily Yucatec Maya from northern Campeche and the state of Yucatan, whose population is currently declining. Landless settlers arriving from diverse Mexican states began establishing ejidos and private landholdings in the 1970s and 1980s to exploit remaining forest resources and practice subsistence agriculture. In the 1990s, there was an influx of Chol and Tzeltal peoples fleeing social unrest in neighboring Chiapas. Table 5-1 shows that the growth rates of the communities in which Chiapan ethnic groups predominate exceed the region's 4 percent growth rate. With the formation of the Calakmul municipality in 1997, a fifth wave of migrants, primarily tourism, service industry, and government workers, has ensued.

Although migration plays a dominant role in the region's population dynamics, current rates of natural increase will affect future natural resource use in the area. Regional fertility rates ranging between 3.9 and 5.2 live births per woman per time unit contrast with the decline in statewide fertility rates cited earlier.[6] If fertility rates remain high, future population growth will move sharply upward due to the fact that 51 percent of the local population was reported to be under fifteen years of age in 1992.[7]

Due to the relatively low population densities in the Calakmul communities, the region has not yet been assigned priority status by government health agencies. Geographic distances, ecological conditions, poverty, and lack of public transportation often restrict the access of locals seeking medical help to the limited number of clinics available in the Calakmul area. Mobile medical units operate in some of the communities but often have difficulty making their rounds due to poor road conditions, insufficient staff, and other logistical problems. Many ejidos have designated one of their members as a rural health assistant to be sent off to attend a government-sponsored short course in basic health care and provided with access to medical supplies, including contraceptives, for distribution in the community. Rural assistants are not trained in promotion of reproductive or general health services; moreover, they are often not able to replenish supplies for lack of money to cover travel expenses. Most of the communities with large indigenous populations have resident *parteras* and *curanderos* (midwives and healers) who often have their own time-tested methods for handling medical problems.

With the formation of the Calakmul county in early 1997, greater financial resources are gradually becoming available to the region for development and infrastructure growth. Known as the "ecological county," Calakmul contains the reserve and 114 communities with a total population of 26,000 inhabitants. The term *ecological county* marks it as distinct from the average county, but there is no real clarity as to the term's specific functions other than its use as a marketing tool to draw funding from the international conservation community. However, the term suggests a valuable opportunity to include both conservation and development objectives in regional planning. It encourages interdisciplinary dialogue and presents an interesting challenge to identify exactly what it means to be an ecological municipality.

Recent infrastructure developments may stimulate increased in-migration in the coming years. Scarcity of water resources in the region has in the past encouraged the out-migration of large numbers of people. A large-scale water project carries water eighty-five kilometers from a shallow lagoon, Laguna de

TABLE 5-1. Growth rates in Calakmul county by ethnic composition of ejido communities

Ethnic Groups as percent of population	Number of ejidos	Inhabitants			CGR (per yr.,%)	T doubling (in years)
		1980	1990	1999		
Tzeltal (more than 30% of population)	4	n.d.	523	918	11	6.16
Chol (more than 80% of population)	22	n.d.	2,753	3,997	7	10.57
Chol (between 30 and 80% of population)	14	n.d.	8,825	10,499	3	19.95
Yucatec Maya (more than 30% of population)	4	n.d.	2,160	2,083	-1	-95.48
Communities with no more than 30% of one/any ethnic group	56	n.d.	7,320	9,102	2	31.30
Communities for which there are no data	14	n.d.	n.d.	n.d.	n.d.	n.d.
Total county population	114	6,858	19,331	24,295	4	19.42

Alvarado, near the border with Guatemala to service twenty-two communities along the more heavily populated southeastern border of the biosphere reserve. The aqueduct is intended for use during the dry season as a complement to already existing water catchment systems in the ejidos. Increased water security will decrease out-migration and may attract more settlers. However, based on population projections calculated on current growth rates, it is estimated that this new system will only be able to service the needs of these communities until 2005. A report suggests that 50 percent of the water carried by the aqueduct is lost due to ruptures in the system (CNA, 1996).

Land markets are emerging due to land tenure changes in the federal constitution pertaining to Article 27 of the agrarian reform law. Land can now be bought and sold within ejidos, with the exception of land covered with standing forest. Due to the large masses of forested land in the Calakmul ejidos, land titling has taken place within deforested urban zones only. Titling of land beyond the respective urban zones to individual ejidatarios in the Calakmul region is not legal; in fact, dividing the land into individual parcels is not legal either but is practiced by the ejidos on an internal basis. Thus, instead of receiving a title to the parcel of land where an ejidatario plants his crops, he receives a *certificado de uso común* (common use certificate), which gives him rights to a specified percentage of his ejido's land. However, whether or not land has been legally titled in these ejidos, there appears to have been an increase in land sales over the past few years. In the ejido Xpujil, for example, more than 100 hectares of land have been sold, only a portion of which is within the urban zone. Buyers are generally investors from outside of the region, purchasing for construction of shops and hotels in anticipation of a growing demand for tourist facilities.

COMMUNITY-LEVEL DYNAMICS

Population dynamics at the community level often fluctuate due to a combination of internal and external forces. Availability of economic opportunities based on resource exploitation within the ejido, internal community conflicts, policies regarding newcomers, and community epidemics such as malaria are among the internal forces causing in- and out-migration. Climatic hardships, government-sponsored infrastructure projects, availability of government credits or programs, and economic opportunities inspired by regional projects such as archeological site excavation and tourism are among the external forces.

Dramatic changes took place in the history of the population of 11 de Mayo. In the mid-1980s, conflict divided the ejido. During a four-year period of internal unrest, families fled the community out of fear, then returned, left, and returned again. In the following years, severe flooding brought about another wave of out-migration due to loss of harvests, the death of domestic animals, and an outbreak of malaria. Since 1990, out-migration seems to have diminished, and the community has experienced a population increase followed by gradual stabilization. Government relief programs following hurricane damage in 1995 and improved road conditions and water catchment devices during these years have helped reduce out-migration and brought about in-migration and stability.

Each ejido community monitors the arrival of newcomers. If an ejido has not yet reached its quota of ejidatarios, the ejido's governing body, the *assemblea*, may or may not choose to give a newcomer rights to the land by making him an ejidatario.[8] In addition to obtaining land rights, ejidatarios also participate in decision making through the *assemblea*. Initially, newcomers are given *poblador* status. They are required to pay an admittance fee to the ejido, provide officials with letters of recommendation from a previous residence, and undergo a waiting period during which the community decides, based on observation, whether or not they will be accepted. *Pobladores* have neither voice nor vote in the *assemblea*. If the quota of ejidatarios for a given ejido has already been met, the newcomer will either wait until a space opens up, borrowing land in the meantime, or will move on looking for land somewhere else.

In the ejido Conhuas, the quota of ejidatarios has not yet been met, and the number of recent arrivals is almost equal to the number of ejidatarios. During the last few years, this ejido has been a magnet for new settlers because it is ringed with archeological sites in which excavation projects provide employment opportunities. Conhuas also has the largest forested area in the region, which attracts *chicleros* during the chicle-harvesting season.

In a participatory exercise designed to determine historical changes, a mixed group of ejidatarios and *pobladores* in Conhuas were asked to develop the matrix shown in Figure 5-2 based on criteria they themselves established and using piles of beans to quantify change over time. Through this matrix and the ensuing discussion, these individuals were given the opportunity to reflect on how quickly the population of their community had grown over the years, and, by looking toward the future, to determine where this growth will lead if it continues at the same pace. As a result of this exercise, the ejidatarios in this group decided to introduce a policy change within the *assemblea* to limit the number of newcomers admitted during the coming years, thus allowing time for current *pobladores* to become accepted as ejidatarios.

In the Calakmul ejidos, land rights are rarely allocated to women. The most common way for a woman to become an ejidataria is if she inherits this status through the death of her husband and only if she has no sons old enough to take on the responsibility. Legally, female *pobladores* can request rights as ejidatarias if the quota is not yet filled, but routinely this is subtly discouraged. Women are sometimes considered to be *gritonas y peleoneras*, basically troublemakers, in the *assemblea* meetings. In 11 de Mayo, 15 percent of ejidatarios are women; in Conhuas 10 percent are women; and in Nueva Vida just 7 percent are women. Access to land rights and participation in decision making at the ejido level are thus conditioned by gender.

A survey of women in 11 de Mayo revealed that the average number of children per family is just over five. Women from Chiapas who speak Chol or

FIGURE 5-2. Historical change matrix of ejido Conhuas

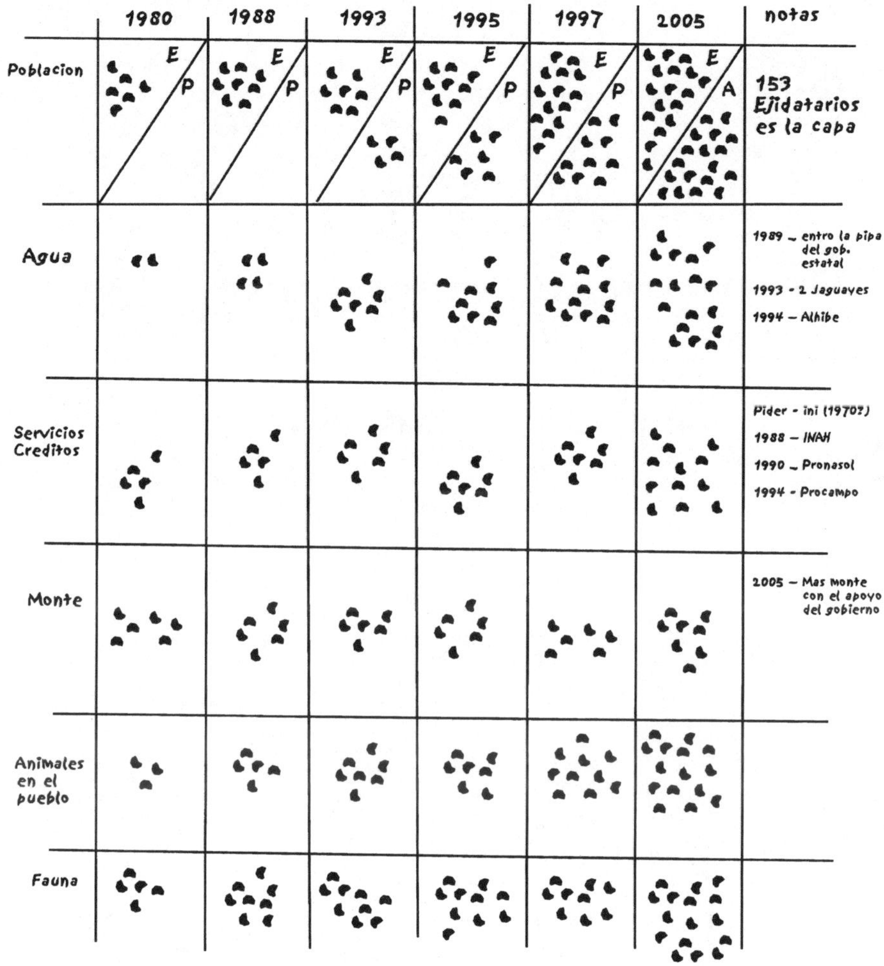

Tzeltal have slightly larger families than women that speak only Spanish. Some of the women surveyed indicated that they had desired fewer children at time of marriage, and 41 percent of them reported having had from one to six more pregnancies than children currently living. Figure 5-3 illustrates the life history of a woman who was one of twelve children, born in Chiapas in 1970, and is currently the wife of a *poblador*. At the age of twelve this woman married and at age fifteen she bore her first child. She now has six children and, because her husband is a *poblador* and must borrow land from an ejidatario to plant his crops, the question arises where their sons will find land when it comes time for them to plant crops.

FIGURE 5-3. Life history of the wife of a *poblador* in ejido 11 de Mayo

Current fertility rates will influence long-term population trends, and future natural resource use will be impacted by choices made today. In 1996, Pronatura initiated a reproductive health project with financial assistance from the U.S. Agency for International Development. Collaboration with Pronatura's reproductive health team in the field was an important part of the case study work on population and environment reported herein. The reproductive health project strives to educate and orient families toward taking responsibility for their own health and demanding and obtaining quality reproductive health care from existing service centers; it also strives to integrate reproductive health issues into community-level discussions.

During a one-year period, Conhuas, with less than 400 inhabitants, received support from eleven different Mexican government agencies for production activities, such as extraction of forest products, agriculture, apiculture, and livestock raising. As previously stated, government support has played an important role in reducing out-migration and helping to stabilize ejido populations since the early 1990s; however, in combination with support from nongovernmental organizations, it has been largely responsible for creating an attitude of opportunism among local residents. Community members can pick and choose, often selecting programs that offer more tangible benefits that answer immediate needs over education-based programs with greater long-term value.

Since the early 1990s and the initiation of former President Salinas's Solidaridad program, which strove to broaden the PRI (Partido Revolucionario Institucional) party's constituency in marginalized rural areas, ejido residents in the Calakmul region have become accustomed to receiving government support. They are slow to take the initiative in solving their own problems because they have learned to wait for a government-sponsored solution. Planning exercises performed with community residents make it clear that the prevailing strategy for attaining community objectives is to solicit community-level authorities, who in turn solicit municipal and state-level authorities, and in some cases nongovernmental organization representatives, for assistance. This phenomenon, combined with the relative socioeconomic isolation of the region and difficult ecological conditions, has created a population largely dependent on outside assistance.

Standing forest in Mexico belongs to the nation and cannot be legally harvested without appropriate studies and authorization from the ministry of environment. When ejidatarios are requested to conserve forested areas on their personal parcels of land or on communal ejido land for which they cannot obtain authorization to harvest timber, the question arises for whom they are conserving. When they put their own labor into reforesting degraded areas, they begin to ask why they are investing labor when they can not legally reap the benefits. This type of insecurity regarding resource ownership undermines trust between political authorities and local residents and acts as a disincentive toward the participation of local residents in conservation. Ejidatarios ask, "Why do we have to take care of the forest if it belongs to the nation? If they want us to take care of it they should pay us!" (Ericson and Maas 1998).[9]

Discussions with local ejidatarios also reveal worries about the presence of foreigners pushing conservation agendas and the notion that these same foreigners may intend to take away land from locals in the future in order to establish their own homes. "The *campesino* begins to think that the nation's

TABLE 5-2. Summary of informant plans for future land and resource use

Activity	Ejido 11 de Mayo	Ejido Nueva Vida	Ejido Conhuas
Forest use and conservation	Preserve part of forested areas for extraction of chicle, wood, guano palm, and cultivation of allspice.	Preserve part of the primary growth forest on plot. Cut down secondary growth vegetation.	Cut down part of the high forest to plant crops. Take care of the forest for future use.
Reforestation and agroforestry	Plant allspice and fruit trees such as guaya. Do agroforestry. Reforest within areas of secondary growth vegetation. Take care of reforested areas.	Reforest the areas of secondary growth. Take care of and plant more trees. Reforest areas previously planted with maize. When wood is needed, take it from preserved forest on plot.	Maintain and expand reforested areas.

Crops	More milpa crops (maize, chihua, beans, camote, yucca), sugar cane, banana trees, grasses (*Zacate estrella*), cover crops (carnavalia and nescafe), and citrus trees.	Maintain milpa crops (maize, chihua, banana, and cover crops) and pasture. After harvesting maize, plant pasture in same area. In lowlands plant rice and in highlands plant maize.	Plant maize, chihua, beans, and banana and citrus trees.
Infrastructure	Use mechanized agriculture methods in areas of secondary growth vegetation, plant pasture, construct water catchment pond, bring in cattle and beehives.	Use mechanized agriculture methods for the milpa and to plant rice. Construct water catchment pond, bring in cattle and beehives.	Plant pasture and construct corrals for goats.
Animals	Cattle and beehives.	Cattle and beehives.	Cattle, goats, and beehives.

land has been sold! One day they [the foreigners] are going to come and live in the forest!" (Ericson and Maas 1998).[10] This line of reasoning stems from the constant presence of foreigners and foreign funding associated with the reserve and conservation of the area's natural resources since the reserve's founding in 1989.

As has already been stated, the Calakmul region is highly marginalized economically. Ejido residents are occupied with meeting immediate demands for survival. They live primarily from subsistence agriculture, which in some ejidos is supplemented by income generated from forest activities such as chicle tapping, timber extraction, and beekeeping. Sustainable harvesting of both chicle and timber has been encouraged by WWF and Pronatura. Beekeeping has also been encouraged due to the principle of forest conservation underlying its practice. The difficulty with these activities, as with sustainable agricultural programs such as organic agriculture and agroforestry, is that income generation is limited due to the relative isolation of the region from markets.

Because most colonists have agricultural preferences, encouraged by government programs like PROCAMPO,[11] and because relatively few precious hardwoods remain in these forests to be harvested, land is more valuable on the market if forests have been removed. Forest clearing for agriculture and cattle ranching is labor intensive and costly. Extensive agriculture is practiced with a few crops such as jalapeno chili and, in some cases, watermelon. However, middlemen, referred to locally as *coyotes*, generally offer low prices to producers, and prices fall even lower in those ejidos that are the most remote. As a result of the overall isolation and poverty of the region, changes in the marketplace as well as the effects of direct and indirect government interventions are often limited (Stedman-Edwards 2000).

Case study results also show that people living in the communities around the reserve have their own set of priorities and define long-term vision in terms other than those of conservationists. Community members prioritize family survival and look to the upcoming harvest or the next series of harvests over a one- to two-year planning period in order to meet immediate needs. In contrast, someone who prioritizes the preservation of biodiversity or perhaps the regeneration of a particular species of tree is probably thinking in terms of twenty to thirty years. Thus, a *campesino* who needs extra cash to pay for agrochemicals and land clearing to plant cash crops may be inclined to fell and sell the last remaining mahogany tree on his plot of land. The conservationist concerned with ecological systems is generally not a resident of the region and does not share the survival strategies of those living from the resources under protection and so often lacks a clear understanding of how to slow their exploitation and eventual depletion.

Table 5-2 indicates that the activity of reforestation is being included in land and resource use plans. Conservation organizations and government-sponsored reforestation programs have been successful in promoting this activity in all three case study communities. Reforestation in these communities is generally practiced with harvesting in mind, despite previously discussed legalities and insecurities about forest resource ownership. The marginalized colonists of the Calakmul ejidos tend to see conservation and replenishment of natural resources in terms of the potential benefits derived from future use of those resources.

Plans to move toward mechanized agriculture and cattle ranching are currently kept in check by financial and ecological restrictions. In some communities, soil conditions and a shortage of permanent water sources prohibit both these activities. Moreover, they require a financial investment greater than that available to the majority of the region's colonists. Although very few ejidatarios can afford to purchase a tractor, many of them pay to have one or two hectares plowed each year by a regional *campesino* organization offering this service.

SUGGESTIONS FOR FURTHER WORK

The following suggestions reflect ideas supported by the six-member land use team in their analysis of the case studies in addition to those presented in discussions held at an institutional level with WWF and Pronatura.

Initiate Planning for Sustainable Development Through the Integration of Conservation and Development Objectives

Currently there is an absence of both land use and development planning in the region. A development plan was drawn up when the Calakmul municipality was formed in 1997; however, development in the region has been haphazard and without consideration for sustainability or conservation of natural resources. Government investment has focused on meeting the immediate demands of settlers for land, roads, and water. A planning mechanism can be developed to encourage the integration of conservation and development objectives in the elaboration of municipal level land use plans, development plans, and biosphere reserve management plans.

Promote Dialogue and Collaboration for Planning Between Actors in the Region

At this time, there is little to no coordination between the parties responsible for elaborating conservation and development plans at the regional level. A forum for discussion, including representatives from local social and commercial organizations in addition to political figures and nongovernmental conservation organizations, could be established to encourage dialogue between parties. Fostering public debate and dialogue on population dynamics and land use at the regional level can be a critically important tool for building consensus on future courses of action and dispelling confusion and misinterpretations that arise from poor communication between government officials and local residents on policy issues. Improvements in planning and reduction of contradictions at national, state, and municipal levels should also be encouraged through greater dialogue and collaboration between governmental and nongovernmental parties with policy, program, and project intervention in the region.

Encourage Discussion and Reflection on Population and Environment Issues at the Community Level

Our case study research indicates that reflection on population and environment issues at the community level can bring about changes in ejido policy, as previously illustrated in the case of Conhuas. Opportunities need to be created for community members to consider the long-term future of the land and resource base within their ejidos with respect to these issues. Capacity building and planning exercises can be facilitated by conservation and development organizations to create these opportunities. It is also equally important for outside facilitators with interests in the region to listen carefully to local residents, enter into discourse with them, and attempt to view conservation and development challenges from their perspective.

Promote Initiatives That Create Positive Incentives for Conservation

Positive incentives can be developed to counteract insecurities regarding resource ownership and to provide recognition for local people willing to work toward conservation goals. An example of this might be a system of registration and public acknowledgment for ejidatarios who reforest their land or protect wildlife habitat. Financial incentives may also be feasible but must be approached with caution given the climate of opportunism already discussed.

Promoting low-impact economic opportunities with the potential to meet both conservation and development objectives, such as apiculture, arts and crafts industries, and carefully planned ecotourism projects, can provide positive incentives and act as environmental education tools. It is crucial, however, that incentives be designed to bring about sizable enough returns to motivate change in resource use. The marginalized socioeconomic reality of the Calakmul region needs to be factored into the conservation equation, and mechanisms should be developed to enable local communities to gain access to larger markets and participate in the national economy.

Establish a System for Population Monitoring

As population dynamics will continue to change rapidly over the coming years in the communities around the reserve, decision makers have requested the development of a multilevel population monitoring system that includes the participation of local residents. Conservationists seek a system that accurately measures trends at both the microlevel of the ejido and the macrolevel of the region. Decision makers are interested to know where and why people settle the way they do and to better understand the determinants of fertility.

With respect to monitoring system design, many issues must be addressed, such as questions of institutional leadership, collaboration of interests, and uses of the information collected. If local people are to be asked to divulge information, the risk exists that data collected may be used to the detriment of informants in the design of top-down management plans. People divulging information about where they live may experience government restrictions and even forced removal from their lands. Alternatively, local people may surmise that such information might bring benefits and so may distort the information accordingly. Local people need to be involved in this monitoring but, unless they can trust that the information will not undermine survival strategies, they may not bc truthful.

Meet Unmet Needs for Public Health and Reproductive Health Care

A reproductive health organization is needed in the region to work in collaboration with government health agencies toward strengthening existing reproductive health services. Initial areas of focus might include capacity building in reproductive health for rural assistants and facilitating the replenishment of reproductive health supplies. The provision of reproductive and family health care assistance can help build trust and confidence with community members, because improvements in health tend to offer rapid returns to the household. This approach is based on the hypothesis that local people will be more open to initiatives that offer long-term gains, such as conservation, once immediate community concerns have been addressed. In addition, resolving conflicts over health care provision may be a step toward resolving a wide array of impediments to planning for economic development and sustainable uses of natural resources.

CONCLUSION

Population and land use are key determinants in the ecological and socioeconomic struggle that is taking place in the Calakmul region. If population growth continues as it has over the past couple of decades, primarily due to dramatic rates of in-migration, the long-term viability of the reserve and its surrounding communities may be at risk unless effective land use planning practices are put in place. At this time in the Calakmul region, there are few structures for empowering local communities to determine their destinies. Caught in a web of dependency, local communities have neither the incentives nor the institutional support to plan for a just and sustainable future. Governments at the federal and state levels play vital roles in building options and incentives for economic development and sustainable land use. At the same time, nongovernmental organizations face great challenges to encourage the creation of institutional structures that increase community participation in decision making and land use planning and ultimately lead to sustainable use of natural resources.

NOTES

1. The names and identities of individuals mentioned throughout this document are fictitious.

2. An ejido is a land grant administered by a group of individuals called ejidatarios who hold usufruct rights accorded to them by the Mexican federal government.

3. In Spanish, this area is referred to as the "municipio de Calakmul"; however, *municipio* translates to the English word *county* rather than municipality.

4. In this context, *indigenous* refers to ethnicity. It is not meant to imply that these people are native to the Calakmul region. Very little is known about the region's inhabitants prior to the arrival of chicle and timber concessionaires in the early 1900s.

5. This ejido was initially established in 1929 in a remote site to the north of where it is currently located. During the 1960s, the people of Conhuas filed a request with the government for permission to move their homes to another site, one that had been allocated to them for forest extraction activities. Although urban areas are generally not allowed on lands allocated for forest extraction activities (known as *ampliaciones forestales*), the urban parts of Conhuas are located along a major road, which enabled greater access to health facilities, electricity, and other types of infrastructure than did the former site.

6. These figures were calculated for the municipalities of Hopelchén and Champotón, which included the communities surrounding the reserve prior to the formation of the Calakmul municipality in 1997 (see CONAPO 1996).

7. At the national level, 36 percent of the population in Mexico is reported to be less than fifteen years of age.

8. When an ejido is established, the agrarian reform office specifies the exact number of ejidatarios who can hold rights to that land. An ejido can request permission to change this quota, as did Nueva Vida when it chose to lower their quota, thus securing larger amounts of land for a smaller number of ejidatarios.

9. "¿Por qué vamos a cuidar la montaña [the wilderness] que tenemos si es de la nación? ¡Si quieren que la cuidemos que nos paguen!"

10. "¡El campesino se pone a pensar que ya está vendida la tierra de la nación! ¡Un día van a venir a vivir en el bosque!"

11. PROCAMPO is part of the Solidaridad program that was instituted under the Salinas administration and later expanded under the Zedillo administration to support rural development. The program provides government subsidies to rural farmers based on the area prepared for the planting of basic crops such as corn and beans.

REFERENCES

Berlanga, Mauro, and Wood, P. 1991. *Las aves de la Reserva de la Biosfera de Calakmul, Campeche, Mexico*. Internal report. Mérida, Mexico: Pronatura Península de Yucatán, A.C.

Boege, Eckart. 1995. The Calakmul Biosphere Reserve, Mexico. Working paper no. 13. Paper presented at international conference, UNESCO South-South Cooperation Programme, Paris.

CNA (Comisión Nacional del Agua). 1996. *Abastecimiento de agua potable en la region de Xpujil, municipios de Hopelchen y Champoton, Campeche*. Internal report. Campeche, Mexico: Gerencia Estatal de Campeche.

CONAPO (Consejo National de Población). 1996. *Poblacion por municipalidad: Estado de Campeche*. Mexico City: Author.

Dominguez, Maria del Rosario, and William J. Folan. 1995. Calakmul, Mexico: Aguadas, bajos, precipitacion y asentamiento en el Petén Campechano. Paper presented at IX Simposio de Investigacions Arqueologicas en Guatemala. Museo Nacional de Arqueologia y Etnologia, Guatemala, July.

Ericson, Jenny. 1997. *Regional assessment: Calakmul population-environment initiative*. Internal report. Pronatura Península de Yucatán, A.C.; World Wildlife Fund; Population-Environment Fellows Program, University of Michigan.

———, Mark Freudenberger, and Eckart Boege. 2001. Population dynamics, migration, and the future of the Calakmul Biosphere Reserve. In *Biological diversity: Balancing interests through adaptive collaborative management*, edited by Louise Buck, Charles C. Geisler, John Schelhas, and Eva Wollenberg. New York: CRC Press, 261–292.

———, and Reyna Sayira Maas. 1998. *La dinamica poblacional en los ejidos alrededor de la Reserva de la Biosfera de Calakmul*. Internal report. Pronatura Península de Yucatán, A.C.; World Wildlife Fund; Population-Environment Fellows Program, University of Michigan.

Haenn, Nora. 2001. *"Biodiversity is diversity in use": Community-based conservation in the Calakmul Biosphere Reserve*. Arlington, VA: America Verde Publications, The Nature Conservancy.

INEGI (Instituto Nacional de Estadística, Geografía e Informática). 1980. Censo general de población y vivienda. Aguascalientes, Mexico.

———. 1991. *Campeche: Resultados definitivos, datos por localidad (integración territorial), XI censo general de población y vivienda, 1990*. Aguascalientes, Mexico.

———. 1996. *Campeche: Resultados definitivos, tabulados basicos del conteo de población y vivienda, 1995*. Aguascalientes, Mexico: Author.

Martínez Salas, Esteban, and Carlos Galindo-Leal. 2002. La vegetación de Calakmul, Campeche, Mexico: Clasificación, descripción y distribución. *Boletín de la Sociedad Botánica de México* 71: 7–32.

Miranda, F., and E. Hernandez-Xolocotzi. 1963. Los tipos de vegetación de México y su clasificación. In *Xolocotzia*, Tomo I. Mexico City: Universidad Autónoma Chapingo.

Partida Bush, Virgilio. 1995. *Migracion interna*. Aguascalientes, Mexico: INEGI.

Stedman-Edwards, Pamela. 2000. Mexico: Calakmul Biosphere Reserve. In *The root causes of biodiversity loss*, edited by Alexander Wood, Pamela Stedman-Edwards, and Johanna Mang. London: Earthscan Publications Ltd., 231–254.

6

Valuing the Maya Forests

E. N. Anderson

Valuing forests is a difficult matter. On the one hand, they have cash value—normally understood to be the cash value of the products of the forest, but now including the value of the forests themselves as ecotourism destinations and genetic preserves. On the other hand, they have spiritual, aesthetic, and "heritage" value placed on them by the people who live in or near them—and, today, often by people who live far from them. In Quintana Roo, all these concerns are matters of intense debate. The Yucatec Maya almost all desire to save their forests. Thus they are becoming involved with ecotourism, with scientific debates about sustainable forestry, and with questions about the income per acre they can derive from harvesting traditional medicinal plants.

Always acutely conscious of the multiple uses of the forest, the Maya are now aware of even more reasons why forests are important. By contrast, "modernization" and "development" schemes imposed from outside have always begun by eliminating the forest—replacing it with cattle, rice, cities, or just bare rock. Today, however, some development agents are aware of the multiple values of the forest. Some Yucatec communities have cooperated with such agents in developing new ways of managing resources.

Mary Elmendorf's work includes a great deal of pioneering effort in the cause of sustainable, local-level development, and Dr. Elmendorf was one of the first to be sensitive to the needs of local rural communities for sound management systems as well as for economic options (Elmendorf 1976). I have written this chapter to speak to that issue, because one of the many lessons

I learned from Dr. Elmendorf's work was that the specific social situation in the local community and its dialogue with the wider world are fundamental in any development initiative.

Anthropologists are fond of taking extreme positions and of attacking others who take extreme positions. In the area of resource management, there is, today, a controversy that need not have happened. On the one hand, the media and some international agencies have caught up with the anthropological thinking of many years ago, which tended to portray traditional or indigenous people as wise managers of their plant and animal resources. The immediate intellectual origins of this view lie in the eufunctional adaptationist school of ecological anthropology, popular in the 1960s and 1970s (see Vayda and McCay 1975 for review and critique). However, in the popular media it has taken on a life of its own, in which we find indigenous peoples portrayed as living "in harmony with nature," using resources with unfailing care and reverence. Critics have caricatured this view with the label "the ecologically noble savage." To my knowledge, no anthropologist now holds this view. (The most recent serious intellectual statement of it I can find is in Hughes 1983; Hughes is a historian.) Even its functionalist-adaptationist ancestor seems to have been largely abandoned (Nyerges 1997).

On the other hand, a strong reaction, or overreaction, has set in. Many writers now portray indigenous peoples as wasters and destroyers who act with no thought of the resources they are using. Such comments range from the reasonable and appropriately qualified to the frankly racist and extremist. An unconsciously humorous example is found in a recent book by the popular science writer Matt Ridley (1997). He spends a chapter vituperatively attacking the very idea that "primitives" can ever manage or conserve anything, and then in the very next chapter he gives several examples of their doing so. He seems to have written this book very quickly.

Of course, all this extremism would not be happening without a political agenda. The "ecologically noble savage" is beloved of New Age writers, environmentalists, and friends of small, oppressed indigenous groups. The "wasteful savage" is the image popular with racists, apologists for logging and agribusiness interests, and, surprisingly, with many biologists. The biologists in question often are interested in creating biological reserves, which occasionally involves dispossessing indigenous peoples from their traditional lands.

In between are the anthropologists and ethnobiologists. Those of us who know something about traditional resource management occupy a vast middle ground that holds room for considerable difference of opinion. Some aspects of this debate have been aired and given full treatment in articles by Michael Alvard (1993, 1995) and Thomas Headland (1997).

In fact, there is a full range from traditional peoples who have almost no concept of management of natural resources (such as the Piro and Machiguenga of the upper Amazon; see Alvard 1993, 1995) to groups who are extremely careful and wise, though not flawless, managers of their resources (such as the

Northwest Coast groups; see e.g., Anderson 1996; Gottesfeld 1994a, 1994b; Nelson 1983).

Bobbi Low (1993) has pointed out that the nonmanagers have very low population densities and live in environments they are not depleting, while the managers are pressing on their resource base and have to take some notice of the sustainability issue. She also notes that groups with clear and definite structures of authority are in a better position to manage than are those without.

The Maya have such clear and definite authority structures, and, as Low would predict, they fall squarely within the class of serious and caring managers. This is not to say they are flawless. The collapse of Classic Maya civilization is widely believed to be due, in part, to excessive pressure on the environment, and this view is supported by clear and unequivocal data from Copán (Abrams et al. 1996). However, it is more relevant to point out that the Maya have been farming their land for 5,000 years without any other major problems (cf. Gómez-Pompa and Kaus 1992). Moreover, evidence from many other parts of Maya territory shows that war, not overexploitation of the environment, was the cause of collapse in those areas (Webster 2002).

Many anthropologists write as if the impingement of wider polities on small, indigenous communities is entirely new (and entirely evil). In the case of the Maya, the state has been a factor for most of the last 2,000 years. By 200 C.E., west central Quintana Roo was densely populated and a center of trade and communication. Then, as now, the effects of new ideas, new ethics, new technologies, and new orders of power must have been complex and demanding. The Maya coped with it.

In western Quintana Roo today, an increasing Maya (and others, including non-Maya Mexican as well as foreign) population is pressing on a fragile resource base. Meanwhile, the outside world is encroaching. The Maya find themselves increasingly involved in debates. Shall the forest be leveled for large-scale agribusiness? Shall it be preserved for ecotourism? Shall it be conserved and managed for sustainable extraction of timber and game? Government agencies, nongovernmental organizations, and Maya communities are interacting in a complex and shifting political and economic universe. This is not a dialogue; it is a polylogue.

The forests of the Yucatan Peninsula are valued by traditional Maya farmers as restorers of soil fertility in the slash-and-burn agricultural cycle of the Maya. The forests also provide medicines, local construction materials, fencing, firewood and charcoal, fiber materials, pesticides, toys, wild fruits and berries, famine foods, game animals, and—last but not least—shade, beauty, and recreation. Today, the modern world values them for production of valuable and utility hardwoods, softwoods for pulp and plywood or for local use, game animals, biodiversity conservation, and ecotourism. Most of these issues and their relationship to conservation are discussed at length in two very important recent books: *Mexican Rural Development and the Plumed Serpent* by Betty Faust (1998), and *Timber, Tourists, and Temples*, edited by Richard Primack and colleagues (1998). In short, the forests produce a wealth of benefits. The U.S. Forest Service talks about multiple uses; the Maya people of Quintana Roo are literally thousands of years ahead of them.

To take just one example, the Maya of western Quintana Roo use over 330 plants for medicinal purposes, most of which come from the forest. Valuing the Maya forest would require serious consideration of all these uses. How many years of life are saved, how many days of work time are saved, by use of those 330 plants to heal sicknesses otherwise hard to treat with local resources?

By contrast, the economic value of the forests was very low until recently. Forests provided mahogany, Spanish cedar (*Cedrela odorata*, which is neither Spanish nor a cedar, but is a tropical American relative of mahogany), and chicle—nothing more. Forests were often stripped of the valuable woods and then leveled for cultivation. Often this was done with no real planning. For instance, I have seen several barren tracts from which forests were cleared in areas where there was not enough rain or available water for farming. All the large-scale cultivation schemes have failed or done poorly, but the forest grows back slowly—if at all—when schemes are abandoned. Because of soil deterioration, increased fire danger, and other factors, large-scale clearings are much less likely to regrow than are Maya milpas. (The foolishness of previous government plans in the Yucatan Peninsula almost surpasses belief, but is thoroughly documented [Kiernan and Freese 1997; Faust 1998; Primack et al. 1998].) Today, most people and agencies involved with the forests recognize that they are valuable for a multitude of reasons. The emphasis is now on conservation. Yet that is not easy to effect, because the policies of years are not easy to change overnight.

The Maya, unlike the Piro and Machiguenga, have an explicit, detailed, highly self-conscious strategy of conserving and managing the environment. It was evidently adequate for the old days of small populations and simple lifestyle—the life of bare subsistence that was still reality for most people in the Chunhuhub area a generation ago. (As recently as 1980, the census showed that a third of the people in the area had no income, i.e., were strictly subsistence farmers.) The old conservation guidelines are no longer adequate because rapid population growth, economic growth, metropolitan Mexican education, and technical change have led both to new pressures on resources and to a breakdown in the cultural traditions.

The old strategy included countless complex procedures, but the main outline and principles are relatively simple. The Maya subsistence economy is, as everyone knows, based on slash-and-burn agriculture and dooryard gardens, with the addition of orchards and groves. (The best published descriptions of Yucatec Maya agriculture, so far, are Terán and Rasmussen 1994 and Faust 1998.) Managing milpas is crucial. This involves more than just letting them grow back. The major further items are as follows.

First, when cutting and weeding the milpa, farmers treat the plants differently according to their uses. Useless trees and brush are cut at ground level, which usually kills them. Trees useful for shade, firewood, construction lumber, and other such industrial uses are cut at waist or breast level, so they quickly regrow. Trees valued for fruit, along with palmettos (valued for thatch and similar uses), are not cut at all; they are left alive. Very large trees are not cut either, not only because they are too big to be worth the effort, but also because they serve as wildlife trees. Being burned, they often develop holes, which are often colonized by wild bees. Then the cut vegetation is dried. It is burned when still damp enough to guarantee a small fire that will not kill the trees left standing. Firebreaks are thus not made. This is different from the practice in drier parts of the peninsula, where burning is more thorough and firebreaks are cut. Milpas without useful trees are dried longer, yielding a cleaner burn. Weeding a growing milpa is a similar process: particularly pernicious weeds are rooted out; most plants are cut to ground level; some are coppiced; and some very useful ones are left to grow. Plants like *kambaljau* (*Dorstenia contrajerva*) and *chak waj* (*Zamia lodigesii*) love to colonize milpas, and are more valuable than many of the crop plants, being valued for medicine and for rat poison, respectively.

Second, when letting milpas regrow, the restoration of soil fertility must be assessed. This is done by observing the soil, but also by observing the growth and type of vegetation. Some plants are markers of fertile soil. Until recently, milpas were allowed to regrow for many years—rarely fewer than eight, and often as many as fifty or more. Today, the range is from two to over fifty.

Third, milpas and their regrowing (successional) stages are also attractive to animals. The Maya of Chunhuhub do not seem to plan milpas intentionally for animal food, as the Maya do in Pich, Campeche (Betty Faust, personal communication). However, they do deliberately leave plants known to be important for wildlife. They are aware that milpas (in production or in regrowth stages) are important to animals, leading to maximization of production. They also use milpas as hunting places. The commonest technique is simply to surprise an animal in the milpa and kill it, but often the Maya will put up tree platforms from which to hunt. Contrary to Jorgensen's data from Xhaasil (Jorgensen 1993, 1994), my data do suggest that "garden hunting" is, or used to be, a conscious strategy that involved increasing game production (as a by-product of milpa cultivation) and then cropping the game.

More generally, any forest operations, such as logging, as well as taking timber for local construction, carefully spare fruit trees. Most important of all are the zapote (chicozapote) trees (*Manilkara zapota*), which have four values: First, and most important, they are the sources of chicle, a major commercial product and local income source. Second, they produce fruit for human food. Third, their fruit is also a major wildlife food, and therefore their presence both increases game and provides the best hunting spots (in fruit season). Fourth, they are valuable timber trees—but it is very rare to see one cut! Because of this hierarchy of uses, they have been favored, and now make up about 15 percent of the large trees in old-growth forests in the area (Plan Forestal, unpublished data). Of course, there is no way to tell what percentage they would be in a totally undisturbed forest, since there is no such forest in Quintana Roo. But it seems reasonable to conclude that they have increased due to human management.

Other commonly spared trees include *wayum* (*Talisia olivaeformis*), a popular fruit tree, and *oox* (ramón, *Brosimum alicastrum*), a source of horse fodder and of a nut used as a famine food.

The other major principle of management is to take nothing except for immediate use and to take as little as possible even for that. There is a definite tendency to economize on wild plants, although people think nothing of cutting down masses of vines and poles when going through forest. On the other hand, they can be very sparing and economical in taking rare but highly valued medicinal plants like *tankasche'* (*Esenbeckia* sp.). With animals, there is a quite different sense: a strong moral charge to preserve them. I have seen a young mother make her children eat butterflies they had killed in play. I have also seen a group of men weeding a cornfield uncover a rabbit nest and carefully cover up the baby rabbits again to protect them—even though rabbits are a major pest of fields. Maya routinely adopt orphan animals, and at least some traditional men feel genuinely obligated to do so if they have shot the parent for food.

More important is the rule against overhunting. Traditional Maya of Quintana Roo do not take many animals at a time, even when the need is strong. My consultant Jacinto Cauich could not resist the temptation to take five pacas one night. Suddenly he saw all the animals of the forest gathering around him, in a silent ring. He was terrified. He heard whistling from above, which he knew to be the Lords of the Forest telling him he would be let off this time, but he had to hunt conservatively from then on or he was doomed. He has never killed more than one animal at a time since. Such stories are common throughout the Maya region, and stories of guardians of the game are widespread (Anderson and Medina Tzuc in press). The traditional Maya woodsman Eleuterio Llanes Pasos has published some of them (Llanes Pasos 1993). Maya hunters (like many hunters worldwide) take pride in killing animals rather than wounding them and thus losing them to death in the forest. These conservationist attitudes are changing, unfortunately. Urban Mexican values—which, until recently, were not protective of animals, and held that a man should kill as much game as he could when he had the chance—have prevailed in some communities. Colleagues and I have observed this in northern Quintana Roo (e.g., Naranjal, personal observation; information from Kurt Heidelberg and Dominique Rissolo). Jeffrey Jorgensen, in a detailed and prolonged study of hunting in east central Quintana Roo, found no significant conservationist behavior (Jorgensen 1993, 1994), though his informants seem not to have been wanton killers either. He has more recently reported that they were concerned about overhunting, but were faced with a "tragedy of the commons" situation (Jorgensen 1998).

In general, then, we see a Maya version of the very general beliefs among Native North Americans (but not Native South Americans) that one should not take more than one immediately needs; one should monitor the resource and save some for others; one should propagate or leave resources when possible; and one should preserve animals unless there is real need to take them. All these beliefs are very widely attested in North American societies and elsewhere (Anderson 1996; Berkes 1999; Colding and Folke 2001). They are not, in themselves, guarantees of conservation. One misses such ideas as taking only male game animals and only in season. (It is hard to imagine how such limits could be enforced.) But they were adequate to preserve stocks of the commoner and more resilient plant and animal resources.

In short, common sense is enough to make the Maya conserve useful trees and other goods within the traditional milpa economy, while strong social pressures backed up by religious beliefs cause people to be cautious about taking animals. With the coming of modern technologies and pressures, these old ways have become inadequate. The benefits of taking wild species have gone up substantially, while the ability to destroy wildlife has also increased.

Commercial logging not only makes certain trees much more valuable—it makes them valuable dead, not alive. Traditional tree use involved harvest of fruit and thatch, coppicing or pollarding of trees for construction and firewood, and thinning of the forest for poles—all uses that keep the trees alive, thus conserving the trees and the forest. Of course, poles and timbers were taken (killing the trees), but trees were used sustainably as far as possible (Faust 1998).

Hunting, by definition, kills the animal. The problem in the case of hunting is that not only population pressure but also the possibility of commercial sale has increased the effective demand for game meat. Moreover, shotguns and rifles are now fairly common, and the traditional conservation ideology was not designed to deal with killing power above the blowgun and snare level; the Maya have had guns for centuries, but until recently the guns were too rare to exterminate game (see also Jorgensen 1998). There is rather little hunting above the subsistence level in the western Zona Maya of central Quintana Roo, but some market hunting does go on. (Jorgensen and I have found that there is much more market hunting and much less conservation ideology in the eastern part of the Zona Maya.) Market hunting is strictly illegal, but the vast majority of the region is remote forest land, where control of hunting by official enforcement agents is simply impossible. People are very much left to their consciences.

Thus, today, more stringent rules are necessary (Jorgensen 1998). Moreover, the new situation demands a much higher level of knowledge of what level of hunting is sustainable. It makes some sort of enforcement structure necessary. People cannot be left to their consciences if the effects of even a few cheaters can be devastating (see Anderson 1996). This obviously creates a dilemma. People must police themselves and each other. Yet a "tragedy of the commons" situation is developing in some areas, making self-policing inadequate. In short, economic forces and an ethic of use-and-destroy are both changing the Maya world in a most unfortunate direction.

At this point, the state consciously enters the picture. In the area under discussion, its most obvious efforts on behalf of conservation have come through the Plan Forestal. This is a plan, now almost ten years old, to manage forests sustainably, according to good forestry concepts. It was formulated with help from Germany. It has recently added a wildlife dimension, thanks to the efforts of a group of exceptionally dedicated wildlife biologists in Chetumal. The story of the Plan Forestal has been told elsewhere, notably by Julia Murphy (1990) and by Michael Kiernan and Curtis Freese (1997) (both of these sources refer to the Plan Piloto Forestal; the plan was so successful that it graduated from pilot stage around 1993). More recently, Hugo Galletti has given an excellent inside account (Galletti 1998; Flachsenberg and Galletti 1998). Suffice it to say here that the plan has the following characteristics:

1. It works with ejidos (collective landholding units) and local ranchers (the two major stakeholders directly involved with land use practices), in what is now often called comanagement. Local decision making is to be paramount, with the government supplying advice and doing the counting.
2. It rationalizes production of timber. This includes keeping statistical records, managing machinery, coordinating activities of the various ejidos, and other managerial functions.
3. Its foresters calculate allowable yield per year, turnover rates, and so on, and continue to monitor logging.
4. It supplies marketing and other advice and assistance (a small training sawmill exists in the local market town).
5. It organizes reforestation efforts.
6. It issues informational bulletins on forestry, forest management, chicle, and so forth.
7. It spun off a subgroup to deal with chicle production.
8. Another spinoff project involves government wildlife biologists, notably A. Ehnis Duhne and Gilberto Avila, who advise ejidos on sustainable hunting levels and encourage them to make rules for hunting (Anderson and Medina Tzuc in press).

Under the Plan Forestal, several ejidos have formulated their own plans for sustainable logging, reforestation, and forest reserves. One, Nohbec, has won the coveted status of recognition by the German government as a supplier of sustainably harvested tropical hardwoods. This gives it favored treatment in the German market. Many other ejidos are rapidly moving in the same direction.

Chunhuhub did not join the Plan Forestal. At first, Chunhuhub's leaders believed they could manage their forest resources without help. When they found, to their cost, that they could not, it was too late; they had virtually no resources left to manage.

In 1991, when I first stayed in the town, ejido leadership was highly responsible about conservation but was locked in disagreement with local and state government. This led to general rejection of government initiatives and also created paralysis, since different levels of government could neutralize each other's initiatives. In 1993 and 1995, when I returned, ejido and *alcaldía* (mayor's office) leadership had gone to more compliant personalities (who had been supported by state government figures). These leaders were cooperative enough, but weak and not widely supported, and were unable to take conservation initiatives. Finally, in 1997, a split among proconservation forces led to the election of an *alcalde* (town mayor) known to be associated with extractive interests and opposed to any sort of regulation. Through all this political change, illegal hunting and logging continued. They remained at a low level, but unfortunately even a low level of activity was enough to destroy Chunhuhub's valuable woods and populations of larger game animals.

By contrast, Naranjal Poniente, near Chunhuhub, worked with the plan. I carried out interviews and observations there. Mahogany (*Swietenia macrophylla*) is logged as a valuable hardwood, and *sak chakah* (*Dendropanax arboreus*) is logged for softwood uses. I found residents to be extremely aware of the need for conservation and vigilant about protecting their forests. In the forest, I found only large trees cut, and many young ones growing. Laura Snook (Snook and Jorgensen 1994; Snook 1998) has pointed out that mahogany grows in light gaps, making some sort of thinning or milpa cultivation necessary if it is to reproduce. Consultants tell me, and my observation confirms, that in the Chunhuhub area these light gaps should be small ones. Naranjal Poniente's old-growth forest is too dense for mahogany reproduction. Thus, logging is actually necessary for the continuation of the mahogany forest (because it is a reserve, protected from agriculture and fire). Reforestation is indeed going on, with tall young mahogany trees flourishing in old logging roads and logging shows (areas where logging was done, logs assembled, trucks parked).

Conservation has also been helped by the plan in Manuel Avila Camacho. This is a small ejido south of Chunhuhub. Like Chunhuhub, it is relatively densely populated and is mostly logged out. However, a large block remains suitable for selective logging. It has been reserved for the purpose, like the old-growth forest of Naranjal Poniente. Moreover, wildlife conservation has become a major issue in Camacho, due to the strong traditional values of the elders. The wildlife biologists associated with the Plan Forestal carried out a survey of wildlife populations and hunting. The ejido, acting on these recommendations, passed rules setting game limits and allowing for new limits as information became available. Most important of all, they resolved to log all game bagged and to make the information public. Thus a running record is available (Anderson and Medina Tzuc in press). Curassows (*Crax rubra*) are protected and are being raised in captivity; this large game bird has been almost extirpated, and the ejido dreams of restoring populations. The leadership for all these rules, and especially for enforcement, has come from senior ejidatarios who still have the full complement of traditional Maya values. They see game as a community resource (not, be it noted, a commons), to be managed and cropped conservatively and sustainably for the good of all. They have been trying hard to stop market hunting, trespassing on leased-out land, and large-scale shooting. They have tried to enforce the protection of curassows. Unfortunately, some of the younger community members have absorbed more metropolitan values. We were offered illegally-shot game for sale at the very entrance of this ejido. Full cooperation remains a goal, not a reality, and game populations are not increasing.

More successful is the story of Tres Garantías (see also Galletti 1998). This huge, thinly populated ejido lies in the far southwest of Quintana Roo. It was the scene of several extremely ill-considered forest clearings for cattle and agriculture. The soil was unsuitable, and much of the area has been taken over by bracken fern and other pernicious weeds that make all use (including reforestation) virtually impossible. Small-scale agriculture pays, badly, in only a few small areas. By contrast, some of the most magnificent rain forest left in Mexico still covers some 20,000 hectares of the ejido. Most of this is now conserved in various types of land use, from absolutely inviolable reserve (400 hectares for wildlife study and protection) to forests open for sustainable logging. Tapirs, crocodiles, and jaguars still roam free. Game is carefully managed, again with the help and monitoring support of wildlife biologists associated with Plan Forestal. There is a small ecotourism development; a forest camp, at the end of a long and virtually impassable jeep track, allows small groups (six at a time preferred) to enjoy something very close to a genuine tropical wilderness experience. Plan Forestal manages reservations and contacts, thus eliminating the otherwise insuperable problems of managing contacts between international tourists and a tiny, isolated community without many modern resources. This logistical and organizational support is critical, and quality local-based ecotourism will probably depend on such contact brokers for a long time to come.

In short, the Plan Forestal actually works, which is very unusual for a government development plan (see the detailed discussions in Kiernan and Freese 1997 and Galletti 1998). There are important lessons in the failure of Chunhuhub to save its hardwoods and wildlife (and perhaps the failure of Xhaasil to save its game animals; Jorgensen 1993, 1998) and the successes of the Plan Forestal. In these cases, the critical variable was not economic self-interest, local control, indigenous rights, or any of the other clichés of cultural ecology and development economics. The important variable was local ability to link social morality with practice (cf. Nyerges 1997).

Chunhuhub failed because of divisiveness within the ejido and because of confrontational tactics against higher levels of government. This was a case in which everyone had a moral position, typically a defensible one; standing on principle, however, cost everyone dearly in the long run. Values enjoining solidarity and unity are what is needed for resource management.

The Plan Forestal has succeeded (so far) because it works with local authority structures. It lends some prestige to local authorities, but, more to the point, it depends on their goodwill. Mutual support means mutual accountability, and the plan is relatively free from the imperious, top-down bureaucracy that has led to ruin in many development schemes. The plan succeeds also because it has been able to draw on the strong indigenous conservation tradition of the Maya.

Particularly instructive is the mutual dependence established between plan personnel and local traditional Maya. The traditionalists, in dealing with young people who have lost the conservationist ethic, need the help of the plan not only to count up trees and animals but also to lend prestige and the possibility of real enforcement capability to their own positions. The plan personnel, in turn, cannot work in an ejido without the full cooperation of the leaders. I have observed the resulting dialogues—among ejido leaders, and between them and plan personnel—in Manuel Avila Camacho and in Tres Garantías (to say nothing of Chunhuhub, where dialogue broke down).

All these factors suggest that, contrary to the implications of much of the recent literature, conservation management can work in the developing world, and indigenous people can work with local governments. However, success does require special circumstances. It requires, specifically, the following:

1. A strong desire, on the part of all parties, to save the goods in question.
2. A strong and widely shared morality of responsibility, such that most people will willingly forgo using a resource if there is any hope at all that others will do so.
3. A management system involving a hierarchy of governmental units, but with accountability and mutual dependence between the levels.
4. Cooperation. Divisive politics are sure to destroy all conservation. This may be the hardest to win; it requires some sort of Durkheimian mechanism to maintain solidarity—if not religion, then some other "collective representation." And even that may not be enough (see the excellent discussion in Terán and Rasmussen 1994). This takes us to difficult questions of trust, cooperation, and mutualism that are not yet resolved theoretically by political science or any other discipline.

The death of most development plans in Quintana Roo (and, I suspect, in the rest of Mexico) has come from the dominance of the top layers over the lower levels (see Faust 1998, and the many excellent case studies in Primack et al. 1998). It should be noted that this is not just a "sin," to be rectified by appointing more moralistic or more left-leaning top dogs. It is an inevitable tendency in administrative systems. All authorities from ancient China (Creel 1974) to the present (Drucker 1992) seem to agree on this. Top-down control means policy is set by people who do not know the conditions on the ground, and who invariably make mistakes, for which they are not accountable and from which they do not suffer. The people on the ground are the ones that pay for the mistakes; but they have no recourse.

Of course, if the top officials are also corrupt or irresponsible, the situation is far worse. This condition seems to have existed in previous forest plans in Quintana Roo (see Kiernan and Freese 1997; and—with appropriate reading between the lines—Flachsenberg and Galletti 1998; Galletti 1998).

Other plans in other regions have failed because of the contrary case: so much was left to the local communities that they were overwhelmed with the responsibility and had no help with things that were beyond their ability to manage (e.g., marketing products in distant markets). Anarchy is, alas, not a viable alternative to top-down authoritarianism. The only viable alternative is good management, of the sort idealized in the relevant literature (see Drucker 1992; this paragraph also draws heavily on the unpublished research of my coworker Kathleen Truman in Yucatan).

This chapter suggests a major reframing of the question about whether traditional (and other) people do or do not manage resources. Of course they do. They do it well or badly according to their success or failure at putting together an effective social system for the purpose.

In the end, classic social theory may be the key to game and forest management in Quintana Roo and elsewhere.

ACKNOWLEDGMENTS

I am deeply indebted to the people of the ejidos mentioned, especially my field assistant Felix Medina Tzuc; also to Betty Faust, who has helped greatly with this chapter.

REFERENCES

Abrams, Elliot M., Anncorinne Freter, David J. Rue, and John D. Wingard. 1996. The role of deforestation in the collapse of the Late Classic Copan Maya state. In *Tropical deforestation: The human dimension*, edited by Leslie E. Sponsel, Thomas N. Headland, and Robert C. Bailey. New York: Columbia University Press, 55–75.

Alvard, Michael. 1993. Testing the "ecologically noble savage" hypothesis: Interspecific prey choice by Piro hunters of Amazonian Peru. *Human Ecology* 21: 355–387.

———. 1995. Intraspecific prey choice by Amazonian hunters. *Current Anthropology* 36: 789–818.

Anderson, E. N. 1996. *Ecologies of the heart*. New York: Oxford University Press.

———, and Felix Medina Tzuc. in press. *Animals and the Maya in Quintana Roo*. Tucson: University of Arizona Press.

Berkes, Fikret. 1999. *Sacred ecology*. Philadelphia: Taylor and Francis.

Colding, Johan, and Carl Folke. 2001. Social taboos: "Invisible" systems of local resource management and biological conservation. *Ecological Applications* 11: 584–600.

Creel, Herrlee G. 1974. *Shen Pu-Hai*. Chicago: University of Chicago Press.

Drucker, Peter. 1992. *The ecological vision*. New Brunswick, NJ: Transaction.

Elmendorf, Mary. 1976. *Nine Maya women*. Cambridge, MA: Schenckman.

Faust, Betty. 1998. *Mexican rural development and the plumed serpent.* Westport, CT: Greenwood Press.

Flachsenberg, Henning, and Hugo Galletti. 1998. Forest management in Quintana Roo, Mexico. In *Timber, tourists, and temples: Conservation and development in the Maya forest of Belize, Guatemala, and Mexico*, edited by Richard B. Primack, David Bray, Hugo A. Galletti, and Ismael Ponciano. Washington, DC: Island Press, 47–60.

Galletti, Hugo. 1998. The Maya forest of Quintana Roo: Thirteen years of conservation and community development. In *Timber, tourists, and temples: Conservation and development in the Maya forest of Belize, Guatemala, and Mexico*, edited by Richard B. Primack, David Bray, Hugo A. Galletti, and Ismael Ponciano. Washington, DC: Island Press, 33–46.

Gómez-Pompa, Arturo, and Andrea Kaus. 1992. Taming the wilderness myth. *BioScience* 42: 271–297.

Gottesfeld, Leslie M. Johnson. 1994a. Aboriginal burning for vegetation management in northwest British Columbia. *Human Ecology* 22: 171–188.

———. 1994b. Conservation, territory, and traditional beliefs: An analysis of Gitksan and Wewt'suwet'en subsistence, northwest British Columbia, Canada. *Human Ecology* 22: 443–465.

Headland, Thomas N. 1997. Revisionism in ecological anthropology. *Current Anthropology* 38: 605–630.

Hughes, J. Donald. 1983. *American Indian ecology.* El Paso: Texas Western Press.

Jorgensen, Jeffrey. 1993. Gardens, wildlife densities, and subsistence hunting by Maya Indians in Quintana Roo, Mexico. Ph.D. dissertation, Department of Forest Resources and Conservation, University of Florida, Gainesville.

———. 1994. La cacería de subsistencia practicada por la gente Maya en Quintana Roo. In *Madera, chicle, caza y milpa: Contribuciones al manejo integral de las selvas de Quintana Roo, Mexico*, edited by Laura K. Snook, and Amanda Barrer de Jorgensen. Chetumal: Ecosur.

———. 1998. The impact of hunting on wildlife in the Maya forest of Mexico. In *Timber, tourists, and temples: Conservation and development in the Maya forest of Belize, Guatemala, and Mexico*, edited by Richard B. Primack, David Bray, Hugo A. Galletti, and Ismael Ponciano. Washington, DC: Island Press, 179–194.

Kiernan, Michael J., and Curtis Freese. 1997. Mexico's Plan Piloto Forestal: The search for balance between socioeconomic and ecological sustainability. In *Harvesting wild species: Implications for biodiversity conservation*, edited by Curtis Freese. Baltimore, MD: Johns Hopkins University Press, 93–131.

Llanes Pasos, Eleuterio. 1993. *Cuentos de cazadores.* Chetumal, Mexico: Government of Quintana Roo.

Low, Bobbi. 1993. Behavioral ecology of conservation in traditional societies. Paper presented at American Anthropological Association annual meeting, Washington, DC.

Murphy, Julia. 1990. Indigenous forest use and development in the "Maya Zone" of Quintana Roo, Mexico. MA thesis, Department of Environmental Studies, York University.

Nelson, Richard K. 1983. *Make prayers to the raven.* Chicago: University of Chicago Press.

Nyerges, A. Endre, ed. 1997. *The ecology of practice.* Amsterdam: Gordon and Breach.

Primack, Richard B., David Bray, Hugo A. Galletti, and Ismael Ponciano, eds. 1998. *Timber, tourists, and temples: Conservation and development in the Maya forest of Belize, Guatemala, and Mexico*. Washington, DC: Island Press.

Ridley, Matt. 1997. *The origins of virtue: Human instincts and the evolution of cooperation*. New York: Viking.

Snook, Laura. 1998. Sustaining harvest of mahogany (*Swietenia macrophylla* King) from Mexico's Yucatan forests: Past, present and future. In *Timber, tourists, and temples: Conservation and development in the Maya forest of Belize, Guatemala, and Mexico*, edited by Richard B. Primack, David Bray, Hugo A. Galletti, and Ismael Ponciano. Washington, DC: Island Press, 61–80.

———, and Amanda Barrera de Jorgenson, eds. 1994. *Madera, chicle, caza y milpa: Contribuciones al manejo integral de las selvas de Quintana Roo, Mexico*. Mérida, Mexico: INIFAP.

Terán, Silvia, and Christian Rasmussen. 1994. *La milpa de los Mayas*. Mérida, Mexico: Government of Yucatan.

Vayda, Andrew, and Bonnie McCay. 1975. New directions in ecology and ecological anthropology. *Annual Reviews in Anthropology* 15: 493–510.

Webster, Daniel. 2002. *The fall of the ancient Maya*. New York: Thames and Hudson.

Maya home garden in Pich, Campeche (photo by Aaron Sumner).

Celebration of local pond as giver of life in Ho'ol, regional fiesta in Campeche (photo by Aaron Sumner).

Don Pedro, a Maya healer, with an orchid saved from a site that was to be bulldozed for a cattle ranch near Pich, Campeche (photo by Aaron Sumner).

Cenote: water and vegetation in the center of Chan Kom, Yucatan (photo by Lindsay Elmendorf and Mary Elmendorf).

Checking the seed corn in a stone house (photo by Ellen Kintz).

Doña Antonia with her envelopes of healing herbs (photo by Lindsay Elmendorf).

Traditional Maya beekeeping (photo by Ellen Kintz).

Public library, Cobá, Quintana Roo (photo by Ellen Kintz).

"Take care of the plants! Don't destroy them!" (photo by Ellen Kintz).

7

The End of Innocence in a Modernizing Maya Community: "There Is No More Timber; Let's Sell the Land!"

Betty B. Faust

> [T]here is nothing in the nature of human thought to prevent it from constructing self-destructive or even world-destroying errors. . . .
>
> —Rappaport 1990, 46

The most commercially valuable timber has been cut in the modernizing Maya community of Pich, Campeche. Selling the land to outsiders is betrayal of traditions, but these traditions have been connected with poverty for generations. This chapter explores why a Yucatec Maya community decided to sell its communally owned lands, sacrificing wild animals to the bulldozers, giving up the land base that once offered security and identity. There were other options, but for most of the community, living with local resources had come to mean being poor. There are new needs, new hopes, new opportunities, and new dependencies: the lure of globalization shaped by a local history of struggles against various forms of exploitation.

LOCAL CONTEXT

Community Origins and History

Pich, Campeche, is a modernizing community of Maya cultural heritage located in the interior of the Yucatan Peninsula of Mexico. Its population has fluctuated between 3,000 and 1,400 during the past twenty years, in response to the availability of paid work. It is located approximately 85 kilometers southeast of the state capitol, Campeche City, and 20 kilometers from the large archeological site of Edzná, with its impressive ancient hydrological system (Matheny et al. 1983).

Pich has long been integrated into large, complex political organizations: successively Classic and Postclassic Maya civilizations, the Spanish Empire, and the modern nation-state of Mexico. Surface archeological evidence indicates that Pich has been a settled community at least since the Late Classic (850–1100 C.E.), when it was probably an administrative center for the agricultural provisioning of the metropolis of Edzná. Archeologists have found that this city and its water management system were under continuous construction from the Preclassic (200 B.C.E.–200 C.E.) through the Late Classic (850–1100 C.E.; Matheny et al. 1983), although research on the many ruins within the communal lands of Pich has not yet been done. After the Classic Collapse of the ninth century, Edzná declined in power, eventually being replaced as a dominant center by the coastal cities of Chak'anputun (or Chan Putún, today's Champotón) and Kam Pech (or Ah Kin Pech, today's Campeche), two major nodes of commerce along the Putun-Acalán-Tixchel trade route (Scholes and Roys 1968). Both these cities were visited by the Grijalva expedition in 1519, and their participation in the Chontal network was documented by the Cortés expedition. The network extended from present-day Veracruz to Belize, and possibly to Honduras (Scholes and Roys 1968). Pich may have been a source of labor tribute and agricultural produce for Kam Pech, approximately 85 kilometers to its northwest. It is likely to have supplied chert for stone tool making, from its numerous outcroppings of white, yellow, and red chert that people still admire and carry home to use as doorstops.

Large ecclesiastical buildings of sixteenth- and seventeenth-century style indicate a relatively large community (for a discussion of sixteenth-century population size and size of ecclesiastical buildings in Mexico, see Kubler 1948; Sanders and Price 1968, 54–55). Historical documents give 1604–1605 as the date for the Franciscan friars' campaign to establish the *Misiones de las Montañas* this area (Scholes and Roys 1968, 69, 277–280, 307, 338, 340). From colonial times to the present, Pichuleños have participated in global markets via exports of dyewood (*Hematoxylon campechianum*), chicle collection (*Manilkara zapota*), sugar production, cattle ranching, and rice production (Faust 1998).

A census of 1815 reports Pich as a town with a surrounding rural area referred to as its district. Today Pich is still the *cabecera* (district center) of a much larger area with a number of other communities in it. In 1815, its population was 65 percent Indian (Maya), 19 percent Spanish and mestizo (mixed-race Spanish and Maya), and 16 percent black (African) and mulatto (mixed race Spanish and black), as reported in Dumond and Dumond (1982, 201). Africans were brought in by the Spanish as workers on sugar plantations in this area, where their resistance to malaria was an advantage. After the Caste War and again after the Mexican Revolution, most of the "whites" (of pure

Spanish descent) departed, leaving property generally in the hands of their mestizo descendents, whose mothers and grandmothers were Maya. Statistics alone indicate something of the well-known Spanish practice of men having both Spanish wives and Indian mistresses. Of a total Indian population of 445, there were thirty more women in town than men, whereas the numbers of men and women listed as Spanish and mestizo were nearly identical (sixty-six and sixty-seven). Outside of this *cabecera*, the population of the surrounding rural district included more even distributions of men and women. These statistics would seem to support oral history that the wealthier Spaniards and mestizos tended to live in town and could afford to maintain mistresses, both black and Maya, whose descendents later mixed.

At the end of the nineteenth century, Chinese workers were brought in to build the railroad from Mérida to Campeche. Some of these workers married local people in the area of Tenabo, and some of their descendents later moved to Pich. The various ethnic groups have not remained distinct but have blended, so that within any one family there is a range of skin color and facial characteristics. Still, this does not mean that Pichuleños do not have preferences. Confidential interviews report that until very recently, having a "white" father, even if he was absent, was considered preferable to having a Maya father living with the family. A white father could potentially give one a Spanish surname, whiter skin, and help with schooling and finding a job eventually. I have been told that even in the worst case of total abandonment by the father, it was still considered better to have a whiter skin, because "the people you have to deal with, buying and selling, and those government people, they all will treat you better if you look white." Even in 1985, the fairest granddaughter of the family I lived with was very much discouraged from marrying a young man with dark skin, referred to as "black." In the twentieth century, the government has provided schools that teach racial equality and pride in the Aztecs as cofounders of Mexico with the Spanish; however, Maya accomplishments and contributions are ignored in the nationalized public schools, and discrimination against "country bumpkins" and "Indians" is still very noticeable among urbanites.

The federal government, in addition to supplying public schools, has supplied other conveniences to communities the size of Pich, resulting in migration out of smaller hamlets to the modernized towns. In addition to a public grade school, today Pich boasts a middle school and a vocational high school. There is also a medical clinic, a drilled well (160 meters) with water piped to individual homes, electricity, paved streets, and a number of government programs with visiting technicians in the areas of agriculture, health, and nutrition. Young people frequently migrate seeking employment, sometimes leaving their children with grandparents until they can establish a city home. Most frequently it is the young men who leave, sometimes returning with a wife from another place. Young schoolteachers and agricultural development workers arrive and sometimes marry local women. The mixing continues.

The Biophysical Environment

Pich is located at the edge of a valley formed by limestone hills, at latitude approximately 19° north, altitude 109 meters above sea level, with the yearly temperature varying between 48° and 15°C and averaging 27°. Average annual rainfall is 1,500 millimeters; average relative atmospheric humidity is 65 percent (Censo Municipal de Campeche 1981).

There is a six- to seven-month dry season, followed by a rainy season. People do not plant corn until the dependable eastern rains begin (formerly in May but more recently in June). Most agricultural work is done from early June through August. These are the months for planting, weeding, and doubling the ripening corn to prevent birds from eating it and rain from rotting it. Harvest for home use begins in early September with the harvest of tender ears for boiling and continues with the harvest of dried corn in a rather leisurely fashion into the new year. Dried corn to sell for cash is generally harvested in December or January. By early September the rains become more unpredictable, coming from sporadic northern storms from then until December. Crops are frequently lost to either drought or flooding. Throughout the Yucatan Peninsula, the soil is generally thin, with many outcroppings of limestone, whose vertical cracks prevent the formation of surface lakes or rivers. The forest throughout the southern and interior portions of the peninsula is Semi-Evergreen Seasonal Forest (Vivo Escoto 1964).

The area around Pich is blessed with better than average conditions: there are deep soils in some large savannas, several presently used in tractor cultivation, as well as extensive areas of medium-height forest with patches of high forest. The forested areas are used for hunting and extraction of wood, stone, and other useful materials. There are also some *ak'al che'* swamp lands where higher concentrations of clay prevent drainage of rainwater for months during the rainy season, providing excellent habitat for many species of mammals, birds, reptiles, and amphibians. The area designated as an agricultural commons consists of 5,160 hectares around the village, and includes several savannas used for tractor cultivation, some fenced cattle pasture, and other areas used for shifting agriculture. The forest commons is separated from the land surrounding the community by areas originally designated as national lands but which have been allotted to colonizing communities and cattle ranchers during the past thirty years. After half an hour's drive on the new highway and another hour on a rough lumber road, one arrives at the northern border of Pich's forest commons, legally granted in 1940 for extraction of the resin of the chicosapote tree, whose resin, called chicle, was used for chewing gum before being replaced by synthetics.

Over 90 percent of this land is within an area recognized as a national priority for conservation of mammals in danger of extinction, with birds, reptiles, and amphibians also threatened (CONABIO 2001). It forms part of the proposed Mesoamerican Biological Corridor, connecting with a nature reserve in Campeche to the north of the Calakmul Biosphere Reserve, which in turn connects to protected areas in Quintana Roo and Guatemala and Belize. Aerial photographs (INEGI 1992, 2000) indicate relatively pristine conditions, with no more than 10 percent low vegetation, much of which is natural, uncultivated savanna, with areas of seasonal swamplands.

The agricultural lands close to the village alternate between *monte bajo* (low forest, traditionally cut and planted after twenty years of fallow, but now often cut after only six or seven years), and milpas (the traditional multicropped cornfields of 1–2 hectares, or 2.2–4.4 acres). In addition, there are two large areas of former savanna (180 and 200 hectares, respectively) that were cleared by government bulldozers in 1983 and 1986 (Faust 1998, 33–50, 145–152), respectively, for cultivation by tractor rather than with the traditional digging stick,[1] and several smaller areas of deep soils cleared by groups of individuals (generally kinsmen) for tractor use (for more information concerning the local environment see Siemens and Puleston 1972; Hammond 1988; Matheny et al. 1983; Faust 1988, 32–38, 1998, 51–54).

The Village Commons: Lands and Owners

The first communal lands were given to the community for agricultural use in 1923. The forest commons was granted as an "extension" in 1940, in response to a local petition requesting the land for use in collecting chicle, to sell for chewing gum. Both grants were part of a national program of land reform initiated by the Mexican Revolution (1910–1921) and amplified by President Lázaro Cárdenas in the 1930s. The forest lands were provided in addition to those for crops and pasture, and their use was controlled by a federal department of forestry. Rights to use both types of common lands were given to all individual men who inhabited the community at that time. In Pich, some of these men are still alive. They and their sons who inherited their position are referred to as the ejido, as is the land itself (Table 7-1). The original number was 216, but it was increased to 286 in the late 1980s (review of ejido documents with the Comisario Ejidal, Reynaldo Cámara, May 24, 1995).

Only ejido members have the right to vote concerning division and privatization of land, acceptance of new members, and logging concessions. Only they receive a share of the fees paid per truckload of logs extracted. In contrast, anyone who lives in the community is allowed to use the ejido lands for their cornfields and for hunting, without payment to the ejido. In ejido organizations, by national law, women have two representatives that can speak and vote in meetings of the members. The women also collectively own a plot that can be used for fruit orchards or other purposes, and they have their own organization. In addition, an ejidatario's widow may participate as representative of a son (or sons) too young to participate. She may become a member in her own right after her husband's death, if accepted by the ejido (interview with Reynaldo Cámara 1995).

TABLE 7-1. Summary of Major Events in the History of the *Ejido* of Pich

Date	Event	Comments
1923	Ejido lands legally given to the community for agricultural uses.	This area surrounds the village
1940	Forestry commons legally given to the community as an extension of the ejido lands.	Distant lands—chicle extraction and hunting
1970s	Paved road from Pich to Campeche to benefit logging of cedar.	Increased timber extraction and hunting
1974	Mechanized rice project and arrival of colonists to form the neighboring community of Bonfil.	Rice replaces hunting, fails in less than 15 years

1983	Government project: First tractor and 180 ha of ejido cleared for mechanized farming of corn.	Replaces hunting, requires money inputs
1984	Mexican government gives Guatemalan refugees nearby "national" land for temporary use.	Replaces use for hunting, milpa
1985	B. Faust begins long-term research.	
1986	Second area of the ejido cleared for mechanized agriculture (200 ha)	Floods, cannot be used
1992	Change in national law allows division of communal agricultural lands.	Continues to forbid division of forestry commons
1994	Logging stopped by SEMARNAP, contested election (first with local candidates from opposition parties).	References to Zapatistas

(Continued)

TABLE 7-1. *(Continued)*

1995	**Logging resumed under contract, with govt. supervision and reforestation, focus on guayacán in south.**	**10-year timber contract with rotation**
2000	**Logging stopped, corn price low, approximately 100 young people migrate out, looking for work, contested election.**	**First national president of opposition party**
2001	Negotiations begin for sale of 21,000 ha in the forestry commons and 1,000 ha in the agricultural ejido, for agribusiness project with local factory employment.	Forest land said to be "degraded"

The number of places in the ejido is set by a federal commission on the basis of its estimate of the carrying capacity of ejido resources. Normally, a father can will his place to only one son. Previously, when a family left the village to live elsewhere, that man's place was given to the son of another ejidatario, one who was not going to inherit his own father's position. Occasionally a place was given to the son of an outsider who had come to live in the community, but this was rare.

According to the 1995 archives in Pich, the community held exclusive legal rights to 65,160 hectares (143,352 acres) under a government-managed communal land tenure system (Faust 1998, 34–41). On September 11, 2001, the Comisario Ejidal (the elected authority responsible for the communal lands), José Rufino Dzib, informed me that government officials reported that new assessments of their boundaries made with GPS (global positioning system) indicate that they have more than 72,000 hectares in their forest commons, plus 5,160 hectares of agricultural lands contiguous to the village. Traveling from the southern limits of the agricultural lands to the northern limits of the forest commons requires about an hour and a half. The first half hour involves using the new extension of highway built in 1998–1999 that goes southeast from Pich to Dzibilchén, connecting to the north-south highway between Hopelchén and the Calakmul Biosphere Reserve. Beyond the highway, it takes about an hour to travel the rough lumber road to the entrance to the forest commons.

Such forest commons in Campeche are generally lands originally thought to be unsuitable for cultivation and are usually quite distant from the village that owns them. Chicle and lumber concessions to private companies were approved by the federal agency responsible for forests. Chicle companies contracted with local cooperatives to buy the resin, boiled and prepared for transport in blocks. Lumber companies contracted with truck owners and their crews to cut logs, paying a fee to the common fund according to the amount of wood extracted. In Pich, these lands have also provided raw materials for small carpentry enterprises, varying in number between four and six over the past twenty years. Two of these enterprises can give official government receipts for selling to city buyers, as well as to their fellow Pichuleños.

Woodcutters normally carry in food supplies and live in the woods for a week or two at a time, cooking and doing laundry for themselves. As a woman, I could not accompany the loggers into the woods, but in 1996 my research assistant, Miguel Medina Yeh, spent five days with them, learning to use a chainsaw and ride on top of the logs. He found that they carry in gasoline, water, beans, tortilla dough, cornmeal, canned fish, salt, dried chilies, coffee, sugar, cookies, and crackers. The fish is eaten on days when no game is taken. These days game is scarcer than it used to be, but still, eating venison, wild turkey, and other favored game is quite frequent in the forest commons. Even in 2002, conversations with loggers and beekeepers indicated that game is frequently taken. In addition to lumber, the men sometimes bring out medicinal plants, vines and bark for twine, poles for house posts, firewood for cooking fuel, and so on.

In the 1970s, a large government project for mechanized rice cultivation brought farmers from other states to form the new community called Alfredo V. Bonfil, 10 kilometers northwest of Pich (Gates 1993). Shortly thereafter, this project expanded into Bolonchen Cahuich, a Maya community located between Bonfil and Pich. In the mid-1980s, Pichuleños who did not have land rights petitioned to participate in the rice project through the formation of a new ejido. It was named Chilam Balam, for the texts written in Maya by local community scribes as a continuation of the pre-Columbian tradition of writing local history in hieroglyphs. In the early 1990s, another group of men without ejido rights in Pich requested the previously designated ejido lands of San Juan Cantemó, a community situated near the center of Pich's forest commons, which had been abandoned gradually during previous decades. Some of those who formed this new group were the grandchildren of those who had previously abandoned the community in order to live in Pich because it had received electricity, water, a clinic, and a school. When the new ejido members petitioned the government for the same accommodations, they were given only the land. They continue living in Pich but use the land as a legal hunting preserve, working as guides for tourist-hunters who arrive through the interventions of an "old friend" in Campeche City, in a plan supervised by SEMARNAT (Secretaría del Medio Ambiente y Recursos Naturales, the Department of Environment and Natural Resources).

Agricultural Practices

The local farming practice has traditionally been shifting cultivation, locally referred to as milpa, and very similar to practices called *swidden* in the Old World. In this area, the traditional practice is to cultivate for two years and allow approximately twenty years of fallow to renew soil fertility. Plots are used cyclically by families over time, and fallowing areas are expected to be respected by others, as belonging to those who had originally cleared them. It has not been considered necessary to ask permission or to register use in the village offices. Only since 1983 have there been permanently cultivated fields, using tractors, herbicides, and fertilizers (see Table 7-1). Since then, individual ejidatarios have formed small groups and prepared other plots for tractor cultivation without government intervention. Fenced cattle pastures were and are locally treated as the private property of the man who fenced them. Fruit trees belong to the person who planted them, and rights of harvest are separable from rights to the surrounding land. These understandings regarding usufruct are similar to those documented in the 1930s in Chan Kom (Redfield and Villa Rojas 1934, 64–67). Before the Mexican Revolution, Pich had a sugar plantation and neighboring cattle haciendas. Earlier, there was commercial extraction of dyewood (referred to as logwood in Belize) in this area. Cattle ranching continues to be very popular, and sheep and goats have been introduced through government projects. Beekeeping depends on the pollen of flowering species in the forest commons, while producing both honey and wax for international markets.

The National Law Changes

In 1992, the federal law changed to permit the division of agricultural ejido lands into privately owned plots and for sale to outsiders, when and if the ejido voted to do so. The ejido also could decide to divide one part and maintain another as common land. However, this did not apply to the forestry commons, which remained under the supervision of what was then SEMARNAP (Secretaría de Medio Ambiente, Recursos Naturales y Pesca). Another change in the agrarian reform law allowed ejidatarios to migrate out of the community without losing their ownership and voting rights in the communal lands. This minimized the probability that any son other than the one designated heir could ever become a member of the ejido. The same can be said for sons of newcomers who have settled there, welcomed into the workforce by the logging entrepreneurs but excluded from voting rights in the ejido. Since it is usually the men excluded from ejido membership who eventually leave, this change in the law is likely to result in increased emigration. The law allows those who migrate to keep control, while denying rights to those who are using the resources and have long-term interest in maintaining them for future use by their children. This accelerates pressures for both deforestation and urban migration, despite the relative scarcity of urban jobs paying at least minimum wage (only $4.00 per day).

ECONOMIC HISTORY OF THE COMMUNITY

Fond Memories of Earlier Days

When the elders of Pich were young (some fifty to eighty years ago), the forest commons were a major source of employment for able-bodied men, providing cash income. The oldest men recall their adventures collecting chicle, including the dangers of falling, snakebite, or attacks by jaguars. Some recounted tales of an insect in those deep woods that was known to deposit its eggs inside a person's skin, so that its larvae would eat the flesh from the inside. There were also leeches in the standing water of the seasonal swamps. The men were flown into and out of the deep forest by the chicle cooperative, as the area is inaccessible during the rainy season when the resin must be collected. One woman worked as a cook for the *chicleros* in the camp, eventually marrying one of them.

Game was plentiful, not only in the high forest of the *chicleros*, but also in the savannas and hills surrounding the community. There were two species of deer (*Odocoileus virginianus* and *Mazama americana*), peccary (*Tayassu tajacu*), armadillo (*Dasypus novemcinctus*), wild turkey (*Agriocharis ocellata*), and great curassow (*Crax rubra*). Tapirs (*Tapirus bairdii*) were commonly encountered in the deep woods of the south, and tales of narrow escapes from jaguars (*Felis onca*) and cougars (*Felis concolor*) are still told, with great drama. Orchids of various species and parrots (*Amazona* spp.) delighted with their beauty, as did many kinds of butterflies. There are fewer of these species now, but people still appreciate them very much. The habits of ants, beetles, termites, wasps, fireflies, and other insects are still well known, in part because people find them interesting.

Local practices and associated religious beliefs were once effective in limiting deer hunting and preventing the cutting of trees on hilltops. Cultivation of native species (together with some adopted exotics) in household gardens, communal management of water in village ponds, and farming practices that allowed ample forest patches between cornfields all contributed to the quality of life during those times. According to oral histories, Pich was similar to other Maya communities described in early ethnographies, like Chan Kom in the 1930s, where Redfield and Villa Rojas reported,

> The obligation not to fell more bush than one needs is a part of the more general obligation never to take from the gods all the yield that is available. When the honey is taken from the hives, a little is left. When the ripe corn is taken from the field, some ears must be offered to the gods before man eats of them. When a deer is slain, certain parts must first be given to the spiritual protectors of the deer before the hunter eats his venison. . . . These offerings return in part and in symbol what is essentially the property of the gods and which is by them ceded to men of pious conduct. (Redfield and Villa Rojas 1934, 116)

In Cobá, Kintz (1990) later found similar traditions of use and respect, including community access to forest resources and respect for wild animals, wild fruit trees, water in caves and rock basins called *sartenejas*, and land for swidden horticulture. Customs condemning overuse have been found in Cobá and Chan Kom, and also in Xocén by Terán and Rasmussen (1994). The Yucatec Maya have generally practiced community access to land, with a division of labor related to resource use by age and gender, and with family ownership of wells, house lots, crops, and fruit orchards, and enjoyment of game and appreciation of many wild species for their beauty and curious habits (e.g., Redfield and Villa Rojas 1934; Villa Rojas [1945] 1978; Elmendorf 1972; Kintz 1990; Faust 1998; Terán and Rasmussen 1994; Anderson, chapter 6, this volume; Kintz and Ritchie, chapter 10, this volume). Beliefs in animal soul mates are common throughout the Maya area and have been analyzed with reference to the symbolic substrate for political action by Gary Gossen (1994).

Before the paved highway was constructed, deer and wild turkeys occasionally wandered into the outskirts of Pich, and hunting was easy. Many recall feeding their families the meat of deer and wild turkeys several times a week, maintaining that on this diet people did not get sick. Village records giving the cause of death indicate otherwise, and more detailed questioning eventually reveals that people did in fact sometimes die of fever, diarrhea, and pneumonia.

Elders also acknowledge that vaccinations have reduced the incidence of diseases like chicken pox, measles, diphtheria, typhoid, and tetanus; however, they still insist that there is much more illness now, although it "does not kill people because of the clinic." A possible explanation of this observation is that the community is now suffering from a relative lack of quality protein, of essential amino acids, formerly provided free by meat from the plentiful game animals. Daltabuit's (1988) data on the diets of three families in Cobá indicate a lack of both protein and iron in that Maya community, at a time when game was becoming relatively scarce.

The Golden Age of Timber

After World War II, the market for chicle gradually disappeared, because synthetic substitutes for chewing gum had been developed. By the 1960s, timber cutting in the forest commons became the major source of cash income, funding a golden age that is still fondly remembered. I asked one village consultant if the paved highway between the state capital and Pich had created the timber business; he replied no, it was the reverse—the timber made the road. Pichuleños built it to get their logs to market more easily. Previously they had hauled logs out with mules, but in the 1960s they struck a deal with the state governor, Carlos Sansores Pérez. The men of Pich provided the labor and even some of the money to build this twenty-kilometer stretch of road, connecting their town with a newly paved highway to the local port and state capitol, Campeche City (Jacobo Martín, personal communication, February 8, 1999; the highway construction is documented for 1968 in Encalada Argaez 1987, 297–298). The main species cut for commercial sale during those years was Spanish cedar (*Cedrela odorata*).[2] Mahogany does not grow easily in this area, according to Pichuleños. A variety of lesser known species of valuable trees were also cut, including *ciricote* (*Cordia dodecandra*), *yaxnik* (*Vitex gaumeri*), *pukte* (*Bucida buceras*), *pich* (*Enterolobium cyclocarpum*), and *granadillo* (*Platimiscium yucatanum*) (scientific identifications by Suárez 1977, 202–203, who documents their extraction for commercial use throughout the peninsula in the nineteenth century). Pichuleños also mention these trees as ones that they have cut in response to requests from buyers in Campeche City.

The 1960s and 1970s are recalled as wonderful times when local entrepreneurs bought lumber trucks, and relatives and friends came from other communities to work in Pich. These original entrepreneurs provided paid work and inspired (by example) other community members to buy more trucks and participate in what is now recalled as the "lumber fiesta." At that time, the community enjoyed three movie theaters (now there are none). There was money for parties and to buy guitars, blenders, stereos, radios, bicycles, and even a few televisions. In 1985 when I first arrived to do fieldwork, community members generally identified five extended families as "rich" and everyone else as "poor," despite acknowledged differences within both groups. The rich families included more than twenty married couples with children of their own, who were descendents of the first entrepreneurs of the 1960s. They were believed to have originally been "poor, like the rest of us," and to have become rich through logging the community's forest.

Loggers had to pay a fee to a fund for community improvements, since they were taking wood that belonged to everyone. Purchases made by the community with these funds included a park with playground equipment, a basketball court, and a dance pavilion by the town pond. There was also support for local baseball and soccer teams and assistance to young people attending high school.

In the early 1980s, the fees paid to the community for the lumber extracted were used to buy a new bus, providing competition for the first bus, owned by one of *esos ricos* (said with a tone of resentment, this perhaps is best translated as "those rich guys" of Pich) who had made his money logging. Both buses left Pich and returned within half an hour of each other, and the rivalry was keen between the communally owned bus and that of the *rico*. Pich was then at the end of the paved road. Its buses served people all along the route, and Guatemalan refugees walked three kilometers from their camp to take the bus in Pich. Both buses generally had all seats full, with a number of passengers standing for the hour-and-a-half trip.

At that time, more than thirty locally owned trucks were actively hauling wood. People complained that some outsiders were also taking advantage of the road, taking out both lumber and deer without permission. It was then believed that there was no way to stop them. This contributed to feelings that the community should get the wood out as quickly as possible, before other people robbed them. Yet people were already lamenting, "*Ya no hay cedro*" (Now there is no cedar). There were such trees in the village and more in the forest commons. What the poor were bewailing was a lack of trees freely available to cut for local use. Trees in house yards were private property, and those still standing in communal lands were too far away to bring home without paying the truckers, who in any event wished to sell these same trees for cash. Spanish cedar had been used for generations to construct furniture for local households. It is easily worked; impervious to termites, mildew, and mold; smells nice; and is attractive. However, its commercial value had by then put it beyond the reach of the village poor, resulting in considerable resentment.

By the 1980s, SEDUE (Secretaría de Desarrollo Urbano y Ecología) began to enforce lumber regulations more strictly in order to allow for forest regeneration, as well as attempting to enforce the prohibition on deer hunting. Both were greatly resented. Since the price of corn was controlled at an unacceptably low level by government agents, growing it for sale was not very attractive to those who had alternatives. Several of the older men told me that the young people (anyone under age fifty) had forgotten how to grow food for sale to the cities, because there was easy money cutting trees. One man known for his hunting skills added to this interpretation, explaining that the government was prohibiting hunting in order to force peasant farmers into growing corn for the cities.

By 1981, the Censo Municipal de Campeche reported that the logging industry of Pich employed 18 percent of the male labor force, but accounted for only 5 percent of the cash income. In contrast, cattle raising employed less than 4 percent of the labor force but produced 67 percent of cash income, reflecting the growth of a local upper class of *ricos*. Meanwhile, beekeeping produced 19 percent of income, despite being a secondary occupation that was not even counted in the occupational categories. Agriculture was the most common occupation, having 67 percent of the workforce, but producing less than 8 percent of cash income! As another indication of the division between rich and poor, agriculture had come to be seen as a default occupation, a last resort for those unable to pay the startup costs for cattle, bees, lumber trucks, and chainsaws. It also continued to provide for family subsistence, a nutritional safety net when city jobs were lost or the prices fall for honey, beef, or lumber. This is a critical livelihood function that is not assessed in the census, since it cannot be measured as cash income.

Some local consultants, however, insisted that the government census figures were wrong, that even in the mid-1980s much more logging was actually being done than was reported. Several loggers with whom I talked in 1986 reported whacking rough passages through the woods for their employers in order to get the trucks to the new highway without being seen in the village, thereby evading community tax obligations. Thus the government statistics probably do underestimate the real amount of logging going on during these years (Faust 1998, 34–40).

In support of this village view, in 1985–1986 I could find little concern among either loggers or owners of logging trucks about the supply of trees. All insisted that there was still plenty of Spanish cedar to cut, and other species as well. Most did not imagine that the timber supply would end; the forests to the south were huge. Some reported driving their trucks to Guatemala and bringing back even bigger logs to sell. There was some vague notion that this was illegal, but there was practically no enforcement, and such piracy was experienced as a grand adventure. Forests were believed to perpetually maintain themselves, providing bountiful crops of valuable wood. Trees grow by themselves (*por sí*), so there is no need to plant them; there are always little ones growing, and they will someday be big. But one eight-year-old boy seemed to understand. Domingo asked me to take his picture on top of a pile of logs before they were hauled to the city, "so that I will remember what these trees were like when I grow up."

The Crisis of 1994–1995

At the end of September 1994, as part of catching me up on what had been happening since I had left, Pichuleño friends repeatedly told me, "There is no work." Then would come the explanation that woodcutting had been stopped, the reasons, the names of people who had left to look for work in cities, and descriptions of efforts to find work. Growing corn for sale generally was understood to give a return for labor expended that did not even meet the official minimum wage (equivalent then to about U.S. $3.00 per day). However, when other work was unavailable, planting milpa at least resulted in food to eat, if the rains came. Thus, subsistence farming was seen as a last resort, and migration to search for paid labor was preferred by the younger men when woodcutting was prohibited. Beekeeping and the sale of honey and wax through a cooperative provided some cash for those too old to easily find paid work as laborers in the cities. Some of the younger men combined it with milpa and raising a few head of cattle pastured on communal lands. Some of the *ricos* kept cattle and farmed inherited shares of small ranches (100–300 hectares) bought by their fathers in 1980 from neighboring Hacienda Solbul. Others traded their lumber trucks for buses or bought taxis for sons who would use them in Campeche City.

In 1985, decreases in the availability of timber were generally denied or mentioned indirectly, but by December 1994 the lack of large trees was openly discussed in relation to a local political crisis: the first local election involving candidates from opposition parties in more than sixty years. This followed the emergence of the Zapatista rebellion in Chiapas in January 1994. Some people said, "Being peaceful does not get us anywhere. We should try shooting things up like the Zapatistas." The 1992 change in federal law allowing the division and sale of ejidos also provoked local conflicts. Interestingly, it was the *ricos* who wanted to keep the communal lands and the *pobres* (poor people) who wanted to sell. These issues affected perceptions and loyalties of the three major candidates for mayor of Pich, who were aligned with national parties. Unaccustomed to open competition, people responded with fear and threats. There was a great deal of discussion in January about who had in fact won the December election and who had paid people to vote for them, and how much. The still-ruling national party backed its candidate's claims to having the most votes and attempted to install him in office. A sizable portion of the populace was involved in various resistance actions, which included attempting to burn the roof of his home and padlocking the town hall. The standoff continued until armed troops were sent in from the state capitol.

By August and September 1995, even some sisters, brothers, and cousins were no longer on speaking terms, in response to accusations concerning fraud in the election of the town's administrative officers. In contrast, the 1985 elections and the January 1986 change of officers had occurred without incident, even though a few friends did inform me that those *ricos* always rigged the elections. I was not present during the 1988 elections and assumption of office; however, I have been told that that year's mayor (elected for a three-year term) had to be relieved by a temporary replacement because he was unable to control a number of conflicts in the town.

On the surface, the conflict was over a contested election; however, the underlying problems were decreasing work opportunities and conflicts over the choice of selling or keeping the communal lands. Both were seriously affected by the overuse of Spanish cedar that had resulted in decreased employment opportunities, decreased cash income for families, and a decrease in money for the community from the remaining logging. The opportunity to divide the community fund came at a time of increasing family poverty, when its division made it possible to recuperate a little of the lost income. Despite this division, the overall situation continued to worsen. According to several community members, the social consequences of the new scarcity of timber included political conflicts, emigration, looting of archeological sites, wife beating, hunger, and "sadness."

In earlier years, people who saved had generally been able to buy one new household appliance a year, for Christmas. Blenders, toasters, radios, tape players, watches, bicycles, and clocks were commonly found. Televisions, stereos, refrigerators, gas stoves, and electric fans required more savings, but a number of them were visible in people's homes, sometimes under thatched roofs. Bathrooms were of less interest, although some invested in this status symbol for the use of guests from the city. Nearly everyone had both electricity and running water in their households, and made use of the local government clinic. The wealthier people traveled to private doctors in Campeche City.

In 1985, participants in the rice project were able to compensate for a gradual loss of income from logging; however, that project ended in the early 1990s. The people's gradually decreasing livelihood was dramatically accentuated in 1994 when the government prohibited logging. At that point, the short-term profit made by the community's *ricos* from the sale of community-owned trees was reevaluated by the community's majority, the self-defined *pobres*. The consequences of overexploitation had become evident to the latter. The past short-term benefits of wage labor were forgotten in the shock that followed this first government prohibition of logging. The prohibition was said to be because the loggers were cutting small trees and not giving them a chance to grow, having already cut the larger ones. However, it is also true that the prohibition and the contract that followed it were associated with changes in national agencies at the end of President Salinas's term and the beginning of President Zedillo's (when SEDUE became SEMARNAP), just as recent efforts to develop agribusiness have occurred within the context of the change from President Zedillo's term to President Fox's (when SEMARNAP became SEMARNAT). Such changes do not necessarily reflect the intentions of the new president but rather perceptions of local government authorities concerning the new regime's expectations and priorities.

After the woodcutting was stopped in 1994, it quickly became obvious that the majority of men could find little or no paid work while "*esos ricos*" were still enjoying the investments made with the profits from the communally owned timber. Newly wealthy, they had invested their profits in private ranches, herds of cattle, and education for their children. Thus, the consequences of overuse were very different for those who became rich than for those who had more minor and more temporary improvements in their lives due to increased cash income. Previously, these differences and the decline in timber had been denied, or were alluded to very indirectly. Most extended families had at least one member, or an "adopted" relative, who benefited to some degree from employment in woodcutting. But by 1995, the advantages had disappeared and the overcutting was public knowledge, openly discussed. Many people wanted to tell me their side of the story and whose fault it really was that there were no longer any trees to cut (Faust 1998, 41–50). Those who continued to reside in Pich, as well as others who returned temporarily from their searches for employment elsewhere, angrily commented on the overcutting of timber, the failure to reforest, and their belief that sufficient money had not been paid to the community by those who became rich in the lumber fiesta.

Overuse occurred despite the fact that there had been government warnings of the impending destruction of the resource, and legal prohibitions, as well as extant indigenous traditions of respect for the forest and knowledge of many species and their interdependencies. SEMARNAP stopped the lumbering with many explanations concerning overuse. However, by Christmas of the following year, and despite their previous prohibitions and pronouncements concerning the need to allow regrowth, local government officials were persuaded by the Pichuleños (with some help from the lumber concessionaire) to resume lumbering. A compromise was reached. A ten-year plan was made for "sustainable" logging of guayacán (*Tabebuia guayacan*), chakah (*Bursera simaruba*), and a little remaining large Spanish cedar. Reforestation of Spanish cedar was begun in a project that provided cash wages for the loggers during much of the dry season. It was paid for by the lumber concessionaire and supervised by SEMARNAP. Renewed logging was enabled by the construction of new roads into the most remote and pristine areas of the forest commons, roads constructed by the men of Pich and paid for by the lumber company, again under supervision of SEMARNAP. This resulted not only in the extraction of timber but also in increased hunting in an area previously rarely visited.

Pichuleños informed me that the guayacán was sold to the Japanese, the chakah was used in Campeche City to produce plywood, and the cedar was used both for plywood and for furniture. Only those trees fifteen inches or more in diameter at breast height could be cut, and the plan specified a different section each year. The goal was to allow the smaller trees to continue growing so that at the end of ten years, the loggers could return to the first section and again have wood to cut. This project lasted only six of its anticipated ten years, until legal logging was again stopped in 2001, this time because the agreed upon restrictions had not been honored, according to the elected head of the ejido, José Rufino Dzib (personal communication, September 11, 2001).

Effects of Forest Use on Community Population

The increases and decreases in wage labor opportunities from lumbering are believed by many members of the community to have affected the size of the local population. The census data available are not very complete but generally confirm the belief that the rise and fall of the lumber industry has had a dramatic impact on the community population. Farriss (1984, 397) has published the 1806 census data, showing a population of 1,838 for Pich. In 1981, when logging was providing 18 percent of the community's paid employment, Pich had 3,000 inhabitants (Censo Municipal de Campeche 1981, 1). However, by 1985, a local census reported a population of only 1,565, a loss of nearly half (Faust 1988, 91; Censo de la Clínica de Pich 1985). By 1990 the population had not recuperated; at 1,478 it was slightly less than in 1985 (INEGI 1990). Unfortunately, no census information was collected in the 1985 or 1990 census concerning the percentage of employment provided by logging. More recently, the number of houses has increased as people have started building along the new access roads, and speculation concerning division of the ejido has accelerated.

The Politics of Stones

The destructive use of communally owned resources by a minority—elite or otherwise—violates traditional Maya beliefs expressed in local talk and in ritual symbolism concerning interdependency, connection, and the special care of those aspects of the world used directly by humans (water, soil, useful plants, game and domestic animals, useful trees, stones). In Pich today, it is generally acknowledged that what has divided the community is differences in wealth. There were differences in wealth before the first highway, although less than now, but what is more important is the perceived loss of a sense of responsibility on the part of the wealthy combined with their recent degradation of the community's timber resources (Faust 1998).

During the 1986 failure of the government-provided water system, people recalled how during droughts the rich used to bring in water in oxcarts. Hence, they expected their rich village mayor to do something to resolve the problem. He waited for the state government to act, narrowly escaping a riot. When the government water trucks finally began to arrive, public sentiment was so strong against him that he decided to take an extended vacation (Faust 1988). There are also frequent references to how the patrons used to keep corn stored to give to their workers, friends, and relatives when crops failed. This obligation of the rich to the poor was generally channeled to specific poor people who were connected to rich families by many-stranded ties—economic, political, and social. However, it is also true that the village has a strong tradition of helping newly arrived strangers, ones not connected by kinship, marriage, or work to any of the patrons. I personally benefited from this tradition on a number of occasions and have watched the village respond to other strangers, including Guatemalan refugees from the nearby camp.

Resentment of the rich was tempered by dependency. The poor could turn to the rich for help in emergencies. They said, "We need even the stones," explaining when I did not understand that the rich are "hard," giving only reluctantly to the needy, but are still there for support, much as hard stones provide stability to tree roots in storms. But that was before the rich began to look to their peers in the city for norms and to see their poor cousins in the village simply as physical laborers to be paid by the day or hour. Desperate Guatemalan refugees settled nearby also provided cheaper labor, further reducing the traditional sense of obligation to local workers.

The combination of all these factors created pressure for emigration. More than 100 young people from the community have left home looking for work in distant cities during the past three years. Migration to the state capitol, Campeche City, and for work on commercial farms had occurred previously but has increased, while migration to more distant cities, such as Cancún and Mérida, is relatively new.

THE DECISION TO SELL THE FOREST COMMONS

Beginning in 1999, a new highway connected Pich with the biosphere reserve and archeological site of Calakmul to the south, as well as to the town of Hopelchén and the archeological site of Uxmal to the north, linking it to the Ruta Maya (Garrett 1989). It is access to this highway that made the forest commons of Pich particularly attractive to potential buyers in the new millennium. It also provided incentives to build a structure to accommodate tourists where the access road to Pich leaves this new highway. The tourist facility has been paid for by a government program and includes plans for Pichuleños to set up small eateries and craft shops.

Suffering from changes in the prices and conditions of their principal products, Pichuleños have decided to sell 22,000 hectares of their commons to an agribusiness firm. Even under the recent changes to the ejido laws, a forest commons should not be divided (José Humberto Castillo, personal communication, January 9, 1995; Victor Alcántara, personal communication, January 12, 1995). However, government permissions have been forthcoming, said to be based upon a legal change in the classification of the land from forest to agricultural. The price promised to the ejidatarios is locally understood to be $1,800 pesos per hectare, or $39,600,000 for 22,000 hectares (including 1,000 hectares of agricultural lands in addition to 21,000 hectares in the forest commons). The author has seen spider monkeys and orchids there, and scat identified by hunters as that of deer and peccary, as well as seeing and hearing wild turkeys and parrots. Aerial photographs for sale by the government agency INEGI (Instituto Nacional de Estadística, Geografía e Informática) indicate that no more than 20 percent of this area is of low vegetation (natural savanna, fallow fields, agricultural fields, or areas burned intentionally by deer hunters to provide good forage for the deer herds; Sabido Zetina 2001).

Some of the young people are concerned about the local environment. In July 2001, twenty young people from Pich helped clean the community's pond with a volunteer group of high school students from Mexican cities, under the supervision of Cuerpos Mexicanos de Conservacion, A.C. (The Mexican Conservation Corps, with financing from Partners for the Americas, as well as local donors). However, sale of nearly a third of the forestry commons by their parents and grandparents will make any return to traditional ways very difficult. Supplying meat by hunting wildlife, as well as use of other nontimber forest products, generally requires an enormous area of forest (for a review of this literature, see Gowdy 1998; Emery and McLain 2001).

SEMARNAT prohibited further cutting of timber in 2001. It is basically the same agency (with changes in name and personnel over time) that has approved the change of use of the soil from forestry to agriculture so that the agribusiness project can proceed. National law prohibits clearing of forest to make agricultural land, but the agribusiness firm and the ejidatarios insist that the land has already been cleared for cornfields throughout the majority of this area. In September 2001, the then director of the Campeche office of SEMARNAT, Ing. Carlos Rijani, gave me his personal assurance that no areas of high forest will be cut for agricultural use and that the country's laws concerning protection of endangered species will be enforced. The mechanisms for deciding which areas will be preserved are not clear; nor is it clear that corridors of natural vegetation will be left between the patches of high forest so that animals can safely migrate.

On September 11, José Rufino Dzib told me that the company has promised full-time employment to 600 workers in two factories associated with dairy farming, with hiring preference for Pichuleños. A newspaper article ("Proyecto agrícola," 2001) previously described the project as costing $23 million U.S. and involving 10,000 hectares of manioc plantations, plus pasture for 24,000 head of dairy cattle and two factories. One will produce supplemental cattle feed from manioc; the other will produce yogurt and dried powder from cow's milk. The legal members of the ejido will receive more than 130,000 pesos apiece from the sale of the land (equivalent to approximately $13,000 U.S. at a rate of exchange that hovered around 9.3 during 2001, rising to 11.3 in late 2003). Considering a daily agricultural wage of 60 pesos in Pich, which does not adequately feed a family of five and still pay for clothing and school supplies, this sum is analogous to winning the lottery. The cash amount is far beyond what any local resident could hope to save in a lifetime. Furthermore, there is the attraction of receiving a cash income from a permanent job in the community, without having to migrate. The advantages to the ejidatarios are thus overwhelming, and the value of the forest is invisible to them now that the timber has been cut and further extraction in the near future has been prohibited. Waiting twenty more years for the reforestation project to produce legal-sized trees is just too difficult. Moreover, the wildlife protection division of SEMARNAT keeps punishing the people for hunting. They feel they would be stupid not to take advantage of this new opportunity, even as they quietly grieve for the trees that will be burned and the animals that will be killed by the chain dragged between two giant bulldozers. They will use the money to buy private ranches, some say. The local healer is trying to grow plants in his yard that he formerly collected wild in the forest, as he believes that the clearing will be total. He took the precaution of moving some orchids shortly after the initial negotiations began.

New market opportunities, "improved" (modern) technologies, and new information have changed values and perceptions, creating deep schisms in the community. Throughout the Maya area, programs for development and natural resource conservation characteristically are in contradiction, and neither has systematically taken advantage of local environmental knowledge. That same knowledge is in danger of extinction as the young people are not interested. Their future lies elsewhere, according to their schoolteachers, the television, their friends, and even their parents.

According to an interview with a member of the ejido, in October 2001 the ejido voted to sell 22,000 hectares to the agribusiness firm Guelpell, in addition to the fifteen private ranches that had already been sold to them by individuals in Pich. They have offered assurances to the community that they will not use areas of high forest where there are monkeys and orchids, nor will they bulldoze ponds. This sale has been approved despite federal designation of the majority of the area as a very high priority zone for biodiversity conservation (CONABIO 2001), not to mention the fact that conversion of land use from forest to agricultural land is generally prohibited by regulations under the national forestry law, and that the destruction of habitat critical to the survival of several endangered species is prohibited by regulations under SEMARNAT (2002).[3]

Only four years ago, even after witnessing a variety of changes in social organization and technology, I described the potential for integrating modernization and traditional local knowledge in the management of forest resources and biodiversity (Faust 1998). However, a large purchase price and the possibility of factory employment near home have proven to be more attractive.

IMPLICATIONS FOR ENVIRONMENTAL AND DEVELOPMENT POLICY

The Maya community of Pich is a modernizing one whose problems reflect global issues of making the modern market system accountable for its effects on the biosphere. The sacrifice of nature to what is seen as progress is similar to that described by Andrews et al. (1974, 1) for Isla Cancún:

> Much of the north arm and adjacent Caribbean front of Cancun has been bulldozed, the *cocales* [coconut groves] are largely a thing of the past, and the mangrove swamps are being filled in. . . . Ironically, a focus of this frantic alteration of the landscape is the strip behind the magnificent stretch of beach north and south of the midden, which until we appeared in 1963, saw no change for over 2,000 years.

As the network of multinational corporations has extended beyond national borders, it has become increasingly difficult to control by law—and increasingly becomes irrational and destructive of the fundamental processes needed to sustain human societies (Korten 1995; Isbister 1993). Capitalism depends on the exercise of law and the control of deceit and theft in the pursuit of individual happiness—otherwise, as Hobbes ([1651] 1958) noted long ago, human society is merely a "war of every one against every one" (p. 108), in which "the life of man [is] solitary, poor, nasty, brutish, and short" (p. 107). The control of externalities affecting communal resources and public goods may be seen as a control of both deceit and theft.

Dependency theorists and other critics of "developmentalism" have repeatedly reported negative ecological impacts in the tropical world resulting from capitalist industrial expansion. Yet, conditions in the former Soviet Union and associated nations, as well as mainland China, make it clear that Marxist or Maoist attempts at modernization have also failed to maintain resources and protect ecosystem services. Environmental conditions in many territories formerly or presently under communist development regimes are very bad, according to reports of travelers, news media, and visiting scientists (for summaries, see Tichle and Welsh 1998 and Wedel 1998). It seems that neither Judeo-Christian religious beliefs (White 1967) nor Western capitalism (Gunder Frank 1967, 1979; Sandoval-Palacios 1982; Tudela 1989, Wallerstein 1976, 1980, 2003) are single causes of our global problem. It is, rather, the expanding exploitation of lands and hands, upon which civilizations have always depended, which is now greatly accelerated by the computerization of both industrial civilization and global commerce. It remains to be seen whether a postindustrial civilization will emerge, based perhaps on solar power, recycling, population control, accountability for pollution, and limits on consumption. It could have some of the positive aspects that are beginning to disappear from Pich, to everyone's dismay: access to managed wilderness areas, respect for other species, joy in nurturing the young and the old. People in Pich fondly recall an intimate community where neighbors were not strangers, gender relations were interdependent and mutually respectful, families produced most of the items they used from abundant local resources, and the ceremonial arts were participatory events rather than spectator sports.

Hopefully, the examination of processes of change in traditional villages, such as Pich, can help us devise a unified social and ecological paradigm that will serve as a basis for political organization and policy. We face increasing signs of a global environmental crisis—when both capitalism and socialism seem to have failed, and religious fanaticism appeals to many of the disadvantaged. Reflection on both the attacks of September 11, 2001, in the United States and the rising violence in the Middle East makes urgent the need to find solutions to conditions that make suicidal terrorism possible as a military strategy. To look at the world from the perspective of poor villagers at the bottom of the system is not comfortable, either morally or intellectually (e.g., Bodley 1982; Marcus and

Fischer 1986). It necessitates that we examine our own categories of thought, our cultural unconscious: the sacred divisions between religion and science, nature and technology, the individual and society, our nation and our species, male and female, the present and the future. It requires making connections between the problems of common property management on the scale of a single community and the management of global commons such as the oceans and the atmosphere (Bromley 1992).

One of the challenges is to comprehend the environmental knowledge represented symbolically in traditional ritual and everyday talk in an indigenous community. This is a first step in understanding some of the linkages between practical reason and the sacred categories of thought that have been seen as binary opposites, the basic dichotomy of Western civilization, set out so clearly in the early teachings of St. Thomas Aquinas and Aristotle (for a general discussion of the problem see Faust 1988).

The dialogic integration of categories in Maya cosmology and ritual noted by B. Tedlock (1982) appear to have functioned in traditional contexts to restrain individual behavior in the interest of the public good (Faust 1998). Symbols and cosmologies channel human emotions, generate political will, and inform individuals concerning their place in the cosmos (Durkheim 1912; Sapir 1949; Gossen 1974; Vogt 1976; Geertz 1983; Turner 1967, 1975). Such traditions may not prevent overuse of resources in the presence of new technologies, markets, and government programs (McCay and Acheson 1987; Bromley 1992). However, they were able to maintain a well-balanced system over many centuries (with occasional disruptions from epidemics, drought, and war).

As Laura Bohannan (Bowen 1964, 297) learned from the Tiv, traditional communities suffer through those catastrophes they cannot control and then, as in her book's title, they "Return to Laughter." They rebuild the familiar system with the people who survived, planting again when the rains come. This, then, is no argument that traditional cultures lived in a perfect ecological heaven in which all processes were always homeostatically controlled by some mystic power of spiritual communication between noble savages and pristine wilderness. On the other hand, I have experienced, like Elmendorf, the remarkably integrated life of plants, animals, and humans in a Maya community.

> As I had looked around me on my first visit to Chan Kom, I had felt a strange feeling of being in a balanced aquarium. In the thatched hut where I ate and chatted, I never saw a fly, never even an ant. In spite of the setting turkey in the corner, the hens and chickens, the innumerable pigs and dogs in the yard, I never once stepped in filth—not even in the dark. Baby chickens ate the fleas off sleeping dogs. This strange cleanliness, this exciting ecological equilibrium continued to amaze me everywhere until my last trip to Chan Kom. When the road opened, when the outsiders started arriving, all at once there were swarms of flies—in the corner store, in my room, in the plaza. I saw three scorpions in one day. Perhaps it

> was simply coincidence, but I felt that the aquarium was already being thrown out of balance. But this original equilibrium was just a part of the larger feeling of harmony with nature, with earth and sky and wind and water, with the feeling that all essentials are really present. (Elmendorf 1976, 121–122)

Clearly, one cannot count these impressions as verifiable data; however, by the time we devise a way to measure such ecosystems, they may no longer exist. In the meantime, we have a great deal of data on what we call the long-term effects of pesticides, herbicides, mechanized agriculture on fragile soils, salinization of irrigated soils, soil erosion related to lumbering, contamination of ground water from agrochemicals, and so on (for one summary, see Bates 2001). In these cases, "long-term" is considered to be ten years or more. There is some evidence that Classic Maya civilization overused milpa soils, suffered from silted-in reservoirs, and cut the majority of forest to supply wood, not only for cooking fuel but also to produce lime for cement. However, this damage occurred over hundreds of years of use, not within the ten or twenty years that it takes to damage soils and forests with modern, industrialized methods.

The evaluation of traditional or indigenous knowledge should not be based on examples of when and where that knowledge did not prevent environmental destruction (Faust and Smardon 2001). If such criteria are to be used, they must in all fairness also be used to evaluate the efficacy of scientific knowledge for prevention of environmental destruction. Said knowledge has not been able to stop nor even appreciably slow the vast majority of accelerating environmental destruction caused by the modern industrial system of commerce in which it is immersed (Korten 1995).

As victims of a progress (see Bodley 1982) that has included many cultural and biological losses, some of Pich's young people have begun to distrust modernization, asking, "Is it too late to go back? Can we retrieve some of what we have lost and still survive economically in this new world?" Most, however, just want out. They are studying to get city jobs, they say. Staying in the village is increasingly seen as a last resort for those who are unable to leave.

EPILOGUE

In the fall of 2002, Pich slowly recuperated from its worst natural disaster in memory. On September 22, Hurricane Isidoro caused more damage than any hurricane in a hundred years, both to the city of Mérida and to many interior communities of Campeche and Yucatan, according to television broadcasters. Pich and several other neighboring communities were badly flooded, and Guelpell's installations gradually emerged from under water in the month of November. The irony is that the engineer in charge of preparing the fields for irrigation filled in the *suuts'*, a natural drain hole for a very wide area. Pichuleños report having tried repeatedly to warn him and being brushed aside. He knew better. He plugged up the *such*, reportedly so that there would be plenty of water for the four or five wells dug to supply the project. Huge John Deere equipment, houses, trucks, the irrigation equipment, and the manioc plants all ended up under several meters of water, which was still there five weeks after the hurricane on November 3, when I visited for the Day of the Dead. I asked the local shaman (*h-men*, in Mayan) if he had called down this rain to punish the agribusiness firm for having destroyed the homes of so many animals. He laughed and said they were being punished all right, but by God and for having filled in the *suuts'*. A very direct and logical consequence! Despite the horrendous destruction of houses, animals, crops, and so forth, people were very grateful for the emergency supplies brought in by helicopter by don Arón, the man understood to be the owner, or manager, of the project. They were also pleased that he was not going to pull up stakes and move elsewhere. They told me with evident relief, "He says he has plenty of money to take care of this and the project will continue." They are still waiting to be paid for the forest commons, and the firm is still waiting for government approval.

The locally elected head of the ejido, the *comisario*, told me on November 9, 2003, that he had been present in Mexico City to witness the signing of the official papers and that they would be receiving their money on December 20, 2003. Individual bank accounts with debit cards had already been established for each member of the ejido. However, on December 12, various ejidatarios told me that they had been told earlier that month that they will not receive their money until sometime in January. This has caused considerable economic hardship as people had bought items on time using up savings to pay a down payment on a vehicle, a major household appliance, a bicycle for a child's Christmas present, with the understanding that they could pay off these purchases on December 20th. Christmas is traditionally a time of major purchases for the entire year. The payment of an extra month of salary, or at least a major gift, is customary for employees in Mexican businesses and government agencies. Typically in rural households, the December harvest helps pay for major purchases. The delay of payment is causing a major problem during the holiday season and may create long-term financial problems for many families unable to make payments on items purchased in anticipation of the payment. Simultaneously with this change of plans has been a change in the director of SEMARNAT for the state of Campeche. It is unknown if this political change has affected the prior arrangements of Guelpell with SEMARNAT regarding the purchase of ejido land in a forestry commons.

ACKNOWLEDGMENTS

This chapter has evolved from three different presentations in professional meetings. The earliest, in 1990, was "We Need Even the Stones," presented in the session titled Conceptions of Poverty in Modernizing Communities, organized by S. Wadley in the Eighty-ninth Annual Meeting of the American

Anthropological Association (New Orleans, LA). The second, in 1997, was "End of Innocence: Over-Used Forests and Under-Used Traditions," in the session, Yucatec Maya Political and Cultural Ecology: Papers in Honor of Mary Elmendorf, organized by E. N. Anderson and B. Faust for the Ninety-sixth Annual Meeting of the American Anthropological Association (Washington, DC). In 1998, parts of the chapter appeared as "Consequences of the Over-Use of Local Resources in a Yucatec Maya Community in Mexico," in the session titled Ressources non-renouvelables: Energies du futur y développement soustenable, organized by Philippe Lefevre-Witier for the Conference Internacionale en Ecologie Humaine et Environnnement: Autonomie et Dependance (Toulouse, France, a precongress of the Eighth World Congress on Ecology, INTECOL 98). The author wishes to thank the discussants, audience, and fellow participants for their criticisms, questions, and suggestions. Miguel Medina Yeh accompanied and interviewed participants in logging activities, providing valuable information. Loggers, community authorities, truck owners, and other members of the community have generously shared their experiences, concerns, and hopes with me over these many years, in addition to providing hospitality. Sue Wadley has provided tremendous support during a recurrent family crisis and a wonderful example of long-term research in India, analyzing processes of change in the context of changing beliefs. Jack Frazier has contributed keen observations on shared visits in the field, patient questioning about the biological impact of the community's behavior over time, and helpful references. Discussions with Willy Folan, Joel Gunn, Gene Anderson, and Ellen Kintz have contributed greatly to my understanding of Maya environmental knowledge and processes of change. In the writing stages, both Jack and Gene have contributed many valuable suggestions for improvements in organization and writing style. I also wish to thank my fellow editors and the authors of this volume, as well as the editorial staff of Greenwood Publishing Group, for their patience with my delays in the publishing process during a period of family crisis. I am deeply grateful to my own family for all their efforts to support my writing and editing, despite their own problems during this period of time.

Research was supported by the Centro de Investigación y Estudios Avanzados del I.P.N., Unidad Mérida (CINVESTAV, Center for Research and Advanced Studies, Mérida Campus) of Mexico and a Smithsonian Small Grant, building on earlier work financed by Wenner-Gren's Small Grant Program and a Shell Foundation Grant for Research in Developing Areas.

NOTES

1. Only the former of these two areas is still being used, since the latter was found to be subject to frequent flooding, sufficient to rot the roots of the corn, which is the main crop of Pichuleños. Income from these mechanized plots has not replaced lumbering as a source of income, as was originally hoped.

2. The generally accepted common name for this species in English is Spanish cedar; however, it is a tree native to Mexico.

3. Government responsibility for protection of the environment has been included in different agencies during different national presidencies: under President de la Madrid (1982–1988) it was Secretaría de Desarrollo Urbano y Ecología (SEDUE, Department of Urban Development and Ecology); under President Salinas (1988–1994) it was Secretaría de Desarrollo y Solidaridad (SEDESOL); under President Zedillo (1994–2000) it was Secretaría del Medio Ambiente, Recursos Naturales y Pesca (SEMARNAP, Department of Environment, Natural Resources and Fisheries); under President Fox (2000–2006) it has become Secretaría del Medio Ambiente y Recursos Naturales (SEMARNAT, Department of Environment and Natural Resources, fisheries having been transferred to the Department of Agriculture).

REFERENCES

Andrews, E. Wyllys, IV, Michael P. Simmons, Elizabeth S. Wing, and E. Wyllys Andrews V. 1974. *Excavation of an early shell midden on Isla Cancun, Quintana Roo, Mexico*. New Orleans, LA: National Geographic Society and Tulane University Program of Research in Yucatan.

Bates, Daniel. 2001. *Human adaptive strategies*, 2nd edition. Boston: Allyn & Bacon.

Bodley, John J. 1982. *Victims of progress*. 3rd ed. Mountain View, CA: Mayfield.

Bowen, Elenore S. 1964. *Return to laughter*. New York: Doubleday. (This book is now known to have been authored by anthropologist Laura Bohannan, during a time when reporting one's personal field experiences was not accepted in the academic community.)

Bromley, D. W., ed. 1992. *Making the commons work: Theory, practice, and policy*. San Francisco, CA: Institute for Contemporary Studies Press.

Censo de la Clínica de Pich. 1985. P. 1. Research report produced internally by staff.

Censo Municipal de Campeche. 1981. Sección de Pich. Research report produced internally by staff.

CONABIO. 2001. Comisión Nacional de Biodiversidad [National Commission on Biodiversity], RTP #151. Available at: www.conabio.gov.mx. Accessed October 15, 2001.

Daltabuit Godas, Magalí. 1988. *Cobá: Estrategías adaptativas de tres familias Mayas*. Mexico City: Universidad Nacional Autónoma de México.

Durkheim, Emile. 1912. *Les formes élémentaires de la vie religieuse*. Paris: Librairie Félix Alcan.

Elmendorff, Mary. 1976. *Nine Mayan women*. New York: Schenkman Books.

Emery, Marla, and Rebecca J. McLain, eds. 2001. *Non-timber forest products: Medicinal herbs, fungi, edible fruits and nuts, and other natural products from the forest*. New York: Food Products (Haworth).

Encalada Argaez, Ricardo. 1987. *Las poblaciones del municipio de Campeche*. Campeche City, Mexico: Ayuntamiento de Campeche.

Farriss, Nancy. 1984. *Maya society under colonial rule: The collective enterprise of survival*. Princeton, NJ: Princeton University Press.

Faust, Betty B. 1988. Cosmology and changing technologies of the Campeche Maya. Ph.D. diss., Syracuse University.

———. 1998. *Mexican rural development and the plumed serpent: Technology and Maya cosmology in the tropical forest of Campeche, Mexico*. Westport, CT: Greenwood.

———, and Richard Smardon. 2001. Introduction and overview, environmental knowledge, rights, and ethics: Co-managing with communities. *Environmental Science and Policy* 4:147–151.

Freidel, David, Linda Schele, and Joy Parker. 1993. *Maya Cosmos: Three thousand years on the shaman's path*. New York: William Morrow.

Garrett, W. E. 1989. La Ruta Maya. *National Geographic* 176 (4): 424–479.

Gates, Marilyn. 1993. *In default: Peasants, the debt crisis, and the agricultural challenge in Mexico*. Boulder, CO: Westview Press.

Geertz, Clifford. 1983. *Local knowledge: Further essays in interpretive anthropology*. New York: Basic Books.

Gossen, Gary. 1974. *Chamulas in the world of the sun*. Cambridge, MA: Harvard University Press.

Gowdy, John M., ed. 1998. *Limited wants, unlimited means: A reader on hunter-gatherer economics and the environment*. Washington, DC: Island Press.

Gunder Frank, André. 1967. *Capitalism and underdevelopment in Latin America*. 3rd ed. New York: Modern Reader.

———. 1979. *Mexican agriculture 1521–1639*. Cambridge: Cambridge University Press.

Hammond, Norman. 1988. *Ancient Maya civilization*. New Brunswick, NJ: Rutgers University Press.

Hobbes, Thomas. [1651] 1958. *Leviathan, parts I and II*. Reprint, Indianapolis: Bobbs-Merrill.

INEGI (Instituto Nacional de Estadística, Geografía e Informática). 1990. XI Censo General de Mexico, Población y Vivienda, 1990. (Resultados Definitivos, Datos por Localidad). Campeche, Cuadro no. 1, Parte A, p. 1.

———. 1992. Aerial photographs, archives, March.

———. 2000. Aerial photographs, archives, January.

Isbister, J. 1993. *Promises not kept: The betrayal of social change in the third world*. 2nd ed. West Hartford, CT: Kumarian Press.

Kintz, Ellen. 1990. *Life under the tropical canopy: Tradition and change among the Yucatec Maya*. New York: Holt, Rinehart and Winston.

Korten, David C. 1995. *When corporations rule the world*. West Hartford, CT: Kumarian Press.

Kubler, George. 1948. *Mexican architecture of the sixteenth century*, 2 vols. New Haven: Yale University Press.

Marcus, George E., and Michael M. J. Fischer. 1986. *Anthropology as cultural critique: An experimental moment in the human sciences*. Chicago: University of Chicago Press.

Matheny, Ray T., Deanne L. Guru, Donald W. Forsythia, and F. Richard Hack. 1983. *Investigations at Edzná, Campeche, Mexico: The hydraulic system*. Papers of the New World Archeological Foundation, no. 46. Provo, UT: Brigham Young University Press.

McCay, Bonnie, and James Acheson, eds. 1987. *The question of the commons*. Tucson: University of Arizona Press.

Proyecto agrícola y ganadero por 23 mdd: Iniciaron operaciones en comunidad de Pich. 2001. *Tribuna de Campeche*, January 21, p. 1.

Rappaport, Roy. 1990. Ecosystems, populations and people. In *The ecosystem approach to anthropology*, edited by Emilio F. Moran. Ann Arbor: University of Michigan Press, 41–72.

Redfield, Robert. 1950. *A village that chose progress: Chan Kom revisited*. Chicago: Chicago University Press.

———, and Alfonso Villa Rojas. 1934. *Chan Kom, a Maya village*. Chicago: University of Chicago Press.

Sabido Zetina, Rafael. 2001. Analysis of aerial photographs of the ejido of Pich, report prepared for Dr. B. Faust. (Sabido is employed by the environmental consulting firm of SIGESE [Sistemas de Información Geográfica y Cartografia Temática].)

Sanders, William T., and Barbara J. Price. 1968. *Mesoamerica: The evolution of a civilization*. New York: Random House.

Sandoval-Palacios, Juan Manuel. 1982. Development of capitalism in Mexican agriculture. Its impact on the humid tropics: The case of the Yohaltun Project in the southeastern state of Campeche. Ph.D. diss., University of California at Los Angeles.

Sapir, Edward. 1949. The unconscious patterning of behavior in society. In *Selected writings of Edward Sapir*, edited by David G. Mandelbaum. Berkeley: University of California Press, 544–559.

Scholes, France V., and Ralph L. Roys. 1968. *The Maya Chontal Indians of Acalan-Tixchel: A contribution to the history and ethnography of the Yucatan Peninsula*. Norman: University of Oklahoma Press.

SEMARNAT (Secretaría del Medio Ambiente y Recursos Naturales). 2002. Available on the SEMARNAT webpage, specifically at: http://sepultura.semarnat.gob.mx/ucamp/data. Accessed October 23, 2002.

Siemens, A. H., and D. E. Puleston. 1972. Ridged fields and associated features in southern Campeche: New perspectives on the lowland Maya. *American Antiquity* 37: 228–239.

Suárez, M. V. 1977. *La evolución económica de Yucatán a través del siglo XIX*, Vol. I. Mexico City: Ediciones de la Universidad de Yucatán.

Terán, Silvia, and Christian Rasmussen. 1994. *La milpa de los Maya*. Mérida, Mexico: Universidad Autónoma de Yucatan.

Tedlock, Barbara. 1982. *Time and the highland Maya*. Albuquerque: University of New Mexico Press.

Tichle, Andrew, and Ian Welsh. 1998. *Environment and society in Eastern Europe*. New York: Addison Wesley.

Tudela, Fernando. 1989. *La modernización forzada del trópico: El caso de Tabasco, Proyecto Integrado del Golfo*. Mexico City: Colegio de México.

Turner, Victor. 1967. *A forest of symbols: Aspects of Ndembu ritual*. Ithaca, NY: Cornell University Press.

———. 1975. *Dramas, fields and metaphors: Symbolic action in human society*. Ithaca, NY: Cornell University Press.

Villa Rojas, Alfonso. [1945] 1978. *Los elegidos de Dios: Etnografía de los Mayas de Quintana Roo*. Serie de Antropología Social, 56. Reprint, Mexico: INI.

Vivo Escoto, Jorge A. 1964. Weather and climate of Mexico and Central America. In *Handbook of Middle American Indians*, Vol. 1, edited by Robert West. Austin: University of Texas Press, 187–215.

Vogt, Evon V. 1993. *Tortillas for the gods: A symbolic analysis of Zinacanteco rituals*, 2nd edition. Norman: University of Oklahoma Press.

Wallerstein, Immanuel. 1976. *The modern world-system: Capitalist agriculture and the origins of the European world-economy in the sixteenth century*. New York: Academic Press.

———. 1980. *The modern world-system: Mercantilism and the consolidation of the European world-economy, 1600–1750*. New York: Academic Press.

———. 2003. *The collapse of the neoliberal offensive*, commentary No. 122 [cited October 1, 2003]. Available from http://fbc.binghamton.edu/122en.htm.

Wedel, Janine R. 1998. *Collision and collusion: The strange case of Western aid to Eastern Europe 1989–1998*. New York: St. Martin's.

White, Lynn, Jr. 1967. The historical roots of the ecological crisis. *Science* 155: 1203–1207.

8

Economic, Political, and Environmental Changes in a Yucatec Maya Village

Grace Bascopé

This chapter presents an ethnographic account of how one village in the central Yucatan copes with economic, political, and environmental changes. Data presented here have been gathered since 1991. Two to four visits have been made to the community every year since then. Time spent in the community totals approximately two months annually. While much of the data have been gathered through participant observation and daily conversations, household censuses were conducted in 1991, 1996, and 1997 (the first two having been conducted by the author and the last under the leadership of Universidad Autónomo de Yucatán student Elias Alcocer Puertos; see Alcocer Puertos 2001). Each year, a representative sample of residents of the community is asked to characterize their corn and other crops and to discuss current economic situations, employment opportunities inside and outside the village, hunting successes, craft sales, and the state of the supply of raw materials necessary to produce these crafts.

Yaxunah is both an ejido (corporately owned farmland) and associated village approximately 20 kilometers from the town of Chan Kom, where Mary Elmendorf has conducted much of her research. Yaxunah shares significant history with Chan Kom. For example, Redfield and Villa Rojas (1934) and Re Cruz (1996) have written accounts of a 1920s conflict when leaders of Chan Kom went to Yaxunah to offer shelter for those who wished to flee the violence between the Liberales and Socialistas. Don José Poot, who still lives in Yaxunah, was a boy of seven at the time and clearly remembers these events. He and his family were among those who went with the Chan Kom patriarchs to temporarily live there in safety. Also, it was to Yaxunah that one of the Chan Kom families came to establish the first Presbyterian church during the religious and political conflicts of the 1930s and 1940s, reported by Redfield (1950), Goldkind (1965, 1966), and Re Cruz (1996). Although today there are no roads between the two communities, the extreme eastern corner of the Yaxunah ejido lands actually touch a corner of the Chan Kom *municipio* (county) as well as the Chan Kom ejido itself.

Yaxunah and Chan Kom have another important similarity. In 1924, the Carnegie Institution of Washington, DC, began a twenty-year investigation in and around Chichén Itzá, and as a part of this project, Chan Kom became a focus of ethnographic fieldwork (see Sullivan 1989 for a discussion of the relationship between the Carnegie Chichén Itzá project and Chan Kom, as well as other communities on the peninsula) and much outside attention (see Redfield 1950). Yaxunah's ejido is itself the site of large pre-Columbian Maya ruins, and researchers conducting investigations there have influenced community life in ways reminiscent of the influences of the Carnegie project on Chan Kom. In fact, the first documentation of archaeological research being carried out in Yaxunah stems from the Carnegie period when investigators from that project did some preliminary reconnoitering in the Yaxunah ruins (see Brainerd 1958).

The population of Yaxunah was reported in the 1991 census as 407, but as of October 1997 there were 536 inhabitants, a 22 percent increase in six years. In contrast, during the same period, the health post serving the region that until recently included Yaxunah only recorded a 1 percent growth rate for its patient universe. Thus, Yaxunah's growth cannot be attributed to regional trends but rather to specific local conditions during the same time period.

At least part of the population increase is due to more widely available health care, leading to a reduction in infant mortality. Pregnant women are given tetanus shots so that their newborns will be protected from neonatal tetanus caused by cutting the umbilical cord with unsterilized instruments. Also, parents have become accustomed to using prepared oral rehydration therapy packets to prevent their small children from dying of dehydration during bouts of diarrhea. These advances, however, are common to many communities in the area and do not fully explain the differential population increase in Yaxunah.

At least one factor, however, sets Yaxunah apart. Over the last decade, because they have had the assurance of a few weeks' employment each year at home with the archaeological project, few men and women have migrated from the village to seek employment on either a temporary or permanent basis (signifying reduced exportation of fertility). Though community-level statistics are scant, those which do exist suggest that fewer men and families migrate from Yaxunah than from neighboring villages (see Re Cruz 1996, for a comparison with Chan Kom) and that the population increase in Yaxunah has been greater than among its neighbors. One neighboring community reports a 25 percent decrease in population during the same time period.

According to my household census, in the last nine years, 5.4 percent of male heads of families have migrated outside the community to seek work on a temporary basis (a period of more than one week's duration), and only one ejidatario (man born in the community with certified rights to communal land use) has moved his family away from the village to pursue permanent employment in Cancún. Nine sons of ejidatarios (themselves having potential future land use rights) have permanently relocated outside the village, while three men not born in the community have moved in and have been granted ejidatario status. Eight daughters of ejidatarios have relocated outside the village for employment purposes. In general, the males seeking temporary employment use their community contacts in Cancún to find work, usually with companies creating gardens in new hotels and other construction sites in the area. The young girls go to work in Mérida, either as housekeepers in hotels or cleaning private homes and doing child care. A few community members find short periods of employment in Pisté, the community closest to Chichén Itzá. As yet no members of the community of Yaxunah have attempted migration to the United States, though they know individuals from a neighboring community who have made such trips on both a temporary and permanent basis.

Men report that the summer archeology work means they can remain in the village and, though this employment is temporary, they realize greater profit because they do not have to pay travel and living expenses elsewhere. One migrating community member explains the numerous hardships posed by traveling to look for work in Cancún: "Not only is my wife frantic, fearing for my safety during my absence, but also, my children cry for me and do not thrive without my care" (every returning migrant has first- and second-hand stories to tell of dangerous, often life-threatening, encounters in Cancún). He explains that while migration provides a little immediate cash, there are no long-term benefits. When he migrates, his house becomes dilapidated; he must sell any cattle or pigs he might own, as his wife cannot gather the food they require; someone must be paid to provide firewood for his family and do some of the work in his cornfields (milpas); and even his dogs, which always accompany and protect him in the cornfields as he clears, plants, weeds, and harvests, and which are trained to help him hunt, will be lost from lack of their master's attention. Certainly a wife does not get rid of her husband's dogs during his absence, but the dogs are reported to become lazy and listless, not to thrive, and potentially to just leave or die if the man is gone too long.

Irrespective of other economic endeavors in which they might be engaged, the men of Yaxunah universally describe themselves as corn farmers. Cornfields are maintained by every family, and corn is the principle food in the village diet. Hurricane Gilbert, which struck in 1988, serves as a benchmark in the minds of Yaxunah farmers. The older adults of Yaxunah report that, until the time of the hurricane, 50 *mecates* of land (which equals 2 hectares, a little less than 5 acres) were sufficient to provide enough corn for a family of five to eight people (depending on their ages) for twelve months, and surplus corn could be sold outside the village. However, since Gilbert there has not been sufficient corn to meet the needs of Yaxunah families, so there is no question of having corn to sell.

Community members report that since the time of Gilbert, the climate seems to be drier and the crops have been sparser. They explain that by 1988 virtually all the *monte alto*, or high forest canopy, was gone from the ejido. With steady population increases, very little of the ejido's land remains unutilized; thus, in the last decade, no fields have been fallowed for more than six years. Villagers acknowledge that for best results after three years of continuous cropping, fields should be allowed to remain fallow for at least twenty years, if full production is to be achieved when the land is put to use again.

Though the villagers are keenly aware that now a 50-*mecate* field will no longer provide enough corn for more than eight to ten months (depending on family size and yield of crop realized), only 12 percent of the men regularly plant larger fields. They cite the lack of availability of quality arable lands; increased expenses of labor for clearing, planting, and weeding; expenses for extra seed, fertilizer, and herbicide (if used); and realistic estimations of their own physical strength as barriers to cultivating more. Planting more, they explain, exposes them to extreme financial risks in case of catastrophic crop failure. They feel it is economically sounder to make up the shortfall in corn by using their time to pursue other economic endeavors, which they hope will allow them to purchase the additional food they need. Yet they report anxiety about the future when even greater population increases will further tax the diminishing fertility of currently cultivated land, and there will be greater pressure to find the money to meet the needs of their families.

On average, one adult male will eat six kilograms of corn per week. As of the 1997 census, corn was sold in the community's government subsidized store for 1.30 pesos per kilogram. Bags of ground cornmeal sold for 2 pesos. Unlike the corn grown in the community, this already ground corn is prepared without calcium carbonate to soften the kernels. This means that during the months corn is purchased, children are deprived of this excellent source of calcium. Worse, perhaps, is the fact that often families do not purchase corn at all but rely on white bread made from refined wheat flour, which can be purchased in the local stores cheaply and easily.

An adult male who had no more homegrown corn would need to spend 7.80 pesos weekly for himself, plus approximately 25–30 pesos more for corn for the rest of his family (based on an average family size of six). Over a period of three months (the amount of time it takes for a new field of corn to grow), this would mean that on average a family would need to purchase approximately at least 400 pesos worth of just corn for themselves.

In 1997, the government program PROCAMPO provided each corn farmer with 500 pesos for each hectare an ejidatario planned to plant (usually two). This money was intended to be used for seed corn, for herbicides, pesticides, fertilizers, and other expenses for the milpa. Nevertheless, as one ejidatario reports, "*uno se fastidia de preocupacciones*" (one grows tired of worrying). One worries, he explains, where the food will come from each day. Further, he notes that where there is not enough corn every day, a man has to plan carefully, to think, to work very hard, and still sometimes there is just no place to get the money. He explains: "A man thinks, if the PROCAMPO money is used for seed corn and other things it takes to make a milpa, then there will be no cash for the three months it takes the fields to grow, and what if there is another crop failure?" He continues: "A man becomes very tired of looking for money for corn every day, and we aren't even thinking about buying other kinds of food, clothes, shoes, medicines, school supplies." He is of the opinion that, chiefly because of the new secondary school in the community, many children and families seem to have had raised hopes and expectations for better education for the community's youth, but recently parents have become more guarded about the prospects of helping their children advance in school. There simply is little money for the supplies needed.

Not only do the men of Yaxunah see the reduction in forest as one of the causes for their drier climate but also they report that the deforestation has been coupled, over the same period, by an estimated 50 percent decrease in deer and peccaries (javelinas), both of which had been significant sources of wildlife protein in the past. Interestingly, deforestation has resulted in a crisis of another kind, the lack of safe, inexpensive roofing materials. Typically, in the past, houses in the community (as elsewhere in the peninsula) have been made of poles and thatched with large palm fronds of *Sabal mexicana*. However, because of overharvesting due to population pressures, these palms have disappeared from the ejido. Long, mature fronds can still be purchased from a community some 20 kilometers away, but at the prohibitive price of 1 peso per frond. This means that, due to the number of palm fronds needed (between 1,000 and 2,000, depending on the size of the house), the roof would cost more than all the materials and labor for the rest of the house combined.

Now houses are most commonly roofed with corrugated cardboard saturated with black petroleum as a preservative. Not only do these cardboard sheets make the houses unbearably hot and foul-smelling as the sun heats them during the day but also during the rainy season the petroleum washes off onto the ground around the houses in the very areas where children play and where chickens and pigs are often found searching for food. Undoubtedly, both children and animals ingest some of this product. Some individuals in the community have built houses of cement block and roofed them with cement, but this is a very expensive undertaking that is not within everyone's means. For block houses, a substitute for roofing with cement has been to use sheets of corrugated asbestos, a product undoubtedly even less healthful than the oil-soaked cardboard on the pole houses.

Though a number of cattle are raised in the ejido, they are not slaughtered there for meat to be consumed within the village. It would be unprofitable to do so. A kilogram of beef costs approximately 26 pesos. A mature cow that weighs

300 kilograms on the hoof produces about 90 kilograms of meat. This represents a potential of 2,340 pesos for beef alone. Community members explain that there is not that much money for food in the village at one time, nor enough refrigeration. The meat would be wasted before it could be bought and eaten, and before any profit could be realized.

A young (a year old or less) beef animal is purchased at approximately 1,200 pesos and then sold to commercial cattle dealers from outside the community after one to two years of care. These cattle are sold at little more than twice their purchasing price. Raising a cow or bull is hard work. As lands are communal, there are no fences in the ejido. Cattle are left in yards within the community during the day and then taken out to feed overnight. To keep them from wandering or doing damage to cornfields, cattle must be tied with ropes. The cattle often step on the tethers, fraying them between their hooves and sharp rocks. The ropes have to be replaced often. Often there are not enough plants on which the cattle can forage, so fodder must be cut and brought to them. This usually takes the form of cutting branches from ramón trees. As the abundance of ramón disappears, men have to look farther and farther from home to find enough food for their animals. Also, as there is no surface water in Yaxunah, water must be drawn from *cenotes* (sinkholes) or *norias* (old colonial wells) twice daily for the cattle. The amount of time and labor required, the fact that to own cattle means a constant, significant energy expenditure that ties a man to home, limiting his ability to travel to seek temporary employment elsewhere, and the small profit realized on top of all this, certainly account for the fact that seldom more than 15 percent of the ejidatarios have owned even one cow since the early 1990s.

Raising pigs is another matter. A weaned piglet costs approximately 70 pesos, though often a woman (pigs are generally considered women's property and responsibility) takes a baby pig from a *comadre* (close friend, ritual kin) and gives another back when one of her full-grown animals has a litter. A pig eats from .5 to 4 kilograms of fodder a day, depending on age, and is raised from one to one and a half years. A full-grown pig can bring approximately 1,300 pesos. Pigs literally represent walking "piggy banks." Though people say that when all the food a pig consumes is considered they only break even, pigs are seen as financial reserves, as they can be sold easily if unforeseen need arises, such as a sudden illness in the family. Pigs are also slaughtered for family celebrations, the *hetz' mek'* (baby-welcoming ceremony), weddings, the anniversary of the death of a family member, and so on, and if the family does not plan to use all the meat in its own celebration, the unused part of the carcass can be sold within the village. Occasionally corn is purchased to feed pigs, but generally, though pigs do receive some of the family's corn crop, it is the least desirable part of the harvest (e.g., the cobs most infested with weevils), a part that might otherwise be wasted. Pigs also eat the flesh of the squashes commonly grown in the milpa with the corn. The squash seeds (*pepitas*) are scraped out and dried for human consumption, but the flesh serves only as pig food. In times when crops fail, pigs have to be sold off, often at low prices, because no one can afford to buy food for the family and the pigs as well. Women complain that they feel sad, that their yards seem empty and sound too quiet when they are not full of pigs. They report they can only be content when they can have their pigs back. They see a lack of pigs in the village as a depressing reminder of hard times.

Yaxunah might arguably be described as an isolated Maya village. For example, all villagers have Maya, as opposed to Spanish, surnames. Only seven adults describe themselves as speaking Spanish as their first language, and none of them were born in the village. In general, all adults report that they speak Maya to each other 100 percent of the time except when someone is in the village (like myself) with whom they must speak Spanish. Only in 1997 did the village get a road, a secondary school, and a clinic, which is staffed approximately three and a half days per week. In response to these new opportunities, twelve sets of parents report that they now speak to their children in Spanish, "so they can get more from the teachers who are sent to the village, as the teachers don't learn Maya."

History, however, has shown that the people of Yaxunah have always been articulated with the outside world and are creative and flexible in looking for ways to obtain necessities not provided by their cornfields. In the 1950s when the population was small, a number of men in the village had rather large herds of cattle (as many as fifty head each). In the 1950s and 1960s, they raised chickens and sold eggs to the Mérida market. Each of these endeavors was wiped out by sudden catastrophic diseases. In the 1960s, villagers made charcoal and cut firewood to be sold around the state (contributing ultimately to the deforestation of the ejido). However, over time, city markets for these commodities were supplanted by the greater availability of fuel oil, butane gas, and electricity. In the 1970s and 1980s, villagers relied on beekeeping to raise needed cash (see Merrill 1984) and went out of the community to seek stints of temporary employment throughout the peninsula. From the late 1980s through the first half of the 1990s, families in Yaxunah employed diverse means to supplement their income. Many still raised bees, though twice the hives were destroyed by hurricanes. Because of a one-time-only government assistance program, more than 80 percent of families owned at least one head of cattle in the early 1990s. Twenty-four families produced citrus crops for the outside market.

In 1995, the corn crop was a 100 percent failure because of another hurricane, Roxanne. In 1996, the crops died due to more than a month with no rain during the critical growing season. For the second year in a row, the crop was a total loss. Not even enough corn was produced either year to provide seed for the following year's planting. In what is fundamentally still a subsistence-level economy, complete crop failure two years in a row places extreme stress on the resources of each family. Because both food and seed for the next year's crop had to be purchased, cattle, pigs, and chickens were sold off.

Drought meant not only that the bees had no flowers from which to harvest pollen but that their malnourished state left them susceptible to the varroa mite, a pest that preys on the immature young of the colony. This combination meant that by the summer of 1998, the total number of beehives in the village was down 83 percent from 1995.

At the same time, the village orange grove not only suffered from drought, thus producing a small crop, but also began to fall victim to *Tristeza*, or citrus decline, a virus for which there is no known cure, but which can be controlled by the expensive process of killing affected trees and removing and sterilizing the contaminated soil. To date, an estimated 7 percent of all orange and other citrus trees have the virus.

Thus, all at once the people of Yaxunah found themselves with not only no field or food crops but also limited ability to obtain cash from the traditional sources of cattle, honey, and citrus. Recuperation from the disastrous events of 1995 and 1996 was made difficult by the *kan kubul* (literally, rain from the yellow cloud) of 1997. Many fields suffered from it, and only a partial crop was obtained by many village farmers. The men described this rain as a hot rain. One farmer explained that he believed it to be seawater that was sucked up into the clouds by a storm or hurricane and poured out on the land. Indeed, the leaves of all the plants in a field over which the *kan kubul* passed looked as though they sustained salt or acid burns.

By remarkable coincidence, in the midst of these agricultural and economic disasters, three years ago a young man moved to Yaxunah who knew how to carve Maya-style figurines for the Pisté tourist industry. He taught the craft to his brother-in-law, and, in an each-one-teach-one fashion, the carving skill spread throughout the community, so that in five months' time fully 85 percent (based on 1996 and 1997 census data) of all families gained income from selling these carvings to wholesalers in various tourist zones around the state.

In the families in which carvings are produced, children over the age of ten carve for an average of fifteen hours a week, the women for thirty, and the men for forty-five (these figures were obtained not from a formal census but by canvassing members of a large number of households about this new woodcarving phenomenon and taking averages based on the information obtained). A number of interesting, if ominous, side effects to this work are occurring.

First, over a period of fifteen months, an average of six serious knife wounds to the hands, upper legs, and the abdomen were recorded per month. Many of these wounds involved loss of function of a digit, tendon and nerve damage, and lingering infections. Such accidents tend to cluster during weekends when the community clinic is not staffed, but when schoolchildren and adults employed in the fields and in wage labor are free to work at home in the company of family members (personal communication with victims, the community nurse, and with a young woman in the community who has had some nursing training and to whom people come for wound care on weekends when the clinic is closed).

Second, eighteen children have either dropped out of secondary school or will not enter it at all in order to dedicate their time to woodcarving. In one family, a twelve-year-old boy is the sole support of four other people (personal communications with the clinic nurse, schoolteachers, and parents of the children involved).

Third, the figures are carved from the wood of the *chakah* tree (*Ursera simarouba*). This is a relatively soft white wood that accepts paint well and does not crack when dried out. Such sudden heavy harvesting is depleting this tree in the ejido. Community members estimate that, at the current rate of harvesting, sound, mature trees will totally disappear within the next twelve months, and only trees with twisted or diseased trunks will remain. Unfortunately, this tree is one of the main sources of flowers for honey production.

Recently, an evangelist came to the village and exhorted the members of both Protestant evangelical churches (who make up 50 percent of the population) to stop carving the figures, saying they are idols and their production can be a stumbling block to the salvation of others. Twelve families were convinced to discontinue production.

One man explained why, though deeply religious, he has not stopped carving:

> I cannot stop making the idols. God sent this to me two years ago when we had nothing left. We are not corn people anymore, we are *chacah* people. I was made of corn; my children are of *chacah*. *Chacah* has fed this whole village for more than two years. I understand the Ten Commandments, but I have to say God sent the figures, and God will find a way to forgive me if I still use them to feed my family. He will show me how to cleanse this sin.

In the midst of these radical alterations in the economy of Yaxunah induced by environmental changes due to climatic events and population pressure on resources, archaeological activity has also influenced the ejido. The Yaxunah site is a huge archaeological complex containing Maya Formative through Postclassic occupation levels, as well as the remains of a colonial cattle hacienda (Alexander 1993; Suhler, Ardren, and Johnstone 1998). Beginning in 1986, and for the next ten years, almost all of the men and boys of Yaxunah worked for a few weeks each summer for a team of archaeologists headed by Dr. David Friedel of Southern Methodist University (SMU), on a project sponsored by the Selz Foundation of New York.

Each week's work crews were filled on a rotation basis, ensuring that all members of the community who wished to work received the same number of days. This not only ensured the same pay for all but also meant that each person was equally trained in how to excavate and consolidate the ruins. The crews were also taught how to protect the site and conserve the artifacts found, and interpretations of the findings were shared with them. Further, elected community officials, not the archaeologists, made the weekly labor assignments and handled any grievances. Every two to three weeks, general assemblies (affectionately known as *mata chismes* or "gossip killers") were held. In them, the archaeologists and all the men of the village had an opportunity to discuss the problems and progress of the dig. This method of equal employment and equal pay was established at the request of villagers because it best fit their tradition of *fajina* or communal labor. In the past, all communitywide projects (e.g., building the school and town hall) were accomplished by this method. This labor arrangement clearly showed the archaeologists to be guests in the ejido.

In 1996, in conjunction with the community, and with various nongovernmental agencies in the Yucatan and throughout Mexico, the SMU/Selz team presented a comprehensive long-term plan to the governing board of the five-nation Mundo Maya Project. The plan called for the continued scientific investigation of the archaeological site and for training and support of community members to become the leaders in development and tourism projects related to the site.

In order to excavate in Mexico, an annual permit must be obtained from the Instituto Nacional de Antropología e Historia (INAH, National Institute of Anthropology and History). At the end of 1996, the permit was denied the SMU/Selz team. Thus their ten years of work in Yaxunah stopped. In the fall of 1996, INAH archaeologists began consolidation of the site. They brought stonemasons from other parts of the state for this reconstruction, ones they had trained on other sites.

INAH archaeologists hired some laborers from the ejido, but not on a rotation basis; every family did not have equal access to this work, nor was the pay equal for those who did find work with the INAH team. Community members protested to the INAH project director. They explained that they were all equally needy because of the crop failure, and that, as ejidatarios, they shared certain democratic values and should have equal access to work on a site located on their own lands. They convinced the archaeologists that they too had been trained as stonemasons and that it would be cheaper for INAH to use their on-site talents. They convinced the archaeologists to accept the rotation system and not import workers in the next field seasons.

One of the chief themes in the large body of Mary Elmendorf's work has been change (see, e.g., Elmendorf 1970, 1972, 1981; Elmendorf and Merrill 1978). She has amply demonstrated to us that change, even in a single Yucatec Maya village, can only be understood if history, the environment, and outside forces (including economics and politics) are considered, and only if villagers themselves are recognized as change agents. In the case of Yaxunah, while community members quite literally have been buffeted by elements outside their control, they have, nevertheless, confronted these losses creatively, finding new avenues of support and even placing conditions on a large bureaucracy working on their ejido lands.

REFERENCES

Alcocer Puertos, E. 2001. Manejo sustentible de recursos naturales y culturales pro parte de una comunidad Maya de Yucatan: El caso de Yaxunah. Thesis, Universidad Autónoma de Yucatán.

Alexander, R. 1993. Colonial archeology of the Parroquia de Yaxcaba, Yucatan, Mexico: An ethno historical and site structure analysis. Ph.D. diss., University of New Mexico at Albuquerque.

Brainerd, G. 1958. The archaeological ceramics of the northern Yucatán. University of California Anthropological Papers no. 19. Berkeley: University of California Press.

Elmendorf, M. 1970. *Nine Mayan women: A community faces change*. New York: Schenkman Publishing.

———. 1972. *Mayan Women and Change*. Cuaderno no. 81. Mexico: CIDOC.

———. 1981. Changing roles of Maya mothers and daughters. In *Women and technological change in developing communities*, edited by R. Dauber and R. Cain. Boulder, CO: Westview Press.

———, and D. Merrill. 1978. Socioeconomic impact of development in Chan Kom. Mérida, Yucatan. Paper presented at the meeting of the Society for Applied Anthropology.

Goldkind, V. 1965. Social stratification in the peasant community: Redfield's Chan Kom reinterpreted. *American Anthropologist* 65: 863–884.

———. 1966. Class conflict and cacique in Chan Kom. *Southwest Journal of Anthropology* 22: 325–345.

Merrill, D. 1984. Mixed subsistence-commercial production system in peasant economy of Yucatan, Mexico. Ph.D. dissertation, Cornell University, Ithaca, New York.

Re Cruz, A. 1996. *The two milpas of Chan Kom: Scenarios of Maya village life*. Albany: State University of New York Press.

Redfield, R. 1950. *A village that chose progress: Chan Kom revisited*. Chicago: University of Chicago Press.

———, and A. Villa Rojas. 1934. *Chan Kom, a Maya village*. Publication no. 448. Washington, DC: Carnegie Institution.

Suhler, C., T. Ardren, and D. Johnstone. 1998. The chronology of Yaxuná: Evidence from excavation and ceramics. *Ancient Mesoamerica* 9: 167–182.

Sullivan, P. 1989. *Unfinished conversations: Mayas and foreigners between two wars*. New York: Alfred A. Knopf.

9

The Landscape of Santa Rita Komchen: The Role of Metaphor in the Ecological History of a Yucatecan Rural Estate

David W. Forrest

During the past five centuries, the landscapes of rural estates such as Santa Rita Komchen have served as an important medium through which the people of Maní, Yucatan, have contested and negotiated the history of their community. Through the landscape, the people of Maní metaphorically link these estates, as well as the set of human environmental interactions that occur on them, to local concepts of the past and the cosmos. These links serve to provide a historical and cosmological context for the place, the environmental interactions that occur there, and the people engaged in these interactions. This chapter uses the ecological history of Santa Rita Komchen as an example of how the people of Maní use these metaphorical links to help them construct community in a context of conflicting ideas about the control and use of resources on the rural estates surrounding the town. This process not only helps to give meaning to human environmental interactions on these estates, but also provides an arena for advocating and contesting the control of land and water resources by different groups in the community.

THE *HETZ' LU'UM* AT SANTA RITA KOMCHEN

In the main corral of Santa Rita Komchen, a rural estate south of Maní, there are five large ceiba trees. Four of them form a square, with the fifth standing along one edge between the two western corners. On one of the trees, *1995* is carved into the bark. The way to the corral passes by the ruins of a chapel and through one of the four large stone gates that provided entrance to its grounds, one from each cardinal direction. The ceibas mark the way to the *noria*, a great well situated upon a large stone platform faced with stones from ancient Maya buildings. The water from this well is the center and source of life for the estate.

In 1994, when I talked to Andrés Parada May,[1] the owner of Santa Rita Komchen, he told me of a commitment he had to fulfill. About ten years earlier, in 1985, five of his cattle at Santa Rita died in one day. He was troubled by this and wondered why it happened. Then the spirit of his dead father came to him and told him to do a *hetz' lu'um* (*hetz'*, seated; *lu'um*, land) in the corral at Santa Rita Komchen.

Don Andrés got a *h-men* (shaman) to do the *hetz lu'um*. They sacrificed a bull and ritually distributed its parts. They put the belly and backbone in a hole in the ground in the center of the corral, and buried the head at the water trough to the west, the tail at the eastern entrance, and a foot in each corner of the corral. With the flesh of this bull, a ritual meal was prepared. They made a *pib* (earth oven) to cook special *wah* (ritual tortillas) and inside each *wah* put a piece of the meat of the sacrificed bull. In this way, the flesh of the sacrificed bull was eaten.

Don Andrés said that the *hetz' lu'um* was successful; only two cattle had died since then. When he did the *hetz' lu'um* in 1985, he promised to do another one ten years later in 1995. "I asked for ten years," he said. In July 1995, when I talked to don Andrés again about this commitment, he was making plans to complete it by doing another *hetz' lu'um* at Santa Rita Komchen in September.

During the past five centuries, the landscape of Maní has served as an important medium through which the history of the community has been contested and negotiated. The many conflicts over the nature of the economy, use of local resources, control of community politics, and religion have been played out, in part, through the creation and interpretation of the landscape. In this chapter, I identify how metaphor plays a role in this process of landscape creation, and I explore how the people of Maní use this process of creating meaning to advocate, contest, and negotiate the control and use of resources on the rural estates surrounding the town. This overview of the history of the landscape of Santa Rita Komchen, a rural estate near Maní, provides insight into the mechanisms used by the people of Yucatecan communities to create a meaningful relationship with natural resources on the rural estate as changing technologies and conditions in the larger economic and political system affect their use of and access to these resources.

THE ANTHROPOLOGY OF LANDSCAPES

During fieldwork in Maní in 1994 and 1995, I often encountered landscapes as complex as that of Santa Rita Komchen. I began to ask questions about the layered purposes and meanings of these landscapes. How and why were the elements of the landscape put there? What were the makers of these elements trying to communicate? How were they used and interpreted by people in the past, and how do they affect the people living there now?

One of the ways that I tried to answer these questions was by collecting data during a type of walking interview that I came to call *rumbos* (Spanish for a road, route, or way). On these walks, I traveled with people of different ages and experiences along routes (paths, roads, and streets) that they knew well to make use of the places and natural resources we encountered to elicit narratives about the environment, such as stories, myths, and practical knowledge about the use of resources. I was amazed at the richness of the narratives told to me on these walks. In these narratives, people mixed together talk about the everyday use and control of resources with talk about the community's past and its place in the cosmos. They might draw on tales of the founding of Maní by the Xiu lineage in the fifteenth century, the Spanish conquest in the 1540s, the abolition of debt peonage after the Mexican Revolution, present-day migrations to work in Cancún or the United States, or aspects of ancient Maya cosmology or medieval European teaching about the nature of an ideal Christian community. What became clear to me was that any one element in the landscape may elicit a multitude of meanings, some relating to its material usefulness and others to its place in local concepts of the past or the nature of the cosmos.

In listening to these narratives, what struck me almost immediately was the way in which the different aspects of the narratives were linked together. People often talked about current events in terms of the past, about history in terms of myth, about local geography in terms of cosmology. I also began to see that many elements of the physical environment were constructed with similar linkages, in which people represented events and concepts from the realm of history and cosmology in terms of forms and layouts in the physical environment.

I realized that the creation of these forms and narratives is structured by the individual's context within the community (a particular interaction with the environment, with other community members, and with the past). "*Cada persona tiene su rumbo*," they say in Maní, meaning that each person has a route, a part of the landscape that he or she knows well, including not only knowledge of material resources but also knowledge of the narratives about that part of the community. It is through their daily activities in the environment that people put local knowledge of history and cosmology into the context of their world.

Like other cultural systems, such as religion, which provide a model of, as well as a model for, society (Geertz 1973, 123), the landscape is a cultural system that provides a model of, as well as a model for, the construction of community. These models make use of metaphor to link two separate domains by using language and concepts appropriate to one domain to provide insight into the nature of the other (Black 1962). By their actions and narratives, people create a metaphorical link between the domain of foreground actuality, the "everyday" realm of daily activities, and the domain of background potentiality, the "beyond-the-everyday" realm of the cosmos, the past, and the future.

This is done as people bring elements of the beyond-the-everyday into the everyday world both in the physical environment, by their use of resources and the creation of certain forms and layouts (physical models), and in the narratives about this space, by the ways they talk about these uses and forms (theoretical models). The landscape provides a greater understanding of both the foreground and background through a process of "mutual implication" (Hirsch and O'Hanlon 1995, 23) with the foreground context of daily life and the background of local knowledge about the past and the cosmos informing each other.[2]

The creation and interpretation of landscape, however, is not the same for everyone in the community, but is structured by discourses shaped by the material and ideological interests of different groups and their positions of power within the community (Moore 1986; Duncan 1990; Barnes and Duncan 1992). One way that these discourses can be presented is as metaphors (Barnes and Duncan 1992, 9). Through the creation of the landscape, people advocate, contest, and negotiate these discourses by referring to powerful metaphors about themselves, their actions, and their place within the community and the cosmos. These metaphors are often alluded to indirectly by the use of metonyms, conventional symbols in the landscape that stand for entities in the world as defined by the metaphor.[3] These metonyms are not always read in the same way by different groups, because the understanding of which metaphor is being referred to occurs within the context of the reader's experience within the community. Dominant groups have often written metonyms in the landscape referring to metaphors that support their position of control in the community. Other groups, however, have sought to promote alternative readings, thereby asserting political power by providing their own interpretation (Moore 1986, 89). It is through these alternative readings that competing discourses can be contested and negotiated in the landscape.

THE LANDSCAPE OF SANTA RITA KOMCHEN

Perhaps in no other aspect of the community does the landscape play a greater role than in providing an arena for advocating and contesting models for the community's interaction with the environment. When local people look at the landscape, they see not only forests, roads, fields, and settlements but also symbols that tell them something about the nature of the world and their place in it. The interpretation of the landscape is one way that people create meaning for their daily interaction with the environment. This ability to promote an interpretation of the meaning of natural resource use and control is a political power, representing an important mechanism in the political ecology of the community.

For example, one of the primary metonyms used in the landscapes of estates such as Santa Rita Komchen is the metonym of the water source as "center." This concept is informed by a reality of everyday existence on the estates of the

Yucatan. With few water sources for cattle or people out away from the towns, the water source of the *casco* (central core) of the estate truly is the center of life for these settlements. Because they are scarce and valuable resources, control over these water sources has often been contested. Throughout the history of the community, various groups have attempted to gain control of water sources and to justify their use and control of this water by linking these water sources to concepts of center in their own cosmology. The concept of the water source as center serves as a common point around which different models of resource use on the rural estate are constructed and different metaphors to justify and support these models are advocated and contested. I would like to briefly describe how the landscape of the *casco* of Santa Rita Komchen provides a powerful example of how alternative readings of metonyms referring to center have played a role in advocating, contesting, and negotiating different models of resource use and control on the rural estate over time.

One of these metaphors posits the estate as *kah tali*, a Maya rural settlement. For centuries, these Maya hamlets have existed in the forest out away from larger urban centers, providing easy access to land for milpa (cornfield) production. As long-term settlements, they were often established at a water source such as a *cenote* (limestone sinkhole). In this metaphor, the water source is a point of settlement around which the forest is used as land to make milpas. Just as in Maya urban spaces where the quincunx pattern of the Maya cosmos (the four cardinal directions plus center) appeared in the design of houses, residential wards, and the town layout (Freidel, Schele, and Parker 1993), links were likely made between the place of the *kah tali* and the space of the Maya cosmos. These metaphorical links placed the water source and the Maya residents of these hamlets in relation to the center of the cosmos. In addition, this link created a connection between traditional Maya resource use, with cycles of forest transformation into milpa and regeneration as forest, and the Maya concept of cyclical time. These ancient concepts are still preserved in the yearly almanac represented by the *xok k'in* (the count of days), in which the weather of the first twelve days of January is used as a prophecy of the ecological conditions of the coming year. In Maní, the *xok k'in* is still part of the traditional Maya milpa farmer's repertory of knowledge about the environment.

In the early colonial period, the term *rancho* came to be used by the Spanish to refer to these unauthorized Maya hamlets (Patch 1993, 52).[4] Such a Maya *rancho* existed at the site of Santa Rita Komchen when Fray Alonso Ponce traveled to Maní in 1588 during his travels in New Spain (1584–1592).[5] This location corresponds to a *kom*, or natural sink in the topography, which probably provided access to the aquifer by way of a small natural well (*chen*), hence the name Komchen.

Soon after their arrival in the Yucatan, the Spanish introduced cattle into the forests where formerly only agricultural production on milpas had taken place. Cattle production was centered on a new type of rural settlement, the *estancia*, a small-scale cattle ranch. As in the Maya *kah tali*, the water source was the center of the colonial *estancia*, with the property often measured out from this water source in four directions (Chardon 1980). Many early *estancias* were centered on natural water sources, such as *cenotes*. Colonial land documents for the northern Yucatan commonly refer to an estate as the ownership of a *cenote* with a particular name and some unspecified amount of land around it (Farriss 1984, 275). The water source made the forest around it desirable as a place to graze cattle. In Maní, where the water table generally lies about 20 to 22 meters below the surface and the number of *cenotes* is limited, these water sources were held at a premium.

The Spanish *estancias*, however, were not limited to naturally occurring sources of water. The technology of metal tools gave the Spanish the ability to dig wells, creating water sources where none had existed before. They could therefore create a "center" for an *estancia* both materially and metonymically by digging a well there. It is important to remember that the control of these water sources was not only about the control of water but also about control of the activities in the lands surrounding them. In the metaphor of the estate as *estancia*, the water source at the center of the *estancia* implies a new use of the forest resources surrounding it. Following the practice in Spain, the Spanish viewed the communal lands surrounding the town (the ejido) as a place for cattle to forage, not for agricultural production as the Maya used it (Chardon 1963). The Spanish in part justified grazing their cattle on the communal forest lands of the town by using the Spanish terms *monte* (forest or unoccupied land) and ejido to refer to this land.[6] This dual use and meaning of the forest, by the Spanish as a place to graze cattle and by the Maya as a place for milpa production, led to the destruction of many milpas by cattle. This created a source of conflict between cattle owners and *milperos* that continued in Maní until the 1980s. In the metaphor of the estate as *estancia*, the Spanish sought to promote a new model of resource use that, with its control of water sources and use of the forest to graze cattle, conflicted with the Maya's traditional use of these resources.

Sometime before 1782, the site called Komchen had become a Spanish *estancia* with the name Santa Rita.[7] The saint's name given to an estate reflects the intimate connection between the estate's existence and its water source. When the Spanish dug a well, the day on which water was reached was considered the day that the well was "christened," with the well and the estate taking the name of the saint whose day was closest to that day, a custom that continues today. Thus, nearly all the estates with wells near Maní have two-part Spanish-Maya names in which the saint's name derives from the name of the well and the Maya name derives from the place name (e.g., Santa Rita: the name of the well; Komchen: *kom*, kettle, *chen*, well, because of a natural sink there).

This hybrid name of the rural estate is a reflection of the hybrid meaning given to these places, by both the Spanish and the Maya. Just as Maya places had become the locations for Spanish commercial production and the ideology supporting it, the Spanish *estancia* had also become a part of the Maya landscape. These early *estancias* were generally constructed on a small scale and so required few salaried workers, such as a few *vaqueros* (cowhands) and the *mayoral* (ranch foreman) (Patch 1993). However, the *estancia* also included tenant farmers called *luneros*, who exchanged one day of work per week (customarily on Monday, *lunes*) or some amount of maize for the rights to use the land of the estates to make their milpas (Farriss 1984, 216). This provided the owners of *estancias* such as Santa Rita Komchen, which had a population of twenty in 1782,[8] with a resident labor force. During most of the colonial period, colonial officials made little distinction between the residents of Maya hamlets (*ranchos*) and those of privately owned *estancias* (both were *fuera de campana*, beyond the church bell). This reflected the early relationship between estate owners and the Maya who lived on their lands, in which the Maya had the freedom to move on and off the *estancias* easily (Farriss 1984, 215). The Maya residents on these estates and their activities furthered the notion of the Mayaness of these settlements.[9]

Changing economic and political circumstances in the Yucatan in the late eighteenth century intensified the struggle between different models for resource use on rural estates and the metaphors used to promote them. Rapid growth in the population of the Yucatan during the late colonial period, together with increasing amounts of trade between the Yucatan and other Spanish possessions, increased the market for Yucatecan henequen fiber, salt, salted meat, lard, maize, beans, and hides (Farriss 1984, 367–369). The resulting increased need for land to produce for the local and overseas market coincided with the main thrust of the Bourbon Reforms (Farriss 1984, 366), which allowed greater privatization of *terrenos baldíos* (vacant lands).[10] As a result, increasing amounts of land were pulled into production on existing estates, and new estates were created to provide for this increasing demand for commercial production (Patch 1993).

This growing market and the Bourbon Reforms provided the catalyst for the rise of the hacienda, an estate based on large-scale commercial production of cattle or market crops (such as sugar or henequen) with a defined territory. This transformation of the rural estate included an expansion in size of these estates, as well as a shift in land use from a rudimentary style of cattle ranching (characteristic of the *estancia*) to the production of cash crops in which staple crops and cattle fodder were grown on the *monte*, which had earlier been used as an open range (Farriss 1984, 370). The result was a dramatic increase in the usurpation of land and water resources by the haciendas. This situation was furthered by conditions after Mexican independence that included an end to the Spanish crown's policies protecting Maya communities, as well as an elimination of the Yucatan's Cuban source of sugar and *aguardiente* (distilled alcohol from cane sugar). Many of the claims for *terrenos baldíos* made during this period centered on water sources.[11] As a result, the hacienda became a new model of resource use and control as the owners of these estates took greater control of water sources and the land around them that formerly had been held in common. For example, in 1782 Maní had three *ranchos de milperías* (large Maya milpa settlements) in the communal lands subject to the town.[12] By 1803, only twenty-one years later, all of these settlements had become haciendas with Spanish owners and had attracted larger populations.[13]

Access to water was an important reason why many Maya lived on these estates, attracted in part because their wells could offer a convenient water supply. The technology of the *noria* (mule-powered water wheel), built to water the cattle of the estate, was also useful to water household gardens clustered around the *casco* (Farriss 1984, 216). John Lloyd Stephens noted this when he visited the hacienda Uxmal in 1840, where the *luneros* living on the estate were obligated to provide labor and maize for the owner "in consideration of their drinking the water of the hacienda" (Stephens [1841] 1993, 240).[14] This control of water by hacienda owners, John Lloyd Stephens stated, "creates a relation with the Indian population which places the proprietor somewhat in the position of a lord under the old feudal system" (Stephens [1841] 1993, 236). The metaphor of the estate as hacienda implies a more intensive use of land, a greater control over the water sources of the community, and a greater domination over the Maya who need access to these resources.

The appropriation of land and water resources by haciendas continued throughout the nineteenth century, reaching a climax during the henequen era (1860–1910), when great henequen estates came to dominate the northern Yucatan. Although a few of Maní's estates grew some henequen, more important for local haciendas was the increased demand for production of foodstuffs to supply the needs of the henequen zone. With its location on the main Maní-Oxkutzcab road and its proximity to the railway at Oxkutzcab (about 6 kilometers to the south), Santa Rita Komchen was in a position to provide foodstuffs to the population of Mérida and the henequen zone.[15] As a result, the production of Santa Rita Komchen during this period, as with most of Maní's haciendas, was geared toward cattle and staple crops rather than henequen.[16]

As the intensification of production continued, many Maya became debt peons on estates such as Santa Rita Komchen, which had a population of thirty in 1910 (Censo Nacional 1910). By the first decades of the twentieth century, with virtually all of the land around Maní held as private property, many Maya had little choice but to move onto these estates, where they often worked under harsh conditions.[17]

During the nineteenth century, as the haciendas intensified their usurpation of land and water resources and increased their exploitation of Maya workers, the forms and layout of the *cascos* and the narratives surrounding them became increasingly supportive of the metaphorical models positing the *casco* as center and the *hacendado* (hacienda owner) as a controller of the center. This landscape represents an attempt by the *hacendados* to strengthen their control over the activities on these estates. The landscape of Santa Rita Komchen thus served as one of the mechanisms by which the *hacendados* sought to justify their increasing control of essential resources. It is during this period that the owners of Santa Rita Komchen and other haciendas near Maní developed the elaborate metonymic representations of water as center that are still found there today.[18]

The metonyms written by the *hacendados* in the *casco* mirror elements that were used by the Franciscans in the sixteenth century when they laid out the center of Maní as a model of a Christian utopia based on medieval notions of Jerusalem as the heavenly city.[19] The main corral of Santa Rita Komchen has four entrances facing the cardinal directions, over which stand arches like those that in the past led to the *atrio* (church plaza) of the parish church. At the rear of the corral near the *casa principal* (the owner's residence) is a high *noria* platform (about 3 meters) connected to large masonry water tanks and troughs. At Santa Rita and other haciendas, the *casa principal* is architecturally and spatially linked to the well and water tanks. In some cases, the *casa principal*, the *noria*, and the water tanks are all built on the same platform. Like those in the church's *atrio*, canals led from the hacienda's water tanks to water the *huerta*, or irrigated garden. The technology used to water the *huerta*, with canals leading from a central basin, was similar to that used in the Moorish gardens of Spain, whose layout was in turn based on Islamic notions of the plan of the Garden of Paradise.[20] The need to irrigate the *huerta* was certainly the practical reason for its placement in the *casco*. However, the link between the place of the *huerta*, irrigated by the well, and the space of the Christian Garden of Eden, irrigated by the four rivers of paradise, which reflected the owner's understanding of the nature of the cosmos, helped to promote the *hacendado*'s control of this water. The metonyms of center in the hacienda's *casco* draw on these multiple understandings of the water source as center.

Overall, the hacienda owners used the architecture and spatial organization of their estates to set themselves apart physically and ideologically from their servants and workers (Bracamonte y Sosa 1990). By linking the *casa principal* to metonyms for cosmological center, the *hacendado* sought to write a message into the landscape about his power and position in the community, with the owner at the center and the Maya workers at the periphery. While the owner of Santa Rita Komchen lived in the masonry *casa principal* connected to the water source, the Maya workers of the estate lived in a number of Maya houses more distant from the well. In addition, a Christian chapel located in the *casco* of Santa Rita Komchen, in which the owners of the estate were entombed, furthered the concept of the core of the estate as a Christian space and the owner as associated with the center. The creation of the hacienda, centered on a Christian well in a Spanish *casco* superimposed on a Maya place, echoes the sixteenth-century Franciscan rewriting of the town of San Miguel with its church as the center of a Christian utopia, on top of the Maya town of Maní.

The Maya, however, were not passive readers of the landscape as advocated by the hacienda owners. Throughout the nineteenth century, the Maya continued to promote alternative readings of the landscape of the hacienda, in which they promoted their access to resources by stressing their links to the place and its water source. The Maya incorporated the new realities of everyday life on these estates into their understanding of the beyond-the-everyday by linking the *casco* of the hacienda to the Maya cosmos. Closely associated with water in both the Maya town and the Spanish rural estate, the *yaxche'* or ceiba tree serves as one of the primary metonyms of center for the ancient and modern Maya. The *yaxche'* is associated with the *Yaxche'il Kab*, the World Tree (Freidel, Schele, and Parker 1993). Ceibas stand next to Maní's *cenote* and, in the corrals of many local ranches, the *ya'ax che'* stands near the well. Like the *cenote* at the center of the Maya town, the well at the center of the estate can be read as a metaphor for the center of the earth. These metaphorical models not only play a role in negotiation of meaning for the Maya of the community but also serve to advocate for their daily access to essential resources such as water and land.

Other metaphors, which interpreted the hacienda as the exploitation of Maya workers and the owner as "the devil," were also promoted by the Maya to express the oppressive conditions they often encountered on the haciendas. These metaphors are still powerful in Maní, so much so that the late-nineteenth-century henequen era is often referred to as "the time of slavery." These alternative readings allowed the people of Maní, both Maya and Spanish Creoles, to create a deep sense of the meaning of everyday conditions by linking the community of the present with the past and the cosmos. A polysemous landscape has been created in which both the Maya and the Spanish Creoles (the poor and the rich, the *campesino* and the *hacendado*, the oppressed and the oppressor) are able to interpret the landscape differently to give alternative meanings to the rural estate and to the models of resource use and control that characterize them. This likely played an important role in controlling activities on the haciendas around Maní during the nineteenth century, whose populations were mostly Maya: around 95 percent Maya in 1806.[21]

The Mexican Revolution brought an end to much of the power held by the owners of haciendas around Maní and changed the relationships between the owners and workers on the estates and between these groups and the land surrounding the town. The abolition of debt peonage was brought about in the 1910s by the Alvarado administration's application of the reforms of the Mexican Revolution (Orosa Díaz 1994, 237).[22] Released from their ties to the haciendas, many Maya left the estates, resulting in the depopulation of many of the haciendas around Maní. Santa Rita Komchen, which had thirty inhabitants in 1910 (Censo Nacional 1910), had only one inhabitant in 1921 (Censo Nacional 1921). Although the number of rural estates (ten) that existed in the municipality in 1921 was the same as in 1910, the populations of these estates had dropped considerably, from 239 or 13 percent of the total population of the municipality in 1910 (Censo Nacional 1910) to 64 or 3 percent of the municipality's population in 1921 (Censo Nacional 1921).

In addition, the seizure of land from the haciendas by the Reforma Agraria (Agrarian Reform) to form the ejido of Maní in the 1930s freed much of the land surrounding the community from control by these estates. Almost two-thirds (over 436 hectares) of the land of Santa Rita Komchen was taken in the 1930s as a "grant" to the ejido of Maní, leaving Santa Rita with 240 hectares. Additional restrictions were placed on the estates in the 1980s with the implementation of enclosure laws which required that cattle be fenced in, thereby ending the centuries-old problem of destruction of milpas by wandering livestock. Santa Rita Komchen, which had over 200 cattle before the enclosure laws, had only about 50 cattle in 1994. Today, Santa Rita and other large estates near Maní are no longer called haciendas (with its connotations of exploitation) but now are called *kah tali* in Maya or *rancho* in Spanish, the colonial Spanish term for Maya hamlets. The use of this name today draws on a long history of viewing the rural estate as a Maya space.[23]

However, even with the nonthreatening name *ranchos*, the large estates such as Santa Rita still contain wealth in natural resources and are still generally controlled by the elite. The quality of the forest on private lands is much better than in the now overused ejido for making milpa, leading many *campesinos* to contract with private landowners to make milpa on their properties. The estates and other private properties, which contain most of the tall trees remaining in the forest around Maní, are also virtually the only source of timber for construction in the community.

While agreeing on the concept of the water source of the *casco* as center, generally the owners and the workers of these estates continue to promote different metaphorical bridges between the *rancho* and the beyond-the-everyday. Most owners of local estates still advocate the reading of the *casco* as a symbol of their wealth and power, while many others in the community continue to read the *casco* as a Maya space. Maya ceremonies such as the *hetz' lu'um* are still carried out there, often by the *rancho*'s overseer without the consent, or in some cases with the express denial, of the owner. These ceremonies serve as metonyms to advocate for the workers' access to resources on these estates by referring to this space as linked more to the metaphors of those who work there than to the metaphors of the owner. A few owners of local *ranchos*, like don Andrés of Santa Rita Komchen, create and interpret several metaphors in the landscape of the *rancho*. By using the resources of Santa Rita Komchen to produce both cattle for the market and milpa for subsistence, and by practicing Maya ceremonies to protect them both, don Andrés negotiates the meaning of his control of the estate's resources through the use of both traditions. These alternative readings of metonyms referring to the *casco* of the *rancho* as center, which have developed over centuries, are still highly important in negotiating the meaning of the relationships between people and resources on the rural estate.

CONCLUSION

Over the past five centuries, the people of Maní have used the landscape as a process through which they have interpreted, contested, and negotiated the meaning of resources and their access. The creation and interpretation of landscape is a human process of using the environment and the narratives about this space to create meaning in the here and now by linking the everyday to the beyond-the-everyday. The example of Santa Rita Komchen shows some long-term trends in how local people have used this process to help them adapt to change brought about by integration of the community into the larger world system, change that has often included innovations in technology and shifts in the economic and political situation that have accentuated differential access by different groups to the resources needed by their families. The Spanish and Creole elite have often used the creation of landscape to justify their control of community resources and their domination of the Maya population. Throughout this oppressive history, the Maya have used the same process to help them not only to maintain a continuity of meaning by creating a link between daily life and their place in the cosmos, but also to advocate for access to basic resources such as water and land, the very necessities of life for traditional Maya. Viewing the landscape as a cultural process provides insight into not only how the landscape has been used as a mechanism of oppression but also how oppressed people have made use of this same process to negotiate the meaning of their place in the community and to advocate for their traditional model of natural resource use. By allowing an "openness of meaning," the cultural process of landscape provides a means of communicating local concepts of history and cosmology, as well as providing an arena for contesting and negotiating the control of essential resources in the present day.

Throughout the community's history, the people of Maní have adapted to change by using the landscape to link together the conditions they face in their daily lives to ways of knowing and acting developed in the past, to provide themselves with models for acting in the present. These linkages have often resulted in the reinterpretation of the past, as well as the new contexts in which people are drawn to new understandings of the nature of the world. By providing a process for the application of past meanings to new contexts, the landscape has been and will continue to be an important tool in the community's adaptation to change.

ACKNOWLEDGMENTS

I owe hearty thanks to the people of Maní for allowing me to experience a small part of their lives during my fieldwork there. I especially wish to thank Bernarda Góngora Mejía, Aurelio Balam Be, Cesar Canul Coot, and Arcadio Parra Mo of Maní for their knowledge of local estates. I am indebted to the directors and staff of the Centro de Apoyo a la Investigación Histórica en Yucatán, the Archivo General del Estado de Yucatán, and the Archivo de la Mitra Emeritense for giving me access to the valuable information in these archives. I also wish to thank Jaclyn Jeffrey for her helpful comments on this chapter.

NOTES

1. A pseudonym.

2. Following the concept of landscape proposed by Hirsch and O'Hanlon (1995), the emphasis here is on the landscape as a cultural process that brings two poles of experience (the foreground and the background) into relation, not on the media (physical environment, narratives, maps, etc.) being used.

3. My use of metaphor and metonymy follows the definitions of these concepts by Lakoff and Johnson (1980).

4. In the later colonial period, the term is also used to refer to *ranchos agrícolas*, commercial ranches operated by the Spanish on communal lands (Bracamonte y Sosa 1993).

5. In the account of Fray Ponce's travels, he noted that between the town of Oxkutzcab and Maní, "At a league [into the journey], a great *ramada* had been made by some *ranchos* in which there were many principal Indians. . . ." (Ponce [1588] 1947, 86). This location is the site of Santa Rita Komchen.

6. Many of the topographical terms, such as *monte bajo* (grazing land and degraded woodland) and *monte alto* (primary or high seconday forest), used by the Spanish in Latin America had been used as land use and landscape classes in the geographical survey of Spain in 1517 by Fernando Colón (Butzer 1992, 548).

7. Visitas Pastorales, Maní 1782. Archivo de la Mitra Emeritense (AME), Mérida, Mexico.

8. Ibid.

9. In addition to those owned by the Spanish, most villages in the seventeenth century had *estancias* that were owned by *cofradías* (religious brotherhoods). The *cofradías* established these *estancias* as cash enterprises and to provide meat and lard for the community's festivals (Patch 1993, 27). The end of ownership of *estancias* by *cofradías rurales* came as a result of the Bourbon Reforms in the late eighteenth century.

10. Following these reforms, the bishop of Yucatan, Fray Luis de Piña y Mazo, decided to begin eliminating *cofradía*-owned *estancias* in 1781 (Farriss 1984, 362; Fernández and Negroe 1994, 26). The sale of these estates, at a time when land was more valuable and more difficult to obtain, was welcomed by the Spanish (Farriss 1984, 371). Those that survived this initial confiscation of *cofradía* estates were confiscated by the state legislature of Yucatan in 1832 (Farriss 1984, 377).

11. Manuscripts, Registro de anotaciones de las denuncias de terrenos baldíos, 1845–1846. Centro de Apoyo a la Investigación Histórica en Yucatán (CAIHY), Mérida, Mexico.

12. These settlements were Ekmaben with a population of eight, Mopila with ninety-seven, and Ikina with fifty-one. Visitas Pastorales, Maní 1782, AME.

13. These haciendas were San Antonio Ekmaben with sixty-seven inhabitants, San Isidro Mopila with sixteen, and Iquilna (Ikina) with ninety. There were also two additional haciendas and six *sitios* within the jurisdiction of Maní in 1803. Visitas Pastorales, Maní 1803, AME.

14. In this case, the obligation included the worker's labor on Mondays and the planting and harvest of 20 *mecates* (0.8 hectares) of maize for the hacienda's owner.

15. This railway line linking Mérida to Peto (by way of Acancéh, Ticul, and Tekax) was completed on September 15, 1900 (Suárez Molina 1977, 2:180). In time, the

development of the railroads led to the decline of many of Maní's roads in favor of the road to Oxkutzcab, which gave the closest access to the railroad.

16. The primary production of Santa Rita Komchen in 1890 was cattle. Poder Ejecutivo, Fomento, Cuadro relativo a las fincas rurales de este partido, Tekax, June 1890. Archivo General del Estado de Yucatán (AGEY), Mérida, Mexico.

17. By 1910, the four haciendas, one *rancho*, and five *parajes* within the municipality of Maní had a combined population of 239, or 13 percent of the total population of the municipality (Censo Nacional 1910).

18. John Lloyd Stephens, writing in the 1840s, gives some of the earliest descriptions of the haciendas' control over water and of the metonymic representations of water and center in the *cascos* (Stephens [1843] 1963; Stephens [1840] 1993).

19. Part of this plan for Maní included blocks whose lots are laid out in a nine-square plan: the floor plan of the Temple of Solomon in Jerusalem.

20. The technology of the *noria* was also brought to Spain by the Moors.

21. Estado que manifiesta el total número de Almas de que se compone esta Parroquia de Maní con distinción de lugares y separación de clases o calidades, 1806, AME.

22. General Salvador Alvarado was governor of Yucatan from 1915 to 1917.

23. The proliferation of small ranches around Maní from the 1940s to the 1960s furthered this concept of the *rancho* of the present day as a Maya space. For the *campesinos* who could obtain these small properties, the possibilities of the *rancho* seemed the way to make enough money to better the way of life of their families in the community. By 1960, there were thirty-three rural settlements in the municipality (Censo General 1960). The enforcement of enclosure laws in the 1980s brought an end to most of these small ranches. By 1990, the number of rural settlements had decreased to four, with a total population of ten (Censo General 1990).

REFERENCES

Barnes, T., and J. Duncan, eds. 1992. *Writing worlds: Discourse, text and metaphor in the representation of landscape*. London: Routledge.

Black, M. 1962. *Models and metaphors*. Ithaca, NY: Cornell University Press.

Bracamonte y Sosa, P. 1990. Sociedades de sirvientes y uso del espacio en las haciendas de Yucatán, 1800–1860. *Historia Mexicana* 40 (1): 53–77.

———. 1993. *Amos y sirvientes: Las haciendas de Yucatán, 1789–1860*. Mérida, Mexico: Universidad Autónoma de Yucatán.

Butzer, K. W. 1992. From Columbus to Acosta: Science, geography, and the New World. *Annals of the Association of American Geographers* 82 (3): 543–565.

Censo General 1960. 1963. *VIII censo general de población, 1960: Estado de Yucatán*. Mexico City: Secretaría de Industria y Comercio, Dirección General de Estadística.

Censo General 1990. 1991. *Resultados definitivos, datos por localidad. XI censo general de población y vivienda, 1990*. Aguascalientes, Mexico: Instituto Nacional de Estadística, Geográfica e Informática.

Censo Nacional 1910. 1918. *División terretorial de los Estados Unidos Mexicanos correspondiente al censo de 1910: Estado de Yucatán*. Mexico City: Secretaría de Agricultura y Fomento.

Censo Nacional 1921. 1928. *1921 censo general de habitantes: Estado de Yucatán*. Mexico City: Departamento de la Estadística Nacional.

Chardon, R. E. 1963. Hacienda and ejido in Yucatán: The example of Santa Ana Cuca. *Annals of the Association of American Geographers* 53: 174–193.

———. 1980. The elusive Spanish League: A problem of measurement in sixteenth-century New Spain. *Hispanic American Historical Review* 60 (2): 294–302.

Duncan, J. S. 1990. *The city as text: The politics of landscape interpretation in the Kandyan Kingdom.* Cambridge: Cambridge University Press.

Farriss, N. M. 1984. *Maya society under colonial rule.* Princeton, NJ: Princeton University Press.

Fernández Repetto, F., and G. Negroe Sierra. 1994. La Fiesta del Santo Cristo de las Ampollas hacia finales del siglo XIX. *Boletín de la Escuela de Ciencias Antropológicas de la Universidad de Yucatán* 114: 23–40.

Freidel, D., L. Schele, and J. Parker. 1993. *Maya cosmos.* New York: William Morrow and Company.

Geertz, C. 1973. *The interpretation of cultures.* New York: Basic Books.

Hirsch, E., and M. O'Hanlon, eds. 1995. *The anthropology of landscape: Perspectives on place and space.* Oxford: Oxford University Press.

Lakoff, G., and M. Johnson. 1980. *Metaphors we live by.* Chicago: University of Chicago Press.

Moore, H. L. 1986. *Space, text and gender: An anthropological study of the Marakwet of Kenya.* Cambridge: Cambridge University Press.

Orosa Díaz, J. 1994. *Historia de Yucatán.* Mérida, Mexico: Universidad Autónoma de Yucatán.

Patch, R. W. 1993. *Maya and Spaniard in Yucatán, 1648–1812.* Stanford, CA: Stanford University Press.

Ponce, A. [1588] 1947. *Viaje a Nueva España.* Reprint: Mexico City: Secretaría de Educación Pública.

Stephens, J. L. [1843] 1963. *Incidents of travel in Yucatán.* 2 Vol. Reprint: New York: Dover Publications.

———. [1841] 1993. *Incidents of travel in Central America, Chiapas, and Yucatán.* Reprint: Washington, DC: Smithsonian Institution Press.

Suárez Molina, V. 1977. *La evolución económica de Yucatán.* 2 Vol. Mexico City: Ediciones de la Universidad de Yucatán.

10

The Transformation of "Paradise": Deep, Social, and Political Ecology Among the Yucatec Maya of Cobá, Quintana Roo, Mexico—a Dialogical Approach

Ellen R. Kintz
Amanda S. Ritchie

There is no better place on this earth than the tropical rain forest of the Yucatec Maya in southern Mexico to exemplify the philosophies of (1) deep ecology, (2) cultural ecology, and (3) political ecology. The Maya, who occupy this exuberant environment, have developed a calculated interface with their natural ecosystem. Sustainable utilization of tropical rain forest resources, a social tapestry designed for communication, negotiation within the social group, and an extraordinary (cyclical) worldview have allowed the Yucatec Maya to maintain "paradise" in the face of an increasingly invasive alien world system. *Paradise* in this context refers to Eden in a biblical sense, a pristine ecological setting. For the southern tropics, paradise is defined as the collection of microenvironments that culminate in the supply of a variety of foods and other useful resources from the tropical rain forest. These microenvironments are articulated by the Maya, and the Maya use the various zones to successfully satisfy various human needs (food, construction, medicine, ritual, etc.). However, the transformation of paradise is a process of cultural mutation initiated by the incorporation of the Maya into the modern nation-state and world system (see Kintz 1990; Re Cruz 1996; among others).

The theories of deep ecology, cultural ecology, and political ecology explicate mutation and transformation. The metamorphosis of community is best revealed through the touch, taste, and smell of life under the tropical canopy.

TRANSFORMATION OF PARADISE

It was a place of delight, bliss, felicity, and happiness, fruitful and bountiful. The tropical sun rose and set on resources natural and social. Howler monkeys howled, frogs sang to announce the rains, jaguars screamed in the night, the orchestra of morning sounds included musical chords of birds, barking dogs, and babies' cries. The children slept wrapped in blanket cocoons in their woven hammocks. The breath of women awakened the hearth fires; slapping hands formed *wahob* (tortillas) and toasted them on the *comal* (griddle). Men packed their *sabucanes* (field bags) with *gordas* (fat tortillas), *posole* (corn mush drink), *coas* (digging tools), and machetes; these items would be carried to milpas (cornfields) hidden in the rain forest. It was paradise.

It was a place of fear, a place of violence and terror, barren, a wasteland. The tropical sun rose and fell on natural and social disorder. There were no sounds of life in the dry, brittle, eroded, drought-plagued land. The birds had flown away. The dogs snarled, maggoty and covered with mange. There were no babies, no children, no breath, no life. It was paradise transformed.

To the uninitiated, the Yucatan Peninsula of southern Mexico appears to be a green homogeneous plain, in the same way that the Arctic seems to be white ice or a snowy homogeneous environment, the American southwest is simply an arid desert zone, and a Micronesian atoll represents a homogeneous oceanic ecozone.[1] Although the Yucatan Peninsula has been described as homogeneous, in fact it has discrete microenvironments, distinctions in vegetation, significant differences in soil types and soil depths, discrete areas of concentrated faunal resources, variation in water sources, and critical distinctions in rainfall patterns. These variations, recognized by the Yucatec Maya in ways comparable to those of indigenous people elsewhere in the world, are managed in complex and calculated ways.

Significant resources managed by the Maya in the knot of biodiversity include the Maya themselves. Human labor functions as a critical variable in the exploitation of the environment as well as in its reconstruction. The Maya are superbly knowledgeable about microbiological zones in their tropical world. They have struggled for more than a thousand years to catalog biological resources, to crystallize the cultural networks necessary to explore and exploit these resources, and to develop an ideology to conceptualize life under the tropical canopy. Their achievements represent a model for the tropical belt that illuminates the interface between ecology and human existence.

This study of the Yucatec Maya is based on twenty-five years of fieldwork in the community of Cobá, Quintana Roo, Mexico. The village, located in the eastern sector of the Yucatan Peninsula, is the site of a large archaeological ruin as well as a small ejido (communal land grant community). Cobá is located southwest of the international tourism center at Cancún, about a three-hour bus trip. Over the last twenty-five years, the eastern sector of the Yucatan has been profoundly transformed from a remote Maya refuge territory into a modernizing state. Tourism facilities have dramatically expanded along the eastern coastal beach zone. Although Cobá is located 45 kilometers inland from the coast, its large Classic Mayan ruins and tropical forest with exuberant bird life have opened the community to an influx of tourists in ever growing numbers.

The village of Cobá, which twenty-five years ago had a population of 100 souls, no electricity, no running water, two stores, and no bus service, has been transformed into a tourism destination with a population of 800 people, electricity, potable water, stores, artisan shops, several hotels, and daily bus service to the coast and other inland destinations. Cobá, which twenty-five years ago was wholly supported by milpa cultivation (cornfield subsistence agriculture), has been transformed into a village where the majority of male workers are employed as wage laborers on the coast. The focus of this study is to document the transformation of the village, changes in use of natural resources, changes in the organization of social networks, and shifts in ideology. These transformations are akin to those occurring in ecozones occupied by rural and indigenous peoples globally and in those zones where isolation is being replaced by integration into a larger world system.

The tropical rain forest and the Maya village have become the site of ecological stress, social disjunction, and struggle for political and economic equity. The village vessel, no longer marginal, hidden, nor protected by the rain forest, has recently been penetrated by the state and, in a sense, violated by the world system. Micro-level changes in the structure of village life mirror the larger disjunctions and struggles polarizing the Maya in the modern world. The transformation of village ecology is apparent in the changing realities of Maya men and women in Cobá. Deer are now rarely found in the forest, and milpa production has declined. Families serve Coca-Cola for breakfast, potato chips for lunch, and watch television while eating dinner. Men are captured by wage labor on the coast and women are engulfed by petty capitalism. Women use birth control, but the population is exploding. The constellation of natural, sociocultural, and political elements reflects the emerging inequity in the Maya village of Cobá, as documented over nearly three decades.

DEEP ECOLOGY: THE 1970s

> The *chuyche'* (pole-and-thatch house) is surrounded and protected by the rain forest.
>
> —conversation at the *rancho* of Demetrio Pol, Cobá, 1975

Nicolás

My father was a *milpero* (corn farmer) and he was one of the earliest settlers in Cobá. More than forty years ago, a few men traveled from Kanxoc and Tixhualactún in Yucatán and arrived at Cobá in the eastern *monte* (forest). The lake zone here was surrounded by *box lu'um* (rich, fertile, black soil), perfectly

> suited for *milpa*. The forest was rich with plant and animal species, huge *pich* (*Enterlobium cyclocarpum*) trees, the giant *cedro* (*Cedrela odorata*), cacao (*Theobroma cacao*), ramón (*Brosimum alicastrum*), *pom* (*Protium copal*), *chechém* (*Metopium brownei*), *chaká* (*Zanthoxylum fagara*), and others. Jaguars (*Felis onca*) and coatis (*Nasua nasua*), roseate spoonbills (*Ajaia ajaja*) and toucans (*Ramphastidae*), the deer, and snakes, among other life forms, filled the forest and entered the *campo* (camp). Zapota (*Manilkara zapota*) trees were slashed for chicle (chewing gum sap). After a few seasons of work, a few whole families traveled from the Yucatan to live in the new *pueblo* (town). My father built a small house: thatch roof, pole walls, dirt floor. He dug a well, bought some cattle, cut his milpa, and grew and harvested his corn. Human and natural resources were tied by interdependence and interaction among families. The impact of the tiny human population on the forest was minimal; humans were enanos, aluxób (dwarfs) beneath the forest canopy. I can remember once when *milperos* accidentally burned the forest in a wildfire and the destruction was a harbinger of what was to come.

Deep ecology, as a theoretical paradigm, is well matched to the Maya conceptualization of the forest ecosystem as an intricate web of plants and animals, including but not elevating human beings. In 1973, the Norwegian radical ecologist Arne Naess established the ecophilosophical framework of deep ecology. He stated that organisms represented "knots in the biospherical net or field of intrinsic relations," and argued for "biospherical egalitarianism," and an "anti-class posture" (Naess 1995). He called for the preservation and promotion of biospherical diversity, the strengthening of local autonomy, and the decentralization of control. Twenty years later, radical deep ecophilosophers such as Zimmerman and colleagues (1993) asserted that the root of contemporary ecological crises is anthropocentrism: positioning humans at the center of the universe. This position, they assert, breeds an arrogance that leads us to treat nature as a commodity; exploitable with the sole purpose of satisfying human appetites, wants, and desires. This leads to ecological disaster, which, according to these scholars, no political reform or revamped capitalist economic development can reverse. Reform and development only address the symptoms of the system, not the penetrating root of the deep ecological crisis inherent in the modern world.

> *María*
>
> I walked for eight leagues (50 kilometers) along a forest path to reach Cobá when I was fourteen and newly married. My husband and I lived in the forest and the frogs sang so loud at night they frightened me. As I rocked in my hammock woven of henequen (sisal), the *guano* (palm, *Sabal japa*) roof formed an intricate design above my head. The sapling pole walls of the small *chuyche* were held together by the strong vines, *anicab* and *akxúx* (family Boraginaceae).

We connected to a natural and social world. The forest provided us with a plenitude of resources which we used to survive. We lived in Cobá, hunting and farming. Grateful for our bountiful harvests, we performed *acciones de gracias* (ritual of thanks) celebrating and sharing our success with our family and our neighbors. Rituals were central to our lives. The *u-hanli-kol* ritual secured our milpas; the *Los Corrales* ritual secured our cattle; *u-hanli-kab*, our bees; and the *lol kah-tali* ritual secured our homes. The bounty of the forest provided herbs for curing illness of the body and the soul. Social relationships and the forest diversity provided our security. We lived on the margins, far from the center, independent, and beyond control.

But I can remember being told by my father that long ago the milpa harvests had failed. People ate wild jicama (*Pachyrhizus erosus*) and tree bark. Families lived in caves. People looked like skeletons and many children died. People wet mad. They were depraved and violent. It was a harbinger of our future.

CULTURAL ECOLOGY: THE 1980s

> My sons make milpa with me, protected by the rain forest.
> —conversation at the isolated *rancho* of Demetrio Pol, 1985

> My son went to work in the city of Campeche. He collected his pay in the *centro* (center of the city) and bought himself a watch. On the long bus ride home to Cobá, he was robbed and his watch was stolen.
> —conversation inside the *solar* (yard area) of doña Manuela Balam Hau

Julian Steward's popularization of cultural ecology in the 1950s sought to predict sociocultural life in terms of the articulation between production processes and habitat. Steward's (1955) strategy stipulated that research should focus on (1) the *interrelationships* between exploitative or productive technology and the environment; (2) the analysis of *behavior patterns* involved in the exploitation of a particular area by means of a particular mode of production; and (3) the extent to which behavior patterns exploit the environment and affect other aspects of culture (ideology).

Steward's ecological theory addressed issues such as labor tied to seasonal and climatic cycles; tools and farm labor associated with food production; human reproductive behavior, population size, density, and change related to mode of economic production and total exploited habitat; human settlement patterns, social organization, and household size related to change in natural biota; and village organization and intercommunity exchange correlated with ecological impact.

Cultural ecology detailed the infrastructure of human networks and human utilization of resources in the environment. In 1993, Murray Bookchin argued that nearly all ecological problems arose from deep-seated social problems. He asserted that the present ecological crisis could not be understood until the problems within society were resolved. For Bookchin, the danger of the ecology movement was that technology, population growth, pollution, or other symptoms were blamed as the causes of ecological crisis. Efforts to ameliorate this crisis were misdirected toward symptoms, producing cosmetic rather than curative results. Bookchin argued that it was the broader market economy, structured around trade for profit and the identification of progress with corporate self-interest, that was the pathological cause leading to ecological disaster (Bookchin 1993). The theory of cultural ecology, which is systemic rather than symptom focused, provides another model to discuss the transformation of the Yucatec Maya.

Nicolás

My father and brother stayed in Chemax when I lived with María and my children in Cobá. I had no one to help me cut and fire my milpa. Milpa production was extremely difficult in the forest. Some years the rains came before we fired our fields and the burn was bad. Tree trunks crisscrossed the field and snakes were plentiful. Some years, it rained so hard when the corn was starting to grow, the low-lying maize became soup and putrefied. Other years, there was no rain and the seeds never germinated. During the years that the milpas gave very few sacks of *elote*, we survived on *macales* (*Dioscorea alata*). It was difficult to harvest the fields alone. My wife helped me weed the fields, but my sons were babies. We needed to hire extra men to assist us but we had no money. So I worked for other *milperos* and, in turn, was paid in corn. Making milpa was better when fathers and sons worked collectively. My work was *solo* (alone) and my family suffered.

My family and I moved out of my father's old house and into a new *chuych'*. Back then, we could take as much land as we wanted for the *solar*. I walled a large area and planted fruit trees: orange, lemon, and avocado. The saplings were like sticks but began to grow. We planted the avocado when the moon was full and the fruit came fat and round. My family was growing and I was working in my milpa and planting fruit trees. At night, I worked as a watchman in the new hotel, but I fell asleep and was fired. So I went to work in Campeche. It was far away and the work was very hard.

María

My husband's family and mine were living in Chemax while Nicolás and I lived in Cobá alone. My husband worked hard in those years, working by himself. I began to have my children, Pedro, and a year later, Nicolás, my firstborn sons. Soon Lola was born and then Berta. Our family was growing. There was so much work. I was raising chickens and working in my small kitchen garden. I grew tomatoes, coriander, onion, and chili. I planted *chile habanero* (*Capsicum* sp.) when there was a full moon and the chili grew fat and round. When we needed it, I collected wild chili (*max ik, Capsicum frutescens*) from in the forest. There was a large bush of *max ic* behind our house, but it is gone now. Sometimes, I went to weed the milpa with my husband and collected *leña* (firewood). My mother never did that.

Nicolás worked with the archaeologists. Sometimes he would leave us to go to Campeche to work surveying the archaeological site of Calakmul. There were six children plus the two of us to feed. Sometimes we were hungry and we dressed in rags.

POLITICAL ECOLOGY: THE 1990s

> Don Demetrio died and Chac Ne, his *rancho*, was abandoned. His sons are working on the coast. They still have their house in the village but they live in Playa Carmen now.
>
> —conversation heard in a local store in Cobá

Political ecology describes the struggle for control over resources and the organization and operation of hierarchy. The theory provides an essential model in the exploration of inequity brought on by the intrusion of state and global forces in small-scale communities. According to Atkinson (1991), the theory of political ecology seeks to explicate a social and economic system which creates a sustainable relationship between nature and society and that deters environmental catastrophe. The Green party of Germany has demanded radical political restructuring of hierarchy to avert ecological deterioration (Bookchin 1982; Kelly 1994; Dobson 1995). Although the keen discourse of Thomas Kuhn (1962) on scientific revolution concluded that when science fails to produce a new paradigm, it is not the end of the world, political ecologists have argued that if we fail to produce a new social configuration, we are likely to be faced with just that—the end of the world (see Atkinson 1991; Bradford 1993; among others).

Intense ecological stress is correlated with intensification of political hierarchy (see Schumann and Partridge 1989; Barkin 1990; Holmberg 1992; Escobar 1995). For the Maya, ecological stress is currently manifest in the reduction of rain forest canopy, the expansion of grasslands, drought, and low-yield harvests. In Cobá, the large cedar and other high rain forest tree cover is now gone. Secondary tropical forest is prevalent, and grasslands for cattle feed are expanding. These changes are correlated with a longer dry season, crop-planting regimes occurring later in the year, and lower yields for crop harvests. In the 1970s, the Maya were harvesting corn that could be stored for up to three years. In the 1990s, the corn harvests are lasting merely three months. These changes in the natural ecology are impacting the sociopolitical order. Families are experiencing disjunction as their level of involvement in wage labor forces them to migrate to the east coast for work. Maya villagers struggle for political and economic equity as state policy and world system intrusions (e.g., tourism) stimulate inequity in the Maya village. These disjunctions and struggles demonstrate the politics of hierarchy, organization of economic inequities, and jolting trauma in the fabric of Maya life.

Nicolás

We are struggling for control over our land now. The ruins of Cobá are located within the boundaries of our ejido lands. Since the 1970s, INAH [Instituto Nacional de Antropología e Historia] has controlled the gate to the ruins. We have never received any payment for use of our land and very few jobs. Now INAH wants to buy the land and set up a new road that would bring in new artisan stores. We fear that the new road would bypass our stores and we will no longer get any tourism trade. We would lose both our land and our ability to gain money from the tourists. The tourist buses come into the village but all they bring is noise. As a community, we have not benefitted from the thousands of people who have visited the site because INAH has not let us sell anything or service tourists inside the ruins until recently. We have been excluded from receiving any money from the tickets sold to view archaeological resources on our own ejido lands. While some people in the village are getting richer, others are getting poorer. The rich are raising cattle and the poor are not able to cut, farm, and harvest enough milpa to support their families. The environment is suffering as a result of these changes, and milpa land is not producing as well as it did in the past. It seems like the end of the world and a *castigo* [punishment] is upon us.

ECOFEMINISM

Ecofeminism, an alternative version of political ecology, argues that there can be no solution to the environmental crisis within a society whose fundamental model of relationships continues to promote domination (see Ruether 1975). Ecofeminism contends that important connections exist between the domination of women and the domination of nature. Recognizing and understanding these connections while uniting the missions of the women's movement with those of the ecological movement provides an alternative to the current patterns of ecological devastation. Ecofeminism promotes a radical reshaping of socioeconomic relations, the fundamental values of the modern industrial world system, and male-biased models of domination (Warren 1993).

Maya women are cognizant of the ecological crisis that their families face. They are the household food providers and are aware that the corn harvests are inadequate to supply their families year-round. They are also aware of the negative impact of this ecological failure on family structure, labor, and opportunity. They are aware of the connection between deterioration of ecological resources and transformation of their position in the family. With ecological deterioration, the women must work harder to supply their families with food and clothing. Maya women have reshaped their family economy with penny capital investments and returns.[2] The larger wage labor market is primarily closed to women, offering negligible opportunities for advancement and no assistance with child care if positions were available.

María

I always kept the money that I made. My husband is working on the coast and when he comes home, he gives me very little money. The reason that he can't give

> me very much money is because it is expensive to live on the coast. The money he makes, he spends in order to live there. I raise my chickens and I sew a little. I also sell ice.
>
> Most women keep and hide their money; if they don't, their husbands may drink it up. I have my fruit trees, *aguacate, limón*, and *naranja* here inside my *solar*. In my garden, I grow *cebolla* (onions), cilantro, *rábanos* (radishes), and chilies. I can sell these or we eat them. Seven of us are living here now. When my husband and son come home from the coast, it makes nine. Mostly, we eat the *chichan pak'alob* (small plants) that I can grow.

Ecofeminism as a theoretical design clarifies the position of Maya women and elucidates their transformation under modernizing conditions on the eastern coast of the Yucatan. Political and economic changes affect the position of Maya women but, more broadly, also affect the status and welfare of their families. If males are removed from the family for wage labor on the eastern coast, family welfare becomes dependent on the service of Maya women as primary heads of households. Women can be empowered or impoverished under these conditions, and ecofeminist theory provides a mechanism to evaluate conditions of domination or defiance.

THEORETICAL ARGUMENTS

The issues that affect the transformation of the Yucatecan Maya system can be analyzed utilizing the theoretical principles of deep ecology, sociocultural ecology, and political ecology. These theories articulate the natural, social, and political complexity of the world in which the Maya live. The issues discussed by ecologists are pointedly relevant to ecological and social problems in the tropical rain forest of the Yucatec Maya, as well as issues facing indigenous and other rural peoples wherever they reside.

Following deep ecophilosophy, development has led to degradation in the Yucatan, disrupting the biospherical net. Land has been wasted and ravaged; short agricultural fallow cycles have restricted fertile soil development; short-fallow tropical rain forest recovery regimes have modified climate conditions; botanical and faunal resources and diversity have become limited by overexploitation and habitat destruction; and human relations to other life forms have been compromised. Recent introduction of petrochemical fertilizers and insecticides by agricultural development agents have permitted the Maya to utilize their land in intensive ways to control insects and to combat other crop plagues. Although this domination of nature appears to be a positive factor in agricultural production in the short run, it has created long-term health hazards for the human population as well as chemically resistant "super" insects. The petrochemicals have increased the cost of maintaining agricultural plots. Productivity per hectare has gone up, but the number of hectares cultivated per family has gone down, while unit costs of production soar. Overall production in kilograms per family has been reduced.

The modern world system has provided opportunities for many Mayan families to become involved in wage labor and the tourism industry on the east coast of the Yucatan Peninsula. This "opportunity" has compromised the independent political position of farmers. The depletion of land, the exploitation of Maya human labor, and excessive tourist development are a burden on the backs of the Maya. The independent farmer, converted to a rural proletariat worker, is captured and consumed by the larger system.

The ecofeminists have a salient point when they argue that economic development encourages and provides greater opportunities for male wage labor than for female employment. Anthropocentric and androcentric development schemes—by their own design—do not offer women equitable access to the development of cash resources (see Behar 1995; Gordon 1995). However, Maya women have been converted to penny capitalists, fostering a dependency on the soft tourist market system.

Human economic development has had a heavy impact on the tropical landscape of the Yucatan. Rain forests are disappearing, and humans continue to build hotels on the east coast of the Yucatan from Cancún to Tulum. Sewage and other forms of garbage and contamination pollute the Cancún resort landscape in large volumes in ways that the smaller traditional populations of Maya *chicleros* (chicle or gum tree harvesters) of twenty years ago never could and never did.

Although Yucatec Maya peasants have been traditionally identified as milpa horticulturalists, they have been involved with larger scale societies for thousands of years. In the 1970s, male and female dialogue in the village of Cobá revolved around making milpa. Male conversation focused on subsistence horticulture; slash-and-burn techniques, area of milpas to be planted, surplus seed or lack of seed, rainfall patterns, weeding regimes, and harvest schedules. Men could recite from memory harvest patterns for decades and their success or lack of success in farming activities. Women concentrated on production activities such as kitchen gardening, animal husbandry, and small-scale exchange. Very few males or females were employed in wage labor, and these jobs were restricted to the village. People held little hard cash.

By the 1980s, male and female dialogue revolved around making milpa, kitchen gardens, and wage labor. The climate of the Yucatan had taken a bad turn; crops were destroyed by hurricanes and, while family size grew, production from the milpas did not. During this transitional time, many men joined the ranks of wage laborers on roads, in construction of resort buildings along the eastern coast, and with teams of archaeologists on a seasonal basis.

In the 1990s, Maya male and female dialogue revolved around wage labor. Males who had made milpa for their entire adult life were abandoning this activity. They were captured by wage labor opportunities and spent much of the week away from their families, living and working on the east coast.

Although some families continued to plant their cornfields, many fields were insufficient to feed the family throughout the year. Corn and beans were frequently bought from the government store or local *tiendas* (stores), but the prices were high and the peso was being severely devalued. The women continued to plant kitchen gardens and tend their fowl and pigs. Many women were engaged in penny capitalism, locally selling *huipiles* (traditional embroidered female dress), their garden products, and other small-scale commodities. Some restaurants opened in the community to serve the growing number of tourists who were visiting Cobá. Success in wage labor and service industries was unevenly distributed from family to family. Wage labor for both men and women was profoundly tied to the boom-and-bust Mexican economy. With this new economic regime, men and women actually had fewer interactive dialogues; the men were absentee husbands and the women as household heads had their hands full.

CONCLUSION

The larger society continues to disturb the balance that the Maya have cultivated with their natural environment, and this loss of control over their own environment, in turn, requires that the Maya depend on the larger society. The situation in Yucatan is not unique. For example, it parallels that of the Guaraní of Paraguay and Brazil (Reed 1997), as well as those of indigenous and other rural peoples elsewhere in the world.

When conquest state economies invade the tropical forests, they draw indigenous peoples into the world system. Rather than resisting the expansion of the invading commercial system, native people like the Guaraní and the Maya interface with it while maintaining their traditional lives. Commercial connections with the larger world have not, to date, destroyed the Guaraní nor the Yucatec Maya. However, neither culture has been wholly assimilated into mainstream Paraguayan, Brazilian, or Mexican society.

However, it was, in part, the diverse economic production of the Guaraní and the Yucatec Maya that was the key to their ethnic survival. Although they entered the international commodity market, they retained their traditional subsistence production. For generations, the Guaraní and the Yucatec Maya adapted a larger system to their own village-based purposes. However, after centuries of negotiation with the larger world, the success and survival of these two groups (and other rural groups like them) within the larger world system is now being compromised and threatened.

The Parable of the Golden Table

María: My father used to tell me the story of the golden table that lay at the bottom of the lagoon in Cobá. He told me that it had been there forever. It was in the lake when the first people arrived here long ago. My father told me that the table was made of pure gold. It was enormous, and the most beautiful thing you have ever seen. It had a radiance. When the night was darkest, you could see the light from the table shine.

Amanda: Has anyone ever seen the table or gone looking for it? Has anyone ever tried to steal it, since surely it would bring someone great wealth? Or hasn't anyone else ever heard that it exists?

Elena: I have heard the story of this golden table before. People in Cobá have told me the story. They know that the table is there.

Nicolás:
Yes, people know the story, and some have tried to bring it up from the depths of the lagoon. But no one has ever moved the table, no one has ever stolen it, and no one has taken its wealth from our community. It's so deep, deep in the heart of the lake, that it is unreachable, and it's very solid because it's made of gold. The golden table is said to have magical powers. Its magic guards the table from those who try to steal it for its great value, its magical powers, and its radiance.

No one, according to the villagers, will ever be able to steal the golden table, and it will always be seated at the base of the lagoon. It is protected, the Maya say, by a huge crocodile, *nohoch lagarto*. According to Tate (1992, 55), Maya iconography depicts the Crocodile-Earth-Water Monster, *Bolon Tz'akab*, "He of Nine or Many Generations." Schele and Miller (1986) identify this monster as the "First Ancestor," associated with ancestor-related sacrificial rituals. This Crocodile-Earth-Water Monster embodies the complex of sacred substances that are the premier necessities for humans: maize, flesh, and blood (see Puleston 1976). Following Taube (1985), the sacrifice of these most sacred of substances is done for the benefit of the community of the living and the dead. The ancient, golden, four-cornered ritual table, covered with golden *elote* (corn in the husk), tortillas, and yellow squash seeds, also reflects the modern-day milpa when the yellow tassel of *nal* (corn) is blowing in the wind. As the Maya recite the parable of the golden table, they are tied to their past, and they are foretelling their collective future, their sacrifice, and their survival.

NOTES

1. *Ecozone* is used here to identify a discrete classification of landform, soil, vegetation, climate, wildlife, and water resources. Its simple classification belies the significant variation of forms contained within the zone.

2. Penny capitalism is taken from the classic study of Panajachel, Guatemala, by Sol Tax (1953) and refers to small, specialized, and diverse production and sales that add small but significant cash income to the family. Whereas Tax referred to penny capitalism in the context of markets in the highlands of Guatemala, Maya women in Cobá are producing and selling garden products and small commodities to their neighbors, stores, local restaurants, and tourists.

REFERENCES

Atkinson, A. 1991. *Principles of political ecology*. London: Belhaven Press.

Barkin, D. 1990. *Distorted development: Mexico in the world economy*. Boulder, CO: Westview Press.

Behar, R. 1995. Introduction: Out of exile. In *Women writing culture*, edited by R. Behar and D. A. Gordon. Berkeley: University of California Press, 1–29.

Bookchin, M. 1982. *The ecology of freedom: The emergence and dissolution of hierarchy*. Palo Alto, CA: Cheshire Books.

———. 1993. What is social ecology? In *Environmental philosophy: From animal rights to radical ecology*, edited by M. E. Zimmerman, J. B. Callicott, G. Sessions, K. J. Warren, and J. Clark. Englewood Cliffs, NJ: Prentice Hall, 354–373.

Bradford, G. 1993. Toward a deep social ecology. In *Environmental philosophy: From animal rights to radical ecology*, edited by M. E. Zimmerman, J. B. Callicott, G. Sessions, K. J. Warren, and J. Clark. Englewood Cliffs, NJ: Prentice Hall, 418–437.

Dobson, A. 1995. *Green political thought*. New York: Routledge.

Escobar, A. 1995. *Encountering development: The making and unmaking of the third world*. Princeton, NJ: Princeton University Press.

Gordon, D. A. 1995. Conclusion. Culture writing women: Inscribing feminist anthropology. In *Women writing culture*, edited by R. Behar and D. A. Gordon. Berkeley: University of California Press.

Holmberg, J. 1992. *Making development sustainable: Redefining institutions, policy, and economics*. Washington, DC: Island Press.

Kelly, P. K. 1994. *Thinking green*. Berkeley, CA: Parallax Press.

Kintz, E. R. 1990. *Life under the tropical canopy: Tradition and change among the Yucatec Maya*. Fort Worth, TX: Holt, Rinehart and Winston.

Kuhn, T. 1962. *The structure of scientific revolutions*. Chicago: University of Chicago Press.

Naess, A. 1995. The deep ecological movement. In *Deep ecology for the 21st century: Readings on the philosophy and practice of the new environmentalism*, edited by G. Sessions. Boston: Shambhala, 64–84.

Puleston, D. 1976. The people of the cayman/crocodile: Riparian agriculture and the origin of aquatic motifs in ancient Maya iconography. In *Aspects of ancient Maya civilization*, edited by F. August de Montequin. St. Paul, MN: Hamline University, 1–26.

Re Cruz, A. 1996. *The two milpas of Chan Kom: Scenarios of a Maya village life*. Albany: State University of New York Press.

Reed, R. 1997. *Forest dwellers, forest protectors: Indigenous models for international development*. Boston: Allyn and Bacon.

Ruether, R. R. 1975. *New woman/new earth: Sexist ideologies and human liberation*. New York: Seabury Press.

Schele, L., and M. E. Miller. 1986. *The blood of kings: Dynasty and ritual in Maya art*. Fort Worth, TX: Kimbell Art Museum.

Schumann, D., and W. Partridge. 1989. *The human ecology of tropical land settlement in Latin America*. Boulder, CO: Westview Press.

Steward, J. 1955. *Theory of culture change*. Urbana: University of Illinois Press.

Tate, C. E. 1992. *Yaxchilan: The design of a Maya ceremonial city*. Austin: University of Texas Press.

Taube, K. 1985. The Classic Maya maize god. In *Fifth Palenque round table, 1983*, edited by M. G. Robertson and V. M. Fields. San Francisco: Precolumbian Art Research Institute, 171–181.

Tax, S. 1953. *Penny capitalism: A Guatemalan Indian economy*. Publication 16. Washington, DC: Smithsonian Institution, Institute of Social Anthropology.

Warren, K. J. 1993. Introduction. In *Environmental philosophy: From animal rights to radical ecology*, edited by M. E. Zimmerman, J. B. Callicot, G. Sessions, K. J. Warren, and J. Clark. Englewood Cliffs, NJ: Prentice Hall, 253–267.

Zimmerman, M. E., J. B. Callicot, G. Sessions, K. J. Warren, and J. Clark, eds. 1993. *Environmental philosophy: From animal rights to radical ecology*. Englewood Cliffs, NJ: Prentice Hall.

11

Women's Human Rights Issues Among the Maya

Susannah Glusker

At the outset of the twenty-first century, the struggle for human rights is a widely respected and recognized field. This chapter honors the vision and dedication that Mary Elmendorf exercised throughout her life, notably in human rights. Dr. Elmendorf, a trailblazer in the field, initiated her work more than sixty years ago and inspired many people of different disciplines and nationalities to focus their energy and talent by contributing to the wider goal of social justice and human well-being.

Elmendorf, from the inception of her career, even before completing her doctoral degree in anthropology, listened to and learned from people. She was—and still is—concerned with issues of equality and equity for women and men, for people of all races, as well as with such basic human needs as drinking water and environmental sanitation. She was one of the Quaker volunteers serving with the American Friends Service Committee (AFSC) and the British Friends Service Council that in 1947 shared the prestigious Nobel Peace Prize for their work in war-torn Europe.

In this chapter, I make an effort to describe the relevance of her work to the recent struggles of the indigenous Maya peoples to live according to their ancestral culture, a polemic that has become much more intense in the last seven years. My research involved following the role of women in this process, as reported by the Mexican press (Glusker 1998b), as well as an analysis of the work and the role of Mariclaire Acosta (1996a, 1996b, 1996c), an important human rights activist in Mexico. In this context, it is important to emphasize that Mariclaire Acosta's work reflects and complements Elmendorf's vision and innovation.

Acosta is an active leader, deeply concerned with defending human rights and improving women's living conditions, issues that are increasingly understood to be integral parts of the discourses on health, resources, and environment. The history of Acosta's involvement illustrates the process of change. She initiated her career working with Amnesty International, then went on to establish her own nongovernmental human rights organization, and then, in response to new political conditions, accepted a position in the government as Mexico's undersecretary for human rights, in the Ministry of Foreign Affairs for three years. She has taught at the Universidad Nacional Autónoma de México (UNAM, National University of Mexico) and worked concurrently as a journalist, founding journals such as *Voices of Mexico* and *Fem*, as well as contributing numerous articles to other publications.

Information published by major newspapers and news journals in Mexico reflects an intense concern by the Mexican populace for human rights, as well as the highly dynamic nature of the process. There is an ongoing struggle for land, water, sanitation, and the right to practice ancestral customs that honor the appropriate use of land. The government and people of Mexico are actively seeking answers to address citizens' concerns for ecological, gender, and human rights issues.

To put the current situation of the Maya peoples in context, the information presented herein is structured within a historical framework of land tenure and general human rights issues in Mexico—particularly regarding women. There is an intimate relation between human rights and environmental conditions, but there is a general lack of detailed information for Yucatan, Chiapas, and other parts of southern Mexico where the Maya live. Hence, reports about Mexico in general are used to indicate the situation relevant to the Maya in this region.

It must be stressed that this chapter cannot be considered finished research with specific results. One primary reason why this work is not definitive is because of the political changes underway in Mexico, especially after the elections of 2000, when the party that had ruled for more than seventy years was "elected out of office." This is an exciting period of searching for appropriate approaches to modify sociopolitical systems and find solutions.

VALUING INDIGENOUS CULTURES IN MEXICO

Mexico, a land of many peoples and cultures, was conquered by the Spanish early in the sixteenth century. Independence from Spain came about 300 years later. The ensuing hundred years were spent searching for a system of government amid power struggles that ended with Porfirio Díaz's dictatorship and the eruption of the Mexican Revolution of 1910.

Once the military phase of the struggle was over, major political and cultural changes were initiated by the new government, which included intellectuals appointed by José Vasconcelos, the minister of education, to rebuild the society using utopian models. One of the most important policies was a concerted effort to identify and revalue that which was indigenous to Mexico: its people and their ethnic traditions.

Manuel Gamio, known as the father of applied anthropology in Mexico, initiated the process by hiring artists, such as Francisco Goitia, to live among the people of Teotihuacán, a major Aztec site north of Mexico City, in order to record what had survived (Luna Arroyo 1958; González Gamio 1987).

The next phase involved a concerted effort to publicly value those traditions. Artists such as Diego Rivera and Jean Charlot joined anthropologists and intellectuals such as Anita Brenner and Frances Toor in documenting the traditions that were still alive after 400 years of subjugation (Brenner 1970 [original 1929]; Wolfe 1939; Toor 1947; Charlot 1963; Glusker 1998a).

The effort was not limited to Mexico City. Charlot was hired by Sylvanus Morley of the Carnegie Institution to join a team of professionals that included Robert Redfield, who was working on one of the most important projects in Yucatan (Anita Brenner 1925–1930). They were excavating and documenting the ruins of the ancient Maya site of Chichén Itzá (Morley 1956). Charlot documented not only the designs that survived on pyramids, but also living descendants and their art with sketches (Anita Brenner, 1925–1930). Redfield, a pioneer in anthropology from the University of Chicago, recorded changing patterns of Maya life.

Gamio's effort to document the vestiges of the past was complemented by Minister of Education José Vasconcelos's program to value the indigenous peoples of Mexico. He hired artists such as Diego Rivera and José Clemente Orozco, among others, to paint murals that would tell the story in a visual fashion to the people of Mexico.

Once the process of recognition was well underway, the Mexican government made a concerted effort to integrate indigenous populations into the mainstream. The Instituto Nacional Indigenista (INI, National Indigenous Institute) was created to develop and execute programs that would increase literacy, promote improved health care, teach Spanish, and support economic development.

Foreign development agencies joined Mexicans to meet the challenge by sponsoring specific programs. Elmendorf, head of the CARE office in Mexico in the 1950s, recalled the excitement of the process. At the same time, she stressed the importance of Alfonso Villa Rojas's work with INI, carried out at their first integrated regional center with headquarters in San Cristóbal de las Casas, Chiapas (Mary Elmendorf, personal communication with reference to the panel honoring Alfonso Villa Rojas at the Society for Applied Anthropology meeting in Cancún, 1994).

Villa Rojas was a Yucatecan Maya schoolteacher who had accepted the invitation to teach at the first school established in the village of Chan Kom. He became intrigued with the tall white people with loud voices that he saw on the bus. They were working with Morley at Chichén Itzá. He walked to the site and watched from the sidelines. He observed until he followed in their footsteps and became an anthropologist himself. Nurse McKay, a member of Morley's team, noticed Villa Rojas and was impressed by the charming, bright young man. She offered him a cold drink and then invited him to join them for a meal. When he refused politely she offered him a gift of food, which he took but later threw away because he was afraid of being poisoned by food that came out of a tin (Anita Brenner, unpublished journals, 1925–1930; Alfonso Villa Rojas, personal communication, 1990).

Villa Rojas was soon hired by Robert Redfield as a field research assistant to observe and record people in the village of Chan Kom. His fieldwork for Redfield led to doctoral studies at the University of Chicago and a subsequent professional career with major contributions about Mayan culture. He was a native Maya who spoke the language. Further, as a schoolteacher he also understood both the traditions and aspirations of his people; thus, he knew how to work toward preserving the culture, while improving their standard of living (Glusker 1994).

Redfield was a pioneer in the methodologies of observation and recording. Villa Rojas, along with other Mexican anthropologists such as Ricardo Pozas and Julio de la Fuente, took their work one step further: through INI they not only documented people's beliefs and traditions but were involved in helping people improve their lives. "The work being done," Elmendorf said, "was so impressive that Arnold Toynbee called it the most advanced applied anthropology in the world" (Mary Elmendorf, personal communication, 1994).

This was the beginning of a tradition in Mexico that fit right in with the philosophy of valuing indigenous culture and helping people improve their lives with literacy and development programs. A new generation of anthropologists followed Gamio in the postrevolution period of Mexico's utopian euphoria. Mexico's finest anthropologists, such as Miguel León Portilla and Gonzálo Aguirre Beltrán, among others, worked with notable Mexican intellectuals such as writer Rosario Castellanos, developing creative programs focused on bringing ethnic groups into the mainstream.

Respecting the cultural values and traditions of each group of people was a primary concern. Through the program, many people were able to learn to speak and understand Spanish. Others became bilingual teachers who worked in villages with young children or at secondary school–level hostels where students lived during the week, returning home on weekends. Personnel from INI worked with federal education and health agencies to develop programs to ameliorate poverty and disease. Although INI programs were successful in some areas and conditions improved, they were not able to eliminate poverty in remote villages, where thousands of people still survive on subsistence farming, with incomes far below the national average.

Elmendorf promoted a CARE self-help program that supplied sewing machines as technical assistance, and this made a major contribution, particularly to women's lives, in many Mexican communities. Women learned to use the machines to sew their children's clothes and their husband's shirts, as well as their own clothes and some additional articles for sale. Although it is hard to improve on hand embroidery as an art that still brings prices as much as ten times higher than machine-sewn articles, introducing sewing machines meant better-made garments to embroider and higher productivity for those who embroidered with the machines.

Another important program that Elmendorf helped implement in the late 1970s was the distribution of CARE midwife kits to graduates of the training programs for traditional midwives. Elmendorf, Villa Rojas, Brigitte Jordan, and others worked with a grant from the Research Institute for the Study of Man (RISM) on a project called Fertility Determinants Among the Maya to help with family planning, midwife training, and primary care clinics. These initiatives were well accepted by INI and the Ministry of Health (Elmendorf 1979).

RELEVANT POLITICAL AND ECONOMIC PROCESSES

Spaniards who conquered Mexico focused on generating wealth by growing profitable crops for two reasons: to meet their quota of gold to send to the king of Spain and to increase their personal wealth. After Mexico gained its independence from Spain, Spaniards born in Mexico, known as criollos, took on the roles of former Spanish landowners. They sought to increase their personal wealth, especially landholdings, even though they no longer sent money to Spain. As a result, many villages were gradually forced to abandon the cultural practices and traditions that renewed and protected the quality of their collectively owned land. Prosperous landowners became politically powerful on local and regional levels. Some even resembled Chinese warlords in their attitude and behavior toward the poor (Brenner [1943] 1976). They are still known as caciques, and they are the ones who lend money to the needy and gradually acquire land in payment for debts.

The political system evolved into a structure of a loose network of caciques, from the village to the regional level, all the way up to members of the state cabinet legislature, governors, and federal officials. Throughout Mexico, most of the land, as well as the general stores and pharmacies in villages, ended up in the hands of the caciques (Alfonso Cruz, personal interview, July 18, 1998). Wealthy oligarchies at the village level were in a position to manipulate most of the population (often in debt to them) in order to be "elected" and thereby formalize their control—a phenomenon that continues today.

ACCESS TO LAND

Mexico's revolution of 1910 brought land reform. The ejido system, as it is known, assigned small plots of land to individuals who worked the land. Ownership (really, the right of use) was transmitted only by inheritance to members of the family: land could not be sold to a third party. While it gives a certain level of subsistence autonomy to peasants, the system has a number of drawbacks.

The ejido is a small parcel of land, sometimes not even enough to grow corn and beans for one family, let alone generate an income to purchase other goods. Some farmers work others' land for a share of the harvest, while still others gradually have taken de facto control of ejido lands in payment for debts acquired during bad harvest years. As certain individuals increase their de facto control of land, the amount of land available for younger members of the community decreases. This forces villagers to seek employment outside their communities. Some have found jobs in cities or villages close by; others have migrated north, crossing the U.S.-Mexican border (Re Cruz 1996).

Land reform called for in the Mexican constitution of 1917 was not fully carried out in many parts of Mexico. Large landowners, known as *latifundistas*, members of the wealthy oligarchies, continued to own large areas of land throughout the country. Moreover, article 27 of the 1917 constitution was amended in January 1992, allowing ejidos to be divided and sold, as well as permitting individuals and corporations to increase the amount of land that they could legally own. This has meant a gradual return to the land tenure conditions that prevailed before the land reform that occurred after the Mexican Revolution of 1910.

As the wealthy increase their landholdings (especially of better quality land for irrigation farming and modern cattle pastures), ethnic communities face a double challenge: they have less and poorer quality land. Further, toxic chemical fertilizers and insecticides have affected the quality of both the land and the water supply (Josefina Mena, personal communication, July 20, 1998).

THE QUESTION OF INDIGENOUS AUTONOMY

Changing economic patterns in villages have led to a growing marginalized population. Today a vast majority of Mexico's indigenous peoples survive by subsistence farming. Acosta defines their situation in terms of racism, referring to *raza*. As opposed to race, which is understood to be just gross physical traits, *raza* refers to a sociocultural identity. Hence, Acosta refers to *raza* prejudice, since most Mexicans are mestizo, a blend of indigenous and European. Her first point is that the indigenous people are descendants of the people who originally lived on the land that was conquered. Second, their culture is significantly different from that of the colonizers. Third, they are economically inferior and politically powerless to affect their circumstances (Acosta 1996a).

Acosta (1996a) has explained that as nations and governments globalize, ethnic groups clamor for their right to exist, to identify as a people with a separate language, culture, religion and most important their own autonomous system of government.

Ethnic groups within Mexico are increasing their demands to elect their own rulers rather than being governed by powerful landowners that rule with impunity and disregard for their traditions. Ethnic autonomy, separateness, or cultural purity is not their goal; what they seek is self-determination for their people and communities.

In her unpublished report for a human rights association in Mexico, Acosta (1996a, n.p.) has summarized the international situation as follows:

> There are 400 million indigenous people in the world, 40 million live in Latin America and of these 15 million in Mexico. Those in Mexico include 56 ethnic groups living on subsistence farming. The illiteracy rate among them is three times higher than Mexico's national average. The 1990 census showed that sixty per cent of those over 12 were unemployed and forty-three per cent of those employed earned less than minimum wage.
>
> Nevertheless these are the people who represent the largest group of victims of the human rights abuses, perpetrated mainly by police, military and even paramilitary groups formed by large local landowners. More than forty per cent of the human rights conflicts involve land ownership of indigenous peoples.

Land is the principal resource required for traditional farmers to meet their basic needs (García Bernal 2001). It is a source of ongoing conflict. Although there are other sorts of conflicts involving basic needs of indigenous peoples, including fishing rights for coastal communities dependent on access to fish and shellfish in both marine and fresh waters, access to land is a basic necessity to nearly all groups.

Considering Mexico's history, one explanation for the situation described above could be that those isolated villages where indigenous people live were never totally conquered by the Spanish. They have not integrated into a mestizo way of life, nor have they modified their culture to follow an economy focused on accumulating wealth. Remarkably, some of these villagers are the same people who are seen as posing a threat to the government of Mexico (García Hernández 2001).

Given their despair and the fact that they had nothing to lose but their dignity, some people in Chiapas and others in Oaxaca and Guerrero organized formally and rebelled (Brenner 1976 [original 1943]). The ensuing occupation by federal troops made their situation worse because even subsistence farming was interrupted (Rosa Rojas 2001; Cabildo and Monge 2001). There are many similarities between the conditions that led to the Mexican Revolution of 1910 and current conflicts between indigenous peoples and the government today, almost 100 years later.

These factors may have triggered the aggression that Acosta identifies as an element of racism. Federal armed forces needed to destroy "the other" in order to be able to recognize themselves (Acosta 1996a). One of the ironies of the present conflict is that although federal soldiers recognize the indigenous people and groups as "visibly different" from themselves and their peers in the army, many of the young soldiers come from indigenous families.

Many of Mexico's indigenous peoples have chosen to identify themselves as distinct ethnic groups, continue to speak their own languages, and live according to their tradition. They want to elect leaders from within their communities as opposed to accepting those that the ruling party selects from among members of local oligarchies. They want the freedom to address their problems within their ancestral system.

THE MAYA, A CASE IN POINT

The Maya people include groups such as Chol, Lacandón, Tzeltal, Tzotzil, and Yucatec Maya who follow their own traditions and speak different languages, although they live relatively close to one another. Most live in small communities where they follow a modified form of their ancestral cultural traditions. Many villagers within these small communities live marginally under conditions of poverty without water and sanitation services, schools, or primary health care clinics.

The Maya communities of Mexico are scattered throughout the southern border states of Campeche, Chiapas, Quintana Roo, Tabasco, and Yucatan. They are divided into disparate groups with languages that are more different than French is from Italian. Quiché, a major Maya group in Guatemala, is as far from Yucatec as German is from Italian, although they evolved from a common proto-Mayan language. Some Maya groups have been more isolated than others, but as a group they are known for being determined and persistent in their pursuit of their lifestyle. They resisted Spanish domination longer than many of the other indigenous peoples of Mexico.

The Maya are generally known as a soft-spoken and artistic people. However, when their patience is tried, they fight for their rights. Their history documents fierce rebellions such as the War of the Castes of the Yucatec Maya in the mid-nineteenth century and the Tzotzil War of Santa Rosa (González Durán 1978; Bricker 1981). Following Acosta's definition, they identify themselves as other than the mainstream of the Mexican nation, while at the same time pursuing changes in the conditions that have led to poverty and disease.

Redfield and Villa Rojas (1934) documented the lifestyle of several Maya villages, including Chan Kom, before the road was built that facilitated access to larger towns, cities, and mainstream culture. Redfield (1950) returned years later to document the changes in *A Village That Chose Progress: Chan Kom Revisited*. Elmendorf (1976) continued the documentation of change in *Nine Mayan Women: A Village Faces Change*, focusing on women's lives and the interaction among villagers when faced with change. She and Deborah Merrill-Sands documented the first migration waves, which provided guidance for Re Cruz's study of the ongoing process of change in the community, including the migration of many younger people to Cancún (Elmendorf and Merrill-Sands 1978; Re Cruz 1996).

Respect for land is a central issue in the conflict between the status quo and the drive to return to ancestral customs; however, it is not the only source of problems. One of the most serious origins of strife, often accompanied by violence, is religion. Most Mexicans, including the Maya, identify formally as Roman Catholics, although many still practice rituals and traditions of their ancestors. An increasing number of Protestant missionaries proselytized throughout Mexico during the twentieth century (Alfonso Cruz, personal interview, July 18, 1998).

Following a Protestant faith can mean a greater sense of community and a formal prohibition of alcoholic beverages. Nonetheless, the process of converting some of the people in the village often results in dividing a community. The enmity between Roman Catholics and Protestants, those choosing change and those remaining Catholic, has led to bitter struggles between them, ongoing violence, and many cases of one group expelling the other from their villages (Redfield 1950; Elmendorf 1972; Mariscal 2001a, 2001b).

Roman Catholics often identify with maintaining the status quo, because converting to any of the Protestant religions means changing their lifestyles. Sometimes there are conflicts between Catholics who are more orthodox and those who are more relaxed in their practice. The power struggle can become political when either group includes ruling landowners, whether members of the oligarchy or those seeking local control (Mariscal 2001a, 2001b; Román 2001).

Women are especially vulnerable in political and religious conflicts because of the contradictions between tradition and modern practices. The gradual disappearance of the custom of sharing all the work, at home as well as in the field, has brought increased burdens for women (Elmendorf 1972), in some cases because their husbands leave the village to seek jobs and in others because the men are ashamed to maintain the tradition in the presence of outsiders (Re Cruz 1996).

Alcohol abuse is a critical factor that contributes to changing patterns as well as increased conflict and pressure on women. It means not only a drain on the family's meager resources but also an increase in physical and verbal abuse (Faust 1998).

MAYA WOMEN VIS-À-VIS THEIR MESTIZO SISTERS

To appreciate the condition of Maya women, it is necessary to understand the overall human rights situation regarding women in general in Mexico and indigenous women in particular. The increasing awareness of gender issues in Mexico is evident in the January 1998 report presented to the United Nations Committee for the Elimination of Discrimination Against Women (CEDAW 1998). The information gathered by a group of Mexican nongovernmental organizations that deal with women's issues and human rights indicates clearly that current conditions for women are far from optimum. Very few changes have registered in the situations that women have had to face for many years, or perhaps centuries, even though Mexico signed the International Agreement to Eliminate Discrimination Against Women in September 1981. Although Mexico established a special agency, the Programa Nacional de la Mujer (National Program for Women) in August 1996, operating funds were not provided until more than two years later.

Women are prime victims of violence at home, on the job, and on the street. This is especially grave in the case of indigenous women, for whom the number of reported cases of harassment, torture, rape, forced temporary and permanent sterilization, as well as induction into prostitution by the military, are increasing. More than 300 women were raped in Chiapas between 1994 and 1998. Nongovernmental organizations report extreme poverty, displacement of people from communities occupied by federal troops, illegal executions, and murder (CEDAW 1998).

On March 12, 1997, the Human Rights Commission stated in a press release that national legislation related to women and girls in Mexico is twenty to twenty-five years behind that in the United States and Western Europe. Many cases of rape by prison guards or police officers have been reported (CEDAW 1998, 8–10). However, "In the State of Chiapas and the State of Mexico the penalty for stealing cattle is more severe than the one for rape" (CEDAW 1998, 11). Juridical statistics show that when the victim is a woman, aggressors are less likely to be sentenced. For every 1,000 male victims, 224 people were sentenced; yet only 126 aggressors were sentenced per 1,000 women victims. Further, in 1995, 160 criminals were sentenced for every 1,000 men murdered, but only 61 killers of women were punished, less than half the proportion for aggressors against men. The figures also show that, on average, women accused of murder receive sentences with a 25 percent greater penalty than men convicted of the same crime (CEDAW 1998, 12).

In 1996, Acosta presented the following data at a conference at the Kellogg Institute of Notre Dame University:

> There are forty-six and a half million women in Mexico, a little over half of the total population. Most of them are in the age group of fifteen to sixty-four years of age (69%), as opposed to those under fifteen (36%) or over sixty-five (4.2%). About 95% of the women in Mexico live with their families, approximately half have had a child by the age of twenty-one; 16% become mothers between the ages of fifteen and nineteen years old. Most urban Mexican mothers have an average of three children; women in rural areas have an average of five within a period of thirteen years. By the age of sixty, more than two thirds of women in Mexico depend economically on their children or other relatives. Fifteen of every hundred women over the age of fifteen are illiterate, two out of three illiterate adults are women, and only two of every three girls between the ages of six and fourteen are enrolled in school. (1996b, 6)

A group of Mexican women's and human rights nongovernmental organizations provided additional information at the 1998 UN session of the

International Federation of Human Rights for CEDAW (1998). The figures show that the number of urban women attending and staying in school is increasing; there are now eighty-two women for every hundred men enrolled in high schools and universities. They also indicate that most of the deaths of women in Mexico between the ages of fifteen and sixty-four are due to breast and uterine cancer, heart attack, diabetes, abortion, and violence (CEDAW 1998).

The report also indicates that 66 percent of women in Mexico have access to birth control, yet there are at least 220,000 deaths per year from poorly performed abortions (CEDAW 1998, 29). The largest number of complaints filed at the National Human Rights Commission was for medical negligence in the care of women and children (CEDAW 1998, 31).

The information and statistics about women who live in rural indigenous villages are even more dramatic. One of every five speaks her mother tongue. Half of the indigenous women in Mexico over the age of fifteen are illiterate; almost a third (32.5 percent) of the girls in school drop out by the age of fourteen, whereas only 17.5 percent of the boys drop out to enter the workforce and help support the family (CEDAW 1998, 27).

The above data must be seen within the greater context of the economic situation throughout Mexico and most of Latin America. The ongoing crisis—at cultural, economic, environmental, social, and political levels—that has affected most of Latin America for more than fifteen years has resulted in a deterioration of most people's standard of living. Currency devaluation, capital flight, debt burdens, and dependence on foreign loans have meant a steady deterioration of real income. In the CEDAW report Acosta said, "Not a single new job has been created since 1982, leaving at least sixteen million people marginalized from the economy" (CEDAW 1998, 15). Acosta's 1996 report presented to the Kellogg Foundation included the following information, according to my notes:

> Consumption per capita of corn, beans and wheat has dropped more than 35% in ten years since their price increased at double the rate of the minimum wage, which has decreased by two thirds since 1982. Malnutrition, coupled with a decline of the health budget to almost half of what it was in 1980, caused the tripling of infant deaths in 1992. Overall spending on education declined from 5.5 to 2.5% of Mexico's Gross Domestic Product.

It is not difficult to appreciate Acosta's term "the feminization of poverty" in a country where 10 million people live within 3 million households headed by women (CEDAW 1998, 17). Women earn one-third of the average income; they contribute economically in one of every three households and are the main support in one of every five; one in every ten households is supported entirely by women (CEDAW 1998, 17). Rural, indigenous, and peasant women are most affected by the crisis because they often work long hours, some as many as sixteen to seventeen hours a day.

Women who migrate to urban centers are subjected to conditions that many countries share. They are paid less than men receive for the same job; as Acosta (CEDAW 1998, 23) points out, "In 1970 women earned 750 for every thousand a man earned. In 1980 it went up to 822 pesos per thousand, yet by 1990 the figure had dropped to 764 pesos for every thousand, and in 1995 the figure had dropped further to 694 pesos per thousand a man earned."

Remarkably, legal support for women has been weakened by recent changes in federal legislation. Sexual harassment is now a punishable offense only when concrete harm can be proved, and there is no provision for multiple offenders (CEDAW 1998, 24). Health and child care benefits were also modified. The 1997 social security law eroded women's rights to day care facilities and retirement benefits. Only 13 percent of women working with social security coverage are entitled to day care for their children. State-sponsored medical care has also been reduced. Pap smears to test for cervical cancer are covered only every two years, and hospitals limit the stay of mothers who have given birth to twelve hours. Medical care is now limited to primary care; laboratory work and X rays are subsidized, but both the worker and the employer must pay part of the costs (CEDAW 1998, 21–22).

Women working in urban environments are also often subjected to humiliating employee regulations such as exhibiting a sanitary napkin used during their menstrual periods to prove they are not pregnant. Some employers require a certificate of sterilization (CEDAW 1998, 18). Other employers, such as banks, terminate women's work contracts by eliminating their positions if they become pregnant. Many employers prefer hiring married men because they are considered to be more stable. The opposite is true of married women, who have a harder time finding employment, especially if they have children. Acosta reported that 27 percent of women economically active in most urban areas earn more than two, and up to five times the minimum daily wage rate. However, in rural areas, only 12.9 percent earn these amounts and almost 30 percent do not have any income at all. Women's workdays are 43 percent longer than those of men because informal services, in which women are routinely involved, are not defined as work (CEDAW 1998, 16).

Acosta's research provides ample statistics about women's working conditions in urban and rural settings. The figures reflect worsening poverty as economic conditions in the country call for cutbacks in employment and expenditure. For example, in the last ten years government expenditure on health dropped to almost half of the amount it was in 1990. The same is true for the education budget (CEDAW 1998, 15). Infant mortality increased threefold in 1992. Conditions in rural areas are even more severe.

The deterioration of basic natural resources and environmental conditions has greatly complicated an already difficult situation. Lack of adequate sanitation services has resulted in contamination at ever deeper levels of the water table (Josefina Mena, personal interview, July 20, 1998). Given the lack of sewage facilities, water is often treated with chemicals, some of which have been found to increase kidney afflictions. The lack of potable water affects the entire community, but women—especially indigenous women—pay this cost disproportionately, as it is they who are primarily responsible for cooking and child care. Hence, there has also been a marked increase in women's workloads, with longer days and greater efforts to survive and provide the basic services needed for maintaining the household. It is estimated that women living in rural areas in Mexico now work more than 100 hours per week without any salary (CEDAW 1998, 15), and a great amount of that effort is in response to lack of basic resources and degraded environmental conditions.

Water problems for Maya women, particularly in the Yucatan, are well known. An early example comes from Elmendorf's concern about childhood morbidity and mortality due to lack of safe water. In the 1950s, she promoted the pilot CARE/AFSC well-drilling project in seven rural villages in Mexico. In this project, both women and men in the communities helped Quaker volunteers provide potable water for their families. President John Kennedy used this project as a case study to help get congressional approval and funding for the Peace Corps. The model of this project is still used by international and national agencies, as well as nongovernmental organizations worldwide, especially for reaching the poorest of the poor (Elmendorf 1977a, 1977b, 1977c, 1980a, 1980b).

HUMAN RIGHTS ACTIVISM

The results of the July 2000 national elections reflected ordinary people's drive to seek change in economic and political conditions in Mexico. People in both small rural communities and urban settings took action. They got organized, voted, and not only participated in the electoral process but stayed engaged, and still take to the streets to make their voices heard (Mariscal and Villalba 2001).

Sometimes the pressure and frustration are overwhelming, and people who live in small communities take the law into their own hands. Acosta (CEDAW 1998, 16) wrote, "Armed violence has started to replace institutional channels as a medium to confront unfulfilled demands that the predominantly peasant population makes. The justice system was riddled with corruption, impunity, and influence peddling as well as racial discrimination against indigenous peoples." She also mentioned the increase of public insecurity and violent crime, with a parallel increase in the impunity of public offenders. "The alarming rate of lynchings of suspicious criminals in the country indicates how mass action is substituting for an inefficient system of police and judicial forces to punish those suspected of delinquency" (CEDAW 1998, 44). Lynching has been reported in the press as late as 2001 (Henriquez 2001).

Participation by ordinary citizens also takes place within an institutional framework. This includes organizing political parties, supervising polls during elections, and refusing to recognize corrupt leaders imposed by local oligarchies in rural communities by working with opposition parties. Through these political processes, people in small communities are trying to improve the infrastructure for housing, water, and other services, such as primary health care and education.

Acosta notes that only 27 percent of Mexico's population lives in rural settings, a shift that changes the nation from predominantly rural to urban (CEDAW 1998, 44). Women are especially active in seeking change; 53 percent of the voters registered in Mexico City in 1997 were women (CEDAW 1998, 44). However, only 30 percent of the women candidates were elected to office. Fewer than 4 percent of small municipalities elected women, and only three women governors have been elected in the last forty-five years. While more than 2 million women are unionized, women represent only 8 percent of the leadership (CEDAW 1998, 44).

> In practice, equality of men and women is mostly a formal right. The dynamics of the economy and society undermine any possibility of affirming the rights of women. Discrimination is . . . a fact of daily life. It is institutionalized. Therefore, it exists in the courts, and in some state laws, as is the case of Chiapas, for example where the penalty for stealing cattle is higher than that for rape. However, as long as current economic policies prevail, women will continue to be at a disadvantage, and no matter how many human rights treaties are signed and ratified by the Mexican state, gross violation of women's basic rights will result. (Acosta 1996b, 13)

An important identifiable element in political violence has been the military. "Violence for political reasons, including sexual violence is increasing, especially in regions where there is intense militarization, due to the emergence of armed guerrillas, such as Chiapas, Oaxaca, Veracruz and Guerrero" (Acosta 1996b, 14). The National Commission for Human Rights in Mexico City studied and recorded cases of violence from 1989 to 1995. They found that those affected included "a total of 31,255 persons, between victims (67%) and their families. The crime most often reported was rape (47.3%), followed by sexual abuse (27%). One in four victims of sexual offenses was under the age of thirteen, and three fourths of them were female" (Acosta 1996b, 15). Acosta (1996b, 16) also reported, "About one in five victims of rape becomes pregnant as a consequence. Abortion is not penalized in these cases. However, once the victim has proven that her pregnancy was caused by rape, there is no established legal procedure for her to obtain a safe abortion in a hospital or clinic."[1]

A joint report prepared and presented by a group of nongovernmental organizations in 1998 described human rights violations that illustrate the situation, for example, in occupied areas in Chiapas:

> As the process of militarization advances, the greater the influence of the military in civilian life and an increased number of complaints of violation of physical integrity and psychology of women, above all in indigenous communities. The military have deteriorated the conditions of health in communities they occupy.

> They pollute rivers, rape women, and control by violence, occupy municipal offices and schools, encourage alcoholism, as well as prostitution and drugs. People cannot come and go freely. They force women to control birth by sterilizing them or inserting devices without their consent, with the discourse that they are bringing social help. In reality they divide families, neighbors and communities.
>
> It is notable that in those places where there is a greater presence of military troops, paramilitary groups act with impunity and malice, as occurred in Acteal, municipality of Chenalhó in Chiapas on the 22 of December in 1998, where 45 indigenous people were murdered—mostly women and children. Pregnant women had their wombs slit. (CEDAW 1998, 46)

Because of the outrage of men and women within the area and throughout the country, unarmed women acted as human barriers to physically expel and prevent further entry of the military and police into their villages. The photograph of the women physically pushing them out was published in major news media, and about a year later Pedro Valtierra, the photographer, received a prestigious award for his work. Some people fled to the hills or to other villages after the massacre. "They left their homes, belongings, and crops behind. They preferred to go hungry rather than accept the so-called humanitarian help offered by the army" (CEDAW 1998, 47).

The parallel official discourse during the worst periods of military activity in Chiapas about democratization and respect for human rights is juxtaposed with a "silent war of systematic and selective violations against human rights. Protesting the struggle against subversion, the Mexican government uses torture, forced disappearances of people and clandestine prisons, paramilitary groups and death squadrons that even attack women, children and senior citizens" (CEDAW 1998, 48).

Harassment is not limited to peasants; it is also used against human rights activists, journalists, and politicians who question or expose violence. Some have been murdered; others, such as Cecilia Rodríguez, have been raped and warned to stop talking, writing, and publishing. In other words, people have been threatened to stop getting the word out. Offices of nongovernmental organizations are trashed, computers and equipment stolen. Highly visible individuals have received threats personally and over the phone. Two students who participated in a witness program sponsored by the Universidad Iberoamericana were imprisoned (Elena Tovar Elena and Deborah Chenillo, personal communication about the witness program to villages in Chiapas, August 13, 2001).

Foreigners have been especially vulnerable to arrest and deportation. They are accused of breaking the law, their immigration documents are questioned, and after intimidation during questioning, they are deported. Foreigners who travel to the areas occupied by the military are required to obtain special visas. Specific cases draw press attention as well as action by human rights organizations. Major human rights activists have been involved, such as Tom Hanson, former head of Pastors for Peace, who, although he had accreditation as an observer, was deported without warning or the opportunity to call the U.S. embassy. He was defended by Acosta's Comisión Mexicana de Defensa y Promoción de los Derechos Humanos, A.C. (Mexican Commission for the Defense and Promotion of Human Rights) and can once again travel to Mexico.

CONCLUSION

Mary Elmendorf taught me to listen and learn, to have a cup of tea and to respond with just a few words that are to the point. Her work focused on Maya women includes observations based on this philosophy. Listening to colleagues and peers, such as Mariclaire Acosta, describe the process of striving to improve human rights sets a tone of hope and confidence in the possibility of change and transformation (Ballinas 2001).

Nobel Prize winner Rigoberta Menchu is among those who participate and make a difference (Castañeda 2001; Enciso 2001). Indigenous peoples throughout the world continue to seek to return to the sacred responsibility of caring for the land that nurtures and sustains their confidence in the conditions necessary to protect agricultural practices, earth, water, fuel, and wildlife for future generations.

Respect for the patterns followed by indigenous peoples for centuries and their desire to protect and provide natural resources for future generations is gaining momentum throughout the world. These peoples are responding to the oppressive fast pace of impersonal industrialization that may feed but does not nurture. Their children and grandchildren seek spirituality in tune with rituals of respect for one another and for their natural habitat.

NOTE

1. The most recent case of this phenomenon to be published in the nationwide press involved Paulina, a fifteen-year-old girl raped in Baja California, whose mother was unable to get any hospital to perform the abortion approved for her by the court. The mother sued for support of her grandchild. The case is still in court at this writing.

REFERENCES

Acosta, Mariclaire. 1996a. Human rights in Mexico: An overview. Report for Comisión Mexicana de Defensa y Promoción de los Derechos Humanos, A.C.

———. 1996b. Overcoming the discrimination of women in Mexico: A task for Sisyphus. Paper presented at the Kellogg Institute at the University of Notre Dame, Indiana, November 9–11.

———. 1996c. Los rostros del racismo [The faces of racism]. Mexico City: Comisión Mexicana de Defensa y Promoción de los Derechos Humanos, A.C., July.

Álvarez Fabela, Martín. 2000. *Acteal de los mártires*. Mexico City: Plaza y Valdés Editores.

Ballinas, Víctor. 2001. Los gobiernos de América Latina deben reconocer el derecho de los indígenas a explotar sus recursos [Latin American governments should recognize the right of indigenous peoples to exploit their resources]. *La Jornada*, June 16, Political Section.

Brenner, Anita. 1925–1930. Unpublished journals. Available online at the Harry Ramsom Research Center of the University of Texas at Austin.

———. 1970 [original 1929]. *Idols behind altars*. Reprint, Boston: Beacon Press.

———. 1976 [original 1943]. *The wind that swept Mexico*. Reprint, Austin: University of Texas Press.

Bricker, Victoria Reifler. 1981. *The Indian Christ, the Indian king: The historical substrate of Maya myth and ritual*. Austin: University of Texas Press.

Cabildo, Miguel,and Raúl Monge. 2001. Investigación de la CNDH: Altos mandos del ejército, implicados en desapariciones forzadas [Investigation by the National Commission of Human Rights: High-level officials in the army implicated in forced disappearances]. *Proceso*, no. 1290. July 22, 9–13.

Castañeda, María del Socorro. 2001. Menchú: El EZLN puso a los indígenas en la agenda mundial [Menchu: The Zapatista National Liberation Army put the indigenous peoples on the world's agenda]. *La Jornada*, June 16, Political Section.

CEDAW (Committee for the Elimination of Discrimination Against Women). 1998. Informe que presentan ONG de mujeres y de derechos humanos mexicanos en coordinación con la Federación Internacional de Derechos Humanos al Comité sobre la Eliminación de la Discriminación contra la Mujer (CEDAW) [Report presented by Mexican women's and human rights nongovernmental organizations in coordination with the International Federation of Human Rights for the Committee for the Elimination of Discrimination Against Women (CEDAW)]. United Nations Eighteenth Period of Sessions, New York, January.

Charlot, Jean. 1963. *The Mexican mural renaissance*. New Haven: Yale University Press.

Elmendorf, Mary L. 1972. *Mayan women and change*. Cuernavaca, Mexico: CIDOC.

———. 1976. *Nine Mayan women: A village faces change*. New York: John Wiley-Schenkman Publishing.

———. 1977a. The changing roles of Mayan women in the home and in the marketplace. In *Proceedings of the first Mexican/Central American symposium on research about women*. Mexico: El Colegio de Mexico.

———. 1977b. The many worlds of women: Mexico. In *Women's roles and status in eight countries*, edited by Janet Giele and Audrey Smock. New York: John Wiley and Sons, 129–172.

———. 1977c. Mexico. In *Village women: Their changing lives and fertility*, edited by Priscilla Reining, Fernando Cámara, Beverly Chiñas, Rosalie Fanale, Sonia Gojman de Millán, Barbara Lenkerd, Iris Shinohara, and Irene Tinker. Washington, DC: American Association for the Advancement of Science, 111–199.

———. 1977d. Public participation and acceptance. In *Environmental impacts of international civil engineering projects and practices*, edited by Charles Gunnerson and John Kalbermatten. New York: American Society of Civil Engineers, 184–201.

———. 1979. Anita: A Maya peasant woman copes. special issue of *Studies In Family Planning*, ed. S. Zeidenstein. New York: The Population Council.

———, and Buckles, Patricia. 1980. *Socio-cultural aspects of water and excreta disposal.* Appropriate Technology for Water Supply and Sanitation, Vol. 5. Washington, DC: World Bank.

———, and Michael McGarry. 1980. What is appropriate technology? A Maya village asks. In *Seven case studies in rural and fringe areas of Latin America.* Appropriate Technology for Water Supply and Sanitation, Vol. 8. Washington, DC: World Bank.

———, and Deborah Merrill-Sands. 1978. Socio-economic impact of development projects on Chan-Kom 1971 to 1976: A preliminary study. Paper presented at Society for Applied Anthropology meeting, Mérida, Mexico.

Enciso L., Angélica. 2001. Ecocidio y etnocidio en zonas indígenas: Menchú. [Ecocide and ethnocide in indigenous areas: Menchú]. *La Jornada*, June 13, Political Section.

Faust, Betty B. 1998. *Mexican rural development and the plumed serpent: Technology and Maya cosmology in the tropical forest of Campeche, Mexico.* Westport, CT: Greenwood.

García Bernal, Cristóbal. 2001. La tribu seri de Sonora recurriría a las armas para recuperar predio. [Seri tribe in Sonora would take up arms to recover land]. *La Jornada*, July 17, State Section.

García Hernández, Arturo. 2001. Los indígenas no son ni han sido amenaza para la seguridad nacional, plantean especialistas. [The indigenous people are not nor have they been a threat to national security; say specialists]. *La Jornada*, July 1, Political Section.

Glusker, Susannah. 1994. From Chan Kom to Chicago: The role of ethnicity, race and gender in the relationships among intellectuals in the United States and Mexico. Paper presented at the Society of Applied Anthropology meeting, Cancún, Mexico.

———. 1998a. *Anita Brenner: A mind of her own.* Austin: University of Texas Press.

———. 1998b. Women networking for peace and survival in Chiapas: Militants, celebrities, academics, survivors, and the Stiletto Heel Brigade. *Sex Roles* 39 (7/8): 539–557.

González Durán, Jorge. 1978. *Los rebeldes de Chan Santa Cruz* [The rebels of Chan Santa Cruz]. Mérida, México: Ayuntamiento de Felipe Carrillo Puerto, Quintana Roo.

González Gamio, Angeles. 1987. *Manuel Gamio: Una lucha sin final.* [Manuel Gamio: A struggle without an end]. Mexico City: Universidad Nacional Autónoma de México.

Henriquez, Elio. 2001. Chocan priístas y perredistas en Zinacantán [Members of the PRI and PRD party clash in Zinacantan]. *La Jornada*, June 11, Political Section.

Luna Arroyo, Antonio. 1958. *Goitia.* Mexico City: Editorial Cultura.

Mariscal, Angeles. 2001a. Denuncian agresiones en municipios autónomos zapatistas; 2 desaparecidos [Aggression in Zapatista autonomous municipalities: Two disappearances]. *La Jornada*, June 18, Political Section.

——— (NOTIMEX). 2001b. Detienen a 3 tojolabales por la expulsión de evangélicos [Three *tojolobales* arrested for expelling evangelicals]. *La Jornada*, June 28, Political Section.

———, and Rodolfo Villalba Sánchez. 2001. Chiapas: Eligen habitantes de San Andrés Larráinzar a sus autoridades [Chiapas: Residents of San Andrés Larráinzar elect their officials]. *La Jornada*, July 17, States Section.

Morley, Sylvanus G. 1956. *The ancient Maya.* Stanford, CA: Stanford University Press.

Re Cruz, Alicia. 1996. *The two milpas of Chan Kom.* Albany, NY: State University of New York Press.

Redfield, Robert. 1950. *A village that chose progress: Chan Kom revisited.* Chicago: University of Chicago Press.

———, and Alfonso Villa Rojas. 1934. *Chan Kom, a Maya village.* Chicago: University of Chicago Press.

Rojas, Rosa. 2001. Rechazan defensa de tres tzeltales violadas por soldados más pesquisas [Defense rejects three Tzeltales raped by soldiers, more arrests]. *La Jornada,* July 17, Political Section.

Román, José Antonio. 2001. La aplicación fría de la ley ante la intolerancia religiosa generaría más pugnas: Fernández Perroni [Straightforward enforcement of the law would generate more friction given religious intolerance, says Fernández Perroni]. *La Jornada,* June 29, Political Section.

Toor, Frances. 1947. *A treasury of Mexican folkways.* New York: Crown.

Wolfe, Bertram D. 1939. *Diego Rivera: His life and times.* New York: Knopf.

12

The "Yucatan Syndrome": Its Relevance to Biological Conservation and Anthropological Activities

John G. Frazier

We—humanity, at least thinking, responsible people—cannot afford to continue with the same mentality that characterized the Iberian adventurers half a millennium ago. Not understanding or appreciating the people and culture that they had come to discover (but also to "pacify," "save," and exploit), the Spanish conquistadores totally misconstrued the simple (and expected) response to their question, "Where do you live?" The term *Yucatan* evidently derives from the reply: "*Matan cub a than*" (i.e., in modern writing "*Ma t'aan, ku y u'ub a t'aan*" [Anderson, personal communication]), meaning, "We do not understand what you are saying" (Restall 1998, 122, 219 n. 81, 220).[1] So, thanks to the cultural insensitivity and domination of the European adventurers, the world has celebrated the We-Do-Not-Understand-What-You-Are-Saying Peninsula ever since they "discovered" it.[2]

Clearly, words are the building blocks of ideas and concepts, and the elements of rhetoric are instrumental in shaping the directions that societies take (e.g., Hymes 1964; Killingsworth and Palmer 1992). But it is not the geographic name that is the problem per se: it is the attitude and the true cost of insensitivity to cultural and social processes that led to the officialization of this appellation. *Yucatan* as a metaphor has dire implications for humanity and the environments on which we depend. Because we live in the postmodern neoliberal age of globalization (PoMoNeLiAGlo),[3] the damage caused by social and political mistakes in one place has tremendous implications for all of us—especially because in this PoMoNeLiAGlo the world's carrying capacity for *Homo sapiens*

sapiens is being challenged by global networks of commerce and consumption. Social, economic, and political impacts now swirl through the "global village," sometimes with amazing speed and horrifying consequences, even on distant societies and their resource bases. There could be no clearer contemporary examples than the 1998 stock market crashes in Asia and then Russia, with profound effects on the economies of Latin America, and consequential aftereffects on their environmental and social programs (Gavin and Hausmann 1999). Even those tremendous impacts pall beside the enormity of the global fallout from the destruction of the Twin Towers of the World Trade Center on September 11, 2001, with the "globalization of terror" and profound impacts on economies and societies worldwide (Iwasaki 2001; World Bank 2001, 2002).

There are many other less known, yet profound effects of global markets. For example, Wright (1990) provides a detailed account of the role that the "global supermarket" has played in relation to agroindustry, which has enormous social and environmental effects on both producers and consumers. The liberalized market in timber products in Japan has resulted in vast deforestation throughout Southeast Asia, with concomitant negative effects—both social and environmental (Dauvergne 1997; Knight 1997).[4] Place (1998) reviews several studies that show similar effects where the environment in one place is influenced by commerce and policy in another. Some students have developed the concept of "shadow ecology" to describe a situation in which ecological impacts in one region shadow, or trace, the course of events in a distant, more economically powerful area (Dauvergne 1997). A comparable concept is the "overdetermined system"; when considered in the context of an "overdetermined community," the term refers to a situation in which forces external to a community, no matter how remote and distant, can have greater importance in driving the community than internal forces (Hackenbert and Benequista 2001).

These concepts are central to the thesis presented here. "A country's shadow ecology is the aggregate environmental impact on resources outside its territory of governmental practices, especially official development assistance (ODA); corporate conduct, investment and technology transfers; and trade, including consumption, export and consumer prices, and import tariffs" (Dauvergne 1997, 2–3). The complexity of the maze of interactions, decisions, interests, and so on that determine where, when, and how an ecological shadow will be cast is daunting, and their unraveling can produce an account that rivals the best of fiction (e.g., Wright 1990; Dewar 1995; Vandermeer and Perfecto 1995; Dauvergne 1997).[5]

THE YUCATAN PENINSULA AS PART OF THE POSTMODERN, NEOLIBERAL, GLOBALIZED WORLD

The Yucatan Peninsula represents many things: the substrate on which lowland Maya culture has flourished for three millennia, a former part of the Spanish

Empire's Nueva España (Restall 1998, 179 n. 5), a breakaway province during Santa Ana's newly liberated Mexican Republic that commissioned the Texas Navy to protect it from the centralist forces of Mexico, a relatively neglected province in the southeast of Mexico (Reed 1964, 29), the northern extension of the much-acclaimed Ruta Maya (Garrett 1989), and the geographic focal point of this volume. Over its various cultural and historic epochs, the peninsula provides clear examples that social issues and environmental problems are interdependent—and often best understood when viewed within a larger geographic context. As with other parts of Mexico, Latin America, and the "New World" (e.g., Moran 1990; Wright 1990; McDonnell and Picket 1993; Dyer and McGoodwin 1994; Johnston 1994; Pye-Smith, Feyerabend, and Sandbrook 1994; Smith 1994; Lean 1995; Painter and Durham 1995; Simonian 1995; Vandermeer and Perfecto 1995; Adam 1998; Place 1998; Decker, Brown, and Siemer 2001; Leff 2001; see also Crumley 1994), the history of the Yucatan Peninsula clearly illustrates that *Homo sapiens* is part of—not apart from—the environment, and that the ways in which people behave and interact have major consequences on the environment, including the availability and distribution of basic resources (e.g., Kintz 1990; Terán and Rasmussen 1994; Faust 1998, 2001; Faust and Bilsborrow 2000; Folen et al. 2000; García 2000). The land of the lowland Maya provides an important setting for examining the issues of political ecology, at diverse levels, from local to global.

Although initially trained as a biologist, over the decades I have come to realize that biological conservation, often described as a vibrant new discipline (e.g., Raven 1987; Meffe and Viederman 1995), is not ultimately about biology. If it is to achieve its goals of "Active management to ensure the survival of the maximum diversity of species and the maintenance of genetic variety within species" (Allaby 1985, 158) or the "protection, maintenance, and restoration of life on this planet" (SCB 1987), in the end biological conservation must be about people as much as about the environment: it must deal with human customs, human relations, human motivations, political institutions, and political structures that drive the interactions between people and their environment. This personal discovery (Frazier 1990, 1994, in press), comparable with those of some other conservation biologists and wildlife managers (e.g., Meffe and Viederman 1995; Vandermeer and Perfecto 1995; Decker, Brown, and Siemer 2001), should be old news to most "social scientists" (Adam 1998 and Brulle 1995a, 2000, 191–192 provide useful syntheses) and so I come, in my naïveté, to beg for help from those whose "study animal" is *Homo sapiens sapiens*. But also, I come to lay a challenge before sister disciplines: anthropology and other "social sciences."

The political and cultural ecology of the Yucatec Maya—as with any other group of human beings on this planet today—are conditioned not just by the local culture, political structure, and environment in which these people live, but are very much under the influence of a homogenizing world culture and mighty global market (e.g., Wright 1990; Johnston 1994; Korten 1995; Utting et al.

1995; Vandermeer and Perfecto 1995; Adam 1998). This has direct and profound implications on the environment and biological resources, which form the resource base of the Yucatec Maya, to say nothing of the effects on their commerce, institutions, and society. Cloistered communities, characteristic of the days when the Spanish "pacified" the peninsula that they called *Yucatan*, have long since disappeared; the days of Gonzalo Guerrero,[6] when lone foreign heroes could waylay unwanted impositions coming from outside the community, have ended. We are in the postmodern, neoliberal age of globalization (PoMoNeLiAGlo), when the individual coffers of hundreds of private corporations each far exceed the national treasuries of all but a score of "sovereign" states (Korten 1995). In fact, perhaps more precise than postmodern, neoliberal age of globalization, these times should be described as "Post-Exuberant" (Brulle 1995a, 57)—certainly, the term is more concise.

NONGOVERNMENTAL ORGANIZATIONS AS AGENTS OF ENVIRONMENTAL CONSERVATION AND COMMUNITY DEVELOPMENT IN THE PoMoNeLiAGlo

With occult—but powerful—factors impinging on societies and their environments, there is an urgent need, first to identify and understand these tremendous forces, and then to develop and implement mechanisms that balance or neutralize their negative and unwanted impacts. Among the various factors at work in the PoMoNeLiAGlo that seemingly compensate for the impacts of the voracious multinational corporations, insatiable consumers, and other manifestations of neoliberalization and globalization, are "benevolent" actions directed at protecting the environment and empowering communities, of which biological conservation and community development are major components. Among the social goals of these programs are the safeguarding of nutritional, physiological, economic, cultural, and spiritual sources of sustenance for the recipient societies; capacity building, civil society building, democracy development, institution building, and women's rights and well-being, among others, are routinely central to the stated objectives; other goals include saving endangered species, protecting environments, and managing for the maintenance of "ecosystem services" (e.g., Friedmann and Rangan 1993; Pye-Smith, Feyerabend, and Sandbrook 1994; Smith 1994; Dewar 1995; Dowie 1995; Lean 1995; Caldecott 1996; Daily 1997; Ghimire and Pimbert 1997; Margoluis and Salafsky 1998; Brulle 2000). Routinely, these initiatives involve nongovernmental organizations (NGOs). Outside the "industrialized North" (e.g., in much of Mexico), bilateral accords between "northern" and "southern" NGOs are commonplace. It is imperative to appreciate that these sorts of activities—even though they may be labeled as conservation or environmental protection—fall within the general realm of development. The mechanisms for transferring information, funds, and other resources, the personal, social, and political links that are involved in establishing and maintaining the interrelations between donors and recipients, and many of the general concepts, attitudes, and procedures are shared between conservation/environmental projects and other initiatives focused on issues such as education, food security, gender relations, and health (e.g., Friedmann and Rangan 1993; Pye-Smith, Feyerabend, and Sandbrook 1994; Smith 1994; Lean 1995; Caldecott 1996; Ghimire and Pimbert 1997; Margoluis and Salafsky 1998).

There are numerous descriptions of the explosive growth of NGOs in Latin America over the past few decades (e.g., Clark 1990, 46 ff.; Dewar 1995; Meyer 1999; Markowitz 2001). In part, this is because these organizations have several benefits over governmental agencies, particularly greater administrative flexibility and responsiveness, as well as relatively lower costs in administration, negotiation, and monitoring (Meyer 1992a, 1992b, 1995, 1996, 1997a, 1997b, 1999). There are many accounts of NGOs—notably small grassroots organizations (or grassroots support organizations, GSOs)—providing invaluable services to marginalized peoples, especially in the Third World (e.g., Breslin 1987; Smith 1994; Lean 1995); NGOs have been called the "building blocks of democracy" and "incubators for alternatives for people who traditionally lack alternatives" (Richy-Vance 1993). NGOs are often categorized as not-for-profit, so Meyer characterizes them as producers of public goods (PPG), commodities that can be shared by many members of society, theoretically without resource depletion or social competition. More often than not, as Meyer explains, Latin American NGOs are producers of international public goods (PIPG), providing services that promote environmental protection, education, health, and other aspects of human well-being. The services and products are enjoyed not only by the citizens of the communities in which the NGOs exist and operate, but on a broader, global level. A key role of PIPGs is the provision of information related to these activities.

Routinely, the activities of implementing NGOs are supported through a form of foreign aid in which resources (equipment, funds, information, concepts, sociopolitical contacts, etc.) are donated (or at least facilitated) by organizations in the "North" to organizations based in the "South." Although the budgets for conservation projects may be modest in comparison to the more conventional types of foreign aid, the processes involved are comparable, if not identical.

Just as in many conventional development initiatives (e.g., Cochrane 1979; Cernea 1991, 2; Metha 2001), specialists from the social sciences are not often a significant component in conservation and environmental activities, or when they are, it is often too late and simply as a social palliative—or even as a deceptive ploy (e.g., Winthrop 1997).[7] To a great extent, these conservation-development actions have been planned, supervised, executed, and evaluated within an intellectual framework constructed by biologists, conservationists, naturalists, resource managers, and economists from the industrialized world (Meyer 1992a, 1992b, 1995, 1996, 1997a, 1997b, 1999; Ghimire and Pimbert 1997; Place 1998, 232; Wedel 1998; Decker, Brown, and Siemer 2001). Admittedly, the social sciences are not invisible to—or totally depreciated by—conservation biologists. Clear examples of the current trend to incorporate aspects from the social sciences are seen in recent publications that are fundamental to academics and decision makers, by providing guidelines for environmental conservation (e.g., Heywood 1995; Mangel et al. 1996; Noss, O'Connell, and Murphy 1997; Primack 1998). Formulated by scores of world-renowned biologists, ecologists, and conservationists, these contemporary "environmental bibles" make it explicitly clear that there is a basic need to take into account social factors, to understand and evaluate social phenomena, and even to carry out studies in the social sciences.

Perhaps in the future, social scientists will be included among the strategic teams of authors who conceptualize, write, and publish these important documents. At that point, one might expect to find a conceptual change in the environmental bibles: instead of referring to "scientific" studies as the foundation, with ancillary help from the social sciences, possibly the social sciences will be accepted within the inner sanctum of scientific studies. But again, these details of names, credits, and social acceptability among scientists and their disciplines are not the fundamental problem; they are merely symptoms.

My concern is that biological conservation, notwithstanding its "natural science" charisma and noble intentions, is a social-political activity, and as such it must be perceived, planned, executed, and evaluated within this context. To ignore or negate this fact is to invite disaster; natural scientists are not trained to identify or understand political and social processes, much less how to develop effective means to deal with them (Brulle 2000, 191–192). Not considering this reality involves the same basic attitude as that which operated when a simple (and expectable) response—"We do not understand what you are saying"—was misinterpreted and transformed into a glorified, formalized, geographic appellation for an entire region.

Several years ago, in an attempt to air concerns about these issues, I gave a presentation to an international group of marine turtle biologists and conservationists, titled "Conserving Sea Turtles and Other Natural Resources: What Ferdinand Marcos and Manuel Noriega Can Teach Us" (Frazier 1994). My goal was to draw attention to a few of the more striking syndromes in conservation activities that are social (i.e., not dealt with, or even considered, under normal biological investigation) and to signal how these are likely to have negative impacts on the desired goals of biological conservation. As far as I have been able to determine, the only clear effect of the talk was to provide evidence of my dementia to a community of professional colleagues and friends with whom I have worked for decades. As several authors have described (e.g., Meffe and Viederman 1995; Decker, Brown, and Siemer 2001), it is not easy for many biologists and conservationists to confer professional priority on "soft" social issues. Yet these conceptual and epistemological deficiencies and disciplinary biases have led to severe criticisms of the environmental movement (e.g., Bookchin 1994; Dowie 1995; Vandermeer and Perfecto 1995, Ghimire and Pimbert 1997; Decker, Brown, and Siemer 2001).

DONORS AND NGOS IN THE PoMoNeLiAGlo

The attitudes and positions of individuals in regard to integrating social-political considerations together with biological conservation and environmental protection warrant evaluation, whether they be biologists, conservationists, or environmentalists on the one hand, or anthropologists, community workers, or sociologists on the other. However, there is an equally profound issue that needs to be understood: the role of key institutions. Recent descriptions of the "environmental movement," particularly in Canada and the United States, have evaluated the behaviors, functions, and effects of those organizations that are ideally positioned to promote fundamental improvements in environmental policies and social conditions. Many of these accounts portray a situation very different from the general dogma and mystique surrounding NGOs, especially concerning the most powerful organizations. Because NGOs represent agents of change, it is imperative to understand these various complexities.

One of the most fundamental points is that NGOs must develop means to acquire and access resources with which they can operate and implement their activities: they must find donors for support. As a consequence, many (if not most) NGOs are driven by their relationship with donors, past, present, or potential, and this is one of the most severe criticisms. Countless authors (e.g., Wright 1990; Meyer 1992a, 1992b, 1995, 1996, 1997a, 1997b, 1999; Dewar 1995; Dowie 1995; Struhsaker 1998; Brulle 2000) have provided detailed explanations on how donors direct the agendas, priorities, programs, and institutional development of the recipients. In many cases, NGOs have adopted strategies of seeking and nurturing political and financial alliances with government agencies and private corporations, even when many of these (potential) donors have checkered social and environmental records (e.g., Pell 1990; Dewar 1995; Dowie 1995). Some authors have described the policies of "mainstream environmental organizations" as a "sell out" (Dore in Place 1998, 231). Not surprisingly, descriptions of how private corporations "influence" (viz. manage and manipulate) environmental groups are not uncommon: Pell (1990) argued that corporate money buys access to environmental groups, insider networks, and thus to agendas of both NGOs and government agencies.

Dewar (1995) gave a detailed account of the environmental movement viewed from Canada, describing the roles of government, private corporations, and NGOs. She found that the maze of interrelationships was so complex that it was a major challenge to tease apart specific links and uncover who was truly responsible for what. She concluded that many NGOs are "PGOs" (private governmental organizations) or "GOPs" (governmental organizations privatized) simply because these organizations are funded—and directly influenced—by governments. In some cases, she found that so-called nongovernmental organizations functioned as extensions of certain government initiatives (often representing the interests of only a few people in government), but without the bureaucracy, accountability, and checks and balances of government. "NGOs heavily dependent on government funding may have sacrificed some of the inherent advantages of being NGO—their independence, their credibility, and their objectivity" (Fox 1987, 13, in Meyer 1996, 456).[8]

NGOS THAT SERVE AS DONORS IN THE PoMoNeLiAGlo

Many NGOs in the "North" serve as donors for supporting activities and organizations in the Third World, and it is important to understand how they function and how effective they are in their own right, before considering the recipients. Clark (1990), Brulle (1994, 1995a, 1995b, 1995c, 1996, 2000), and Brulle and Brown (1997) provided detailed analyses of the environmental movement, particularly in the United States, with a special focus on the NGOs specializing in this field. Brulle (1994, 1995b, 2000) explained that the environmental movement offers the potential as a major political and social factor to empower society and produce an improved social order.

Although there has been a clear evolution of the movement and its key agents (Brulle 1996), numerous authors (e.g., Clark 1990; Ibister 1993; Meyer 1992a, 1992b, 1995, 1996, 1997a, 1997b, 1999; Brulle 1995a, 2000; Dewar 1995; Dowie 1995; Wedel 1998) have explained that despite the dominant beliefs, NGOs are frequently top-down in structure and nonparticipatory in function. Brulle (1995a, 2000) concludes that in structure and function many of the more powerful environmental NGOs are oligarchic—not democratic or even "representative." Nepotism and favoritism are usual (Clark 1990, 59; Brulle 1995a, 2000; Dewar 1995; Dowie 1995): "They are subject to favoritism and particularism" (Meyer 1997b, 132). This theme is emphasized by Dowie (1995) over and over again throughout his book, arguing that the conservation, or preservation, movement was primarily for white men, an elitist club characterized by being "genteel, white and very polite, . . . safe, respectable, very polite" (p. 3). This is consistent with the detailed analysis of Brulle (2000). Not only is this structure contradictory to what most NGOs preach (thus, hypocritical), but it is inefficient in terms of developing and making the most out of the personnel in the organization (Clark 1990, 58).

The strategy to court power and social acceptability has been given such priority that many NGOs have not only lost, but in some instances have cast aside or even suppressed, an awareness of fundamental social and grassroots problems—even though these issues are basic to conservation issues. One of the most profound criticisms of the environmental movement in Canada and the United States is that, despite enjoying generous political and financial support, it is irrelevant—not pertinent to the needs of society (Bookchin 1994; Dewar 1995; Dowie 1995).[9]

Like Dowie (1995), Brulle (1995a, 2000) argues that the environmental movement in the United States has failed in many respects—that between rhetoric and concrete action there is a deep chasm. He extends the investigation deeper into the social causes, identifying the socialization process, with concomitant deep cultural and economic roots, as a primary cause of environmental problems. As Brulle explains, profound social change will be required to resolve fundamental environmental problems, so, to be effective, NGOs must provide true leadership roles, not just platitudes and socially agreeable images. But instead of dealing with the major issues at hand and addressing the root problems, many of the more powerful U.S.-based conservation organizations are described as having taken on the customs of their sociopolitical superiors: in addition to consolidating their own social and political power, the compulsion to establish and maintain a socially agreeable image has become paramount, even if the basic objectives in environmental protection and biological conservation are never met or are actually contradicted (Miller et al. 1994; Dowie 1995).

In their desperate drive to win respectability and access to power, mainstream leaders of environmental NGOs politely pursue a course of accommodation and capitulation with elected officials, regulators, and even polluters and wildlife traffickers. This almost reflexive insistence on compromise has "pushed a once-effective movement to the brink of irrelevance" (Dowie 1995, 6; see also Bookchin 1994).[10] To consolidate power and outcompete other NGOs for donations and access to power, some NGOs have adopted strategies that are more in line with the cutthroat corporate world than with charitable organizations espousing ethics, democracy, and other noble humanitarian goals (Thornton and Currey 1991; Dewar 1995; Dowie 1995). Conflict of interest, willful blindness about contradictory situations, misrepresentation, nepotism and favoritism, manipulation, collusion, and conspiracy are a few of the characteristics that flow through Dewar's (1995) detailed account. One of her most disturbing conclusions is that certain environmental NGOs have been adept players in using "a cloak of green" to promote policies that are in the best interests of transnational corporations, particularly the globalization of trade. The result has been the fortifying of transnational corporations with the concomitant debilitation of sovereign governments: routinely, this is in direct opposition to local environmental needs.

In declaring that the environmental movement in the United States is "courting irrelevance," Dowie (1995) offered three fundamental causes: a naive (or convenient) reliance on authority and good faith in the government; a hegemonic drive to concentrate power, despite alienating and undermining the grassroots of their own movement; and the incapacity (or lack of desire) to estimate "the fury of their antagonists." He acknowledged that the air, water, and land "are certainly in better shape than they would have been had the movement never existed," providing examples of true leadership by certain politicians and civil servants. However, Dowie (1995, x) was adamant that the environment "would be in far better condition had environmental leaders been bolder; more diverse in class, race, and gender; less compromising in battle; and less gentlemanly in their day-to-day dealings with adversaries."

Since institutions are primary agents of change, a tremendous responsibility for resolving root environmental issues lies with the environmental NGOs. However, numerous authors (e.g., Bookchin 1994; Dowie 1995; Brulle 1995a, 2000) argue that, in the main, protagonists and organizations in the United States are not addressing the fundamental issues. These authors demonstrate that in many instances, the measures taken are cosmetic, short-sighted, or even self-serving. Certainly, Dewar's (1995) description painted an extremely troublesome image of both donor and recipient NGOs, where deception and manipulation were portrayed as the norms for consolidating power, but transparency and democratic procedures were unusual, to say nothing of the dissonance between the stated objectives—noble and altruistic—and the actual behavior, best described as crass self-interest under the guise of charity.

RECIPIENT NGOS IN THE PoMoNeLiAGlo

How does the situation regarding NGOs—particularly those in the United States and Canada—relate to people in the Third World, such as those whose ancestors responded "*Ma t'aan, ku y u'ub a t'aan*" half a millennium ago? Just as corporations are multinational, extending tentacles over the world, exploiting resources and impacting social and political decisions, conservation and other development activities are exported and guided in a Newtonian fashion: they emanate from where there is an abundance of material resources and information to where there is a scarcity of these elements.[11] As mentioned earlier, biological conservation and environmental protection ventures regularly share the same basic characteristics with activities categorized as development, foreign aid, and institution building. When this flow of resources, information, and political contacts is influenced—or driven—by much larger social, political, and economic forces, the best of intentions may not be adequate to ensure that the actions will really be effective and beneficial to the majority of people in the recipient society. This is the case with any sort of activity that falls under the categories of development, foreign aid, or institution building, as shown by a large and diverse literature (e.g., Cochrane 1979; Blase 1986; Clark 1990; Cernea 1991; Isbister 1993; Dewar 1995; Dowie 1995; Wedel 1998; Meyer 1999).

Northern-based NGOs, just like the northern-based transnationals that tower above them, have tremendous impact on the planet's environment and rural societies by influencing the ways in which diverse peoples and institutions around the world interact with their environs. This is not only because of the direct actions taken by these NGOs but is also due to the ways in which they influence—consciously or unconsciously—their sister organizations in the Second and Third Worlds.

Often the controlling influence of donors is not totally intended but results from inadequate procedures. For example, the issue of how "aid" is directed to recipients is critical. To make possible and facilitate the transfer of resources and information, a target for reception must be identified and incorporated into a scheme. Obviously, it is much simpler—and administratively much more effective—to focus the point of reception rather than to try to direct resources to diverse, dispersed, heterogeneous members of a society. How is a focal point of this nature identified and selected? There must be a modicum of communication and confidence between the exporter (donor) and the importer (recipient). The recipient must speak the donor's language, must be sufficiently sensitive to the donor's culture to gain and maintain their confidence. With these criteria, the recipient can ensure a flow of resources and information (Clark 1990, 55; Frazier 1994).

However, these criteria to establish and maintain a bond of confidence between donor and recipient need not take into account the details of how the latter is related to the society in which he/she is immersed. In other words, the activities, concepts, and priorities of the importer may or may not be pertinent to the needs and expectations of the society in which the importer lives. The problems that the agent, or NGO, identifies and actions that it promotes may or may not be congruous with the majority of society (Frazier 1994). Myriad authors (e.g., Clark 1990, 5; Meyer 1992a, 1992b, 1995, 1996, 1997a, 1997b, 1999; Dewar 1995) have described how NGOs in the Third World are normally driven by the demands of donors, which are not necessarily relevant to the societies in which the recipients exist and work.

In large "foreign aid" activities, this phenomenon can take on remarkable proportions, with donors dictating totally irrelevant, even counterproductive, priorities (Wedel 1998). Indeed, the actions and policies of donors can disrupt and destroy justifiable institutions serving a clientele in the recipient country (Meyer 1992a; Wedel 1998). Both Meyer (1992a) and Wedel (1998) argue that it is not rare for institutions irrelevant to the needs of society to be specifically created because of the demands and influence of donors. These authors explain in great detail that it is normal for the recipient in these transactions to represent the elite[12] and not necessarily the "grassroots" or even the majority of the so-called recipient society. Agents are often driven by what looks good or proper to them, or what makes them feel important, criteria that may be contradictory to the needs of the recipient society (Scott 1998).

In the end, recipient NGOs are responsible only to their donors and not to the general public (Clark 1990, 46 ff.); "externally funded NGOs in developing countries are primarily serving the international donor community rather than the domestic beneficiary" (Meyer 1992b, 1123). The problem is especially grave when wealthy donors, pushing primarily development agendas for financial and political reasons (e.g., World Bank, USAID, CIDA), have redirected priorities and actions in conservation and development programs—especially in the Third World (Isbister 1993; Dewar 1995; Struhsaker 1998).

There is nothing new or unique about the way that NGOs acting as brokers of resources react, and diverse examples of similar anomalous relationships with other sorts of agents abound. Half a millennium ago, certain Maya elites became adept at building alliances with the Spanish conquistadores at the cost of other Maya groups and the common Maya people; indeed, some "Maya conquistadores" may have been far more severe in their treatment of the Maya than were their Iberian allies (Restall 1998). Another case in point, albeit extreme, comes from the shores of Venezuela. A local NGO, well connected to northern counterparts, arranged the staged filming of a dolphin kill in order to create a high-visibility, national scandal and thereby attract support from foreign sponsors in the activist conservation community (Salvatierra 1995a, 1995b). Accounts of recent fiascoes in Central Europe, where billions of dollars in "foreign aid" yielded results counterproductive to the stated goals, detrimental to the recipients as well as the donors (Wedel 1998), drive even farther home the point that simply transferring and receiving resources does not necessarily solve social and environmental problems—not infrequently, it complicates them!

Given the basic faults that have been identified in the more powerful environmental and conservation organizations in Canada and the United States (e.g., Dewar 1995; Dowie 1995; Brulle 2000), it should be no surprise to find recipient NGOs in other countries that are irrelevant to their own societies. Because the structure, function, priorities, and behavior of recipient NGOs is routinely influenced, or even directed, by their donors, it is more than likely that the recipients will acquire, or be directed to adopt, the same characteristics as their patrons, the donors. If there are fundamental lacunae in the social organization and function of the donor organizations, one can expect little better in the recipients who will be encouraged to emulate their sociopolitical "superiors."

The exception to the generality of donors driving agendas is found when there is a clear understanding of priorities and processes. When the objectives of a donor are to listen, learn, and then help try to strengthen a recipient, with development and dignity, agendas are more likely to be relevant to solving the issues at hand (Breslin 1987; Smith 1994). "Transfer of money, infusion of capital and growth do not necessarily foster development. At the heart of the issue is a change in attitudes, self respect, self esteem, dignity, identity, autonomy, independence, outside respect, creating a space for participation and change, and self determination" (Richy-Vance 1993, 123).

External agents often play a critical role in facilitating empowerment of marginalized communities, implementation of conservation plans, and effective environmental protection and community development. These agents can be local and need not necessarily be international or national. "But the danger in depending on external agents is that grassroots struggles for livelihood may become mere exemplars of larger societal processes that have become ideolized [*sic*]" (Friedmann and Rangan 1993, 18). Moreover, it is a major challenge for most donor organizations to sacrifice the brandishing of their image and other fast track measures of success in favor of small-scale, inconspicuous improvements of marginalized peoples who have no political or financial power with which to reward the donor.

DEMYSTIFYING NGOS IN THE PoMoNeLiAGlo

The myth that NGOs always represent the grassroots has been severely challenged: as a rule, the people they provide services for are those who have access to political and financial resources. Both Meyer (1992a, 1992b, 1995, 1996, 1997a, 1997b, 1999) and Wedel (1998) explain that it is usual for local elites to benefit most from the programs of NGOs and "foreign aid." In challenging the dogma surrounding these organizations, Meyer is very clear in her description of not-for-profit NGOs, explaining that they "have the reputation of reaching the poor and of promoting local participation because of their small size and political independence; and they are considered to be low cost and innovative," but "[t]here may be more mystique than substantive advantages in these reputed characteristics of NGOs" (1992b, 1118–1119). Dowie (1995) repeatedly explains that although there is dedication, passion, and vitality in the roots of the environmental movement, neither political power nor "real money" routinely reach the lower levels: "most funding is siphoned off by international NGOs and southern professionals before it hits the grass [grassroots organizations and groups]" (Meyer 1995, 1278).[13] Clark (1990, 46 ff.) comes to similar conclusions, based on NGOs all around the world. Indeed, tribal peoples—the so-called Fourth World—have routinely been victims of development and "foreign aid" initiatives carried out by both governments and NGOs (Bodley 1988, 1990; Johnston 1994).

Meyer (1995, 1279) reports that specialists "lament the proliferation of such illegitimate, opportunistic NGOs." She goes on to explain, "While the vocabulary of entrepreneurship, with its profit-seeking capitalists, is little used in the NGO literature, a recognition of the dynamic entrepreneurial nature of the response to funding for environmental protection is critical." Because of the way the major NGOs have operated, monopolizing political power, social contacts, media coverage, and financial support, environmentalism has remained narrowly defined. Although the politically powerful NGOs can take credit for many environmental accomplishments, they must also share the blame for diverse environmental setbacks—a responsibility that is very rarely honored (Dowie 1995). Clark (1990, 46 ff.) also explains that it is usual for NGOs to focus on self-promotion and fund-raising, with little effort toward objective evaluation or learning from experience: there are numerous accounts of dissonance between perceived or projected success and true performance.

Even without the relationship between donors and recipients, there are basic aspects of NGOs that need to be critically and objectively assessed (Clark 1990). Markowitz (2001) explains that there is a remarkable paucity of anthropological information in this arena; tellingly, she suggests that this may be, in part, because most research is donor driven.

The issue of evaluation is yet another Pandora's box. To begin with, objective assessments of conservation and development work are not easy to find. Even so, it has been well established that there is no single recipe for measuring success in NGOs, conservation, and community development: each case must be treated individually (e.g., Richy-Vance 1993; Pye-Smith, Feyerabend, and Sandbrook 1994). Moreover, there is a constant risk that the criteria, scales, and measures used will depend on who is in control and what their interests are (Lélé and Norgaard 1996); measures and criteria of success are subject to political pressures and often are changed to accommodate modified agendas (Struhsaker 1998).

This is not to say that there are not NGOs or even donors who shine brilliantly with their clarity of purpose and integrity (e.g., Breslin 1987; Thornton and Currey 1991; Friedmann and Rangan 1993; Richy-Vance 1993; Meyer 1995), but this depends on fundamental objectives of the organization, and it is not uncommon for establishments categorized as not-for-profit and charitable to have agendas very different from their publicly stated objectives (Miller et al. 1994; Dewar 1995).

As Clark (1990, 50 ff.) and Meyer (1997a, 1127–1128) explain, NGOs are both political and economic entities, and it is difficult to separate the economics from the politics. Markowitz (2001) explains the risks and challenges of doing ethnographies of NGOs, and the fact that this requires working within a web of relationships that are inherently unstable, as well as using "multilocal research strategies," because of the strength of the transnational processes involved. There is also a problem in assuming residential and institutional stability of the organization or people under review when in fact no such stability exists.

After a series of detailed studies, Meyer (1997b, 133) concludes that often the reputation of NGOs has been overstated and "[t]hey are not a panacea and cannot replace government." She goes on to warn that "participation may be merely perfunctory as entrenched local interests maintain control of most incoming benefits" (p. 143). Hence, "[s]kepticism is in order as donors begin another round of institution building" (p. 134). Other authors have come to similar conclusions: the complexity of understanding NGOs and the many items on their agendas is daunting (Markowitz 2001). As Richy-Vance (1993, 28) explained, *NGO* is an unwieldy term that defines a vast diversity of organizations, based on what they are *not*.

Analyses of NGOs, conservation, and development activities are replete with commentaries, from indirect to very detailed, about the issues discussed above—the social and environmental dangers are enormous (e.g., Clark 1990; Thornton and Currey 1991; Meyer 1992b, 1995; Dewar 1995; Dowie 1995; Wedel 1998; Brulle 2000). The concentration of resources and information, even if the amount is not particularly significant to the exporting/donor organization, may represent an incomparable treasure to the recipient. With this, the imbalance in resources and disparity in economic, social, and political influence is exacerbated in the region where the importation is made. Empires in the Third World can be quickly built and fortified with relatively small amounts of money and other resources. The impacts on the local society and, by consequence, the effects on the local environment can be devastating (Meyer 1992a, 1992b, 1995, 1996, 1997a, 1997b, 1999; Korten 1995; Wedel 1998).

In this respect, the allegory of Ferdinand Marcos and Manuel Noriega as political giants—*cum* monsters—should be perfectly clear: they were created and nurtured by targeted "foreign aid" and "development" projects from U.S. donors who were driving political agendas. The careers of these two social pariahs led to enormous social and environmental costs, borne not only by their respective homelands but by many other people and areas under their sway. There is a great deal that these two historic figures can teach us about biological conservation, development, and "foreign aid," and their histories stand as poignant metaphors from which conservation biologists and field anthropologists should learn about the social processes that routinely drive biological conservation, development, and "foreign aid" (Frazier 1994).

ENVIRONMENTALISM, NGOS, AND POLITICAL ECOLOGY IN THE LAND OF THE MAYA

Focusing on the We-Do-Not-Understand-What-You-Are-Saying Peninsula, there appear to be no studies (anthropological, economical, sociological, etc.) of either the indigenous NGOs or the role of large external organizations as donors and exporters to the region. Nonetheless, it is remarkable that descriptions of conservation, development, "foreign aid," and institution building from other regions strike notes of recognition for people familiar with conservation and development activities on the Peninsula.

It would be deceitful to pretend that I am expert in the cultural anthropology or political ecology of Yucatec Maya, or versed in the theoretical underpinnings of these disciplines. However, decades spent working in rustic situations with rural peoples around the world, many of them indigenous, have convinced me through "folk learning" that grave cultural, environmental, political, and social problems derive from a complex and powerful synergy, and many of the more powerful forces are not evident from the sociopolitical or geographic levels at which many field anthropologists and biologists prefer to work. Nearly eight years working in the Centro de Investigaciones y de Estudios Avanzados, INP, Mérida, Mexico, as a researcher/professor/conservation biologist interacting with villagers, fishers, students, other professors and researchers, state and federal governmental agencies, NGOs, and other members of society, has convinced me that numerous descriptions of conservation and development activities in other parts of the world are very relevant to the Yucatan Peninsula.

During the last decade, as elsewhere in Latin America (Meyer 1992a, 1992b, 1995, 1996, 1997a, 1997b, 1999; Richy-Vance 1993; Markowitz 2001), there has been a rapid proliferation of NGOs in the Yucatan; they come in multiple sizes and forms, representing a diversity of interests, with environmental and social issues central to most of them. Some of these NGOs are intimately linked to larger institutions (particularly in the United States and Mexico City); others have been created to meet the agendas of international donors such as the United Nations Development Program (UNDP) (Hartasánchez 1997); while still others are small and seem to have developed from self-defined needs to address specific issues in a restricted geographic area.

This growth of NGOs in the Yucatan, especially since the end of the 1980s, has been accompanied by an inflow of much larger sums of money, technology, information, and other resources, allegedly to help the local people and to protect their environment. In some cases, "partnerships" have been forged between donors and recipient NGOs, in which one importer has exclusive rights to an exporter's resources (e.g., Pronatura Península de Yucatán and The Nature Conservancy, J. Quiroz, personal communication). As expected, this has resulted in the concentration of resources and power. Following the example of their more powerful northern role models, local organizations strive for political endorsement, which produces intense competition and noncooperation between local NGOs (cf. Clark 1990, 56).[14] Frequently, the small, indigenous grassroots organizations have been trampled to the bottom of the heap by those that have access to greater power, political connections, or media coverage.

Remarkably, the role and participation of the Yucatec Maya, the "original" and numerically dominant inhabitants of the peninsula, are inconspicuous in the planning, execution, and evaluation of the projects that are devised and carried out, ostensibly for their benefit: there is, however, no lack of people from Mexico City (or elsewhere, such as Europe and the United States) in positions of authority and power in local conservation NGOs and governmental agencies. For all the knowledge, tradition, and geographic seniority that the Yucatec Maya may have, in the main their most consequential role for the global conservation and development brotherhood is as charismatic symbols of romantic, bygone Classic times: images of Maya ancestors, their glyphs, temples, gods, and icons adorn the publications of many Yucatecan NGOs, but there is little effective communication between decision makers and other members of the empowered sectors of society on the one hand and the living Maya on the other—a situation that is little different from that which existed between the Mayan ancestors and the Spanish conquistadores.

Accounts of abuse of resources for personal gain, along with blatant favoritism, are rife in local NGOs in the Yucatan Peninsula, but it is extraordinary to hear of even token demands for accountability, much less sanctions. Following the models of powerful First World NGOs, programs are ordinarily reported as raving successes, while mistakes and failures never occur, or so it would appear (cf. Clark 1990, 46). Success is routinely measured in terms of the quantity of resources received and controlled by the organization and the numbers of political contacts and high-profile, media-focused events, with emphasis on short-term goals rather than long-term objectives. Obviously, this is inconsistent with the recommended ways to measure success in conservation and development activities (e.g., Margoluis and Salafsky 1998).

Given the way in which "foreign aid" routinely works, with donors driving agendas in the recipient countries (Meyer 1992a, 1992b, 1995, 1996, 1997a, 1997b, 1999; Struhsaker 1998; Wedel 1998), and given the way in which many powerful NGOs in Canada and the United States are structured and function, with autocratic, nondemocratic procedures and focused on political and financial gain (Brulle 1995a, 2000; Dewar 1995; Dowie 1995), it should not be surprising that the less powerful organizations on the We-Do-Not-Understand-What-You-Are-Saying Peninsula simply follow suit.

Hence, although there have been no detailed studies of the NGO phenomenon in the Yucatan, the patterns of institutional structure, development, organization, administration, and interactions mirror what has been documented elsewhere. What is clearly missing is an independent, honest assessment of how these organizations are actually affecting the social situation and environment in the Yucatan.

CONCLUSION AND APPEAL

The contributions that NGOs have made to environmental conservation and development are many and varied. They "have helped force changes in the way multilaterals [multinational corporations] do business and have increased their accountability to stake holders" (Rich 1994 in Meyer 1997b, 123). Donors have enhanced institutional innovation by making new ideas and resources available to political entrepreneurs (Meyer 1992a, 642). NGOs have great strength in their heterogeneity and flexibility (Meyer 1992b, 1119). They have provided "space" and mechanisms for participation at the "forgotten end of the social scale," enhancing the development of civic responsibility and democracy (Richy-Vance 1993). Organizations of this nature can respond to and nurture the emotional, spiritual, and religious[15] needs that are routinely ignored by countless developmental initiatives but are known to be essential for community development and environmental protection (Richy-Vance 1993; Smith 1994; Lean 1995). They have been instrumental in providing much-needed foundations for self-respect, local pride, and a sense of community and place, through providing favorable conditions and appropriate forms of support for local initiatives to take root (Pye-Smith, Feyerabend, and Sandbrook 1994). Despite his stinging critique of the environmental movement in the United States, Dowie (1995) does admit that NGOs have made important contributions. But, at the same time, it would be naive and irresponsible to pretend that everything is perfect, and free of problems—to accept the myth that NGOs are a panacea: "Ideology cannot be allowed to blind us to the realities of existing environmental struggles" (Friedmann and Rangan 1993, 19).

One is accustomed to hear the analogy that ecologists and environmentalists are the physicians of the environment (e.g., Norton 1988). Imagine the state that modern medicine would be in if medical practitioners promoted a routine of keeping the populace content by refusing to discuss the basic problems of a patient's health and instead maintaining an image of compassion by handing out sweets and other addictive substances!

As unpleasant as it may be, a responsible practitioner has no choice but to identify the source of a problem and work toward its cure in collaboration with the sufferer. When the organism that is unwell is a social organization, and the ills discovered are social and political, practitioners show an overwhelming reticence to deal openly with the issues—even when the malady involves clear risks to society and the environment. Although there is an extensive literature on diverse aspects of institution building and the responsibilities of institutions, much of it available for more than twenty-five years (see Cochrane 1979; Blase 1986; Cernea 1991; Miller et al. 1994; Lélé and Norgaard 1996, 363), it is consistently underutilized—if not ignored.

Admittedly, the allocation of benefits and provision of environmental IPGs (international public goods) is not expected to be either optimal or fair—indeed, there are no clear mechanisms to evaluate "fairness" (Meyer 1995, 1286). However, as Dowie (1995) and Brulle (2000) explain, to be relevant, environmentalism must embody justice, with representation and participation of the masses. Indeed, if the scientists and scientific procedures involved in environmental work are to have broad acceptance, they must relate to social realities and values (Lélé and Norgaard 1996).

"Several insightful anthropologists have shown that, although aid agencies tend to promise neutral technical solutions, they nonetheless reflect political ideologies that have important unanticipated consequences for recipients" (Wedel 1998, 10; see also Wright 1990). Moreover, there are serious criticisms that conservation biology and science itself cannot be held up as positivist, totally neutral and value-free (Norton 1988; Nader 1996), despite the fact that the practitioners of these disciplines typically evade the issue of value judgments (Lélé and Norgaard 1996; Adam 1998). It cannot just be assumed that the environmental movement, conservation activities, and community development in the Yucatan (or anywhere else), just because they are appealing and well presented, are actually meeting their stated objectives—nor indeed that their objectives are realistic or acceptable to the communities being directly affected. Nor is it responsible to ignore institutional problems and pretend that they will just go away. For example, there are indications that even the more carefully conceived and executed small grassroots programs in the Yucatan Peninsula have basic problems (Hartasánchez 1997)—to say nothing of those NGOs that are perpetually in the media, constantly claiming myriad and incredible accomplishments.

At the root of all this is a much more complex issue. Rarely have environmentalists "challenged the fundamental canons of western civilization or the economic orthodoxy of welfare capitalism—the ecologically destructive system that gives the nation's resources away to any corporation with the desire and technology to develop them" (Dowie 1995, 28). Economic growth is simply accepted, nearly on the same level as gravity, by most people in the

PoMoNeLiAGlo, or Postexuberant Age, despite the fact that this clearly is a fantasy and contradicts the fundaments of long-term access to basic resources (Frazier 1997; Czech 2000). In the same light, the global impact of commoditization has had immense and devastating effects on diverse environments (Manno 2000). Several authors who have dared to explore the roots of contemporary problems (e.g., Clark 1990, 150 ff.; Brulle 1995a, 2000; Vandermeer and Perfecto 1995; Adam 1998; Czech 2000; Manno 2000) call for a new pragmatism, with not only structural adjustments but structural transformations to redefine development activities. Léle and Norgaard (1996) point out the fundamental—and hypocritical—problem of conservationists relying on technologies that are "unsustainable," and Czech (2000) explains why the canons of contemporary economic systems are at odds with environmental protection. Certainly, the high fashions in conservation and development, such as "sustainable development," are more a political convenience or deceit than a boon (Frazier 1997). Until there are fundamental changes in the way the dominant society functions, the issues discussed here will continue to be but symptoms, subject to social "Band-Aids" of dubious long-term benefit. Clearly, NGOs are in many respects part of the dominant society, despite the location of their headquarters.

So, what am I both begging and challenging my colleagues in the social sciences to do? You are professionals, trained in the identification, description, and evaluation of human social processes, political structures, social interactions, communication, symbolism, cultural matrices, and the sources of human motivation that illuminate the questions of why people do what they do. As Durrenberger (1990, 82) explained, "these matters of allocation and distribution effects—political matters—are often what anthropologists are most competent to talk about. . . . It is not that anthropologists or social scientists are the only ones to have noticed this, but that it is part of our subject matter, part of the data we try to understand while it is not part of the subject matter of biologists." At one level, one could say that social scientists understand about the accelerators, steering wheels, and brakes of societies. But that simile is inadequate. You are the scientists who understand the thrust systems, gyrostabilizers, orientation systems, range, useful life, and costs of vehicles much, much more complex—and far less predictable—than rockets.

There is an urgent need for social scientists to be more active in biological conservation and environmental protection, particularly in the root problems of social organization and motivation; there is an unhealthy reticence—or perhaps taboo—for social scientists to incorporate environmental issues as priority matters for investigation (Beck 1992, 1995), and this does no good for either the social sciences or the environment on which societies depend.[16] The investigations of political ecology that are carried out in the Yucatan must contemplate major forces impacting resource availability and distribution. To understand these questions today, in the PoMoNeLiAGlo, attention must be given to the institutions involved at diverse levels; it is just not adequate to know how much effort goes into planting a milpa and how much it will yield in a good year. There is a crying need to understand and evaluate those institutions that claim special relevance to the issues of environmental conservation and community development, particularly donor and recipient NGOs. Social scientists should go "beyond the traditional cultural ecology approaches" to incorporate the role of public policy processes and their actors (Orbach and Johnson 1988, 9). "What we need at this critical juncture is an effort to transcend the impasse created by debates around abstracted and ambiguous categories such as environment, development, and sustainability and the romanticized notions of popular environmental action. In order to move beyond the totalizing tendencies of grand structural theories, we need to develop middle-range explanatory frameworks that integrate global processes with local environmental action and reveal the particular outcomes experienced by peoples and communities living with localities and regions" (Friedmann and Rangan 1993, 11). Furthermore, these studies must be disseminated to international, multidisciplinary groups involved in diverse aspects of these activities (e.g., Faust and Smardon 2001).

The Yucatan Peninsula provides prime examples of persistent imperialism. Not only is the geographic name a monument to cultural ignorance and domination, but the contemporary situation illustrates clear cases of imperialism in the developmental and environmental arenas. Libraries are replete with volumes on the cultural and economic conquest of the Americas by European—and Amerindian—conquistadores, and discourses on past and present effects of these historic events are endless. What is often missing is explicit analyses of the ways in which these issues are played out under contemporary systems. As Markowitz (2001) and others concerned with the interaction between environment and society have lamented, there is a paucity of information on how the agents of change, particularly NGOs, are dealing with these complex, interdisciplinary issues.

Communications and interactions within conventional discipline-bounded professional groupings and organizations in anthropology and the other "social sciences" will do little to solve the root problems. There are countless professional societies, journals, and other academic vehicles and actions directly involved with biological conservation and, with few exceptions, these are designed and run by "natural scientists." Don't get me wrong—some of my most respected friends are biologists! But we are not trained in rocket science, much less in the complexities of social science. Anthropologists, sociologists, and other social scientists investigate questions of human social life, beliefs, and behavior—but all too rarely analyze those effects in relation with ecological processes, biological diversity, and "environmental health," which are so critical for humans and their societies.

Just as I, an outsider, have invaded your professional space, you must do the same. Social scientists must go to meetings about biological conservation; let them know the relevance of your work. Publish in their journals.[17] Publicly debate their conclusions, illuminate their unspoken assumptions, test their taboos, shake their intellectual structure so that in collaboration a stronger edifice can be designed and built.

Why should you do this? If you are interested in the people you study, you must be concerned about their resource base and the environment in which they live. Environmental protection is not just a luxury of First World elites—it is necessary for the resource base on which all societies depend (Allaby 1985; Myers 1992; Heywood 1995; Vandermeer and Perfecto 1995; Daily 1997; Baskin 1997). Like it or not, we are all part of the PoMoNeLiAGlo—postmodern, neoliberal age of globalization. Pretending to be Gonzalo Guerrero among the Yucatec Maya when unwanted impositions of the outside world could be deterred with wizardry will do little to help the people who are your study subjects—the intended beneficiaries of our intellectual efforts.

ACKNOWLEDGMENTS

Earlier drafts of this chapter were improved by comments and criticisms from Gene Anderson, Liz Barnaby, Robert Brulle, Megan Gallagher, Angela García, Carrie Meyer, and Brian Miller; discussions, information, and suggestions provided by Betty Faust, and her vast library, were invaluable in shaping, enhancing, and correcting many of the arguments.

NOTES

1. Although there are variations in the interpretation of the origin of the term *Yucatan* (for example, Gates' [1937] 1978, 2 translation of Friar Diego de Landa's classic 1566 account of the early conquest of the Peninsula stated that when early Spanish explorers asked the name of the land, the Maya responded, "*Ci uthan*," meaning, "they say it," while Pragden's [1975, 31] translation from the same source claimed that it came from "*Ciuthan*," meaning "they say so"), there is consistency in the explanation that the term is not Maya and derives from a misunderstanding of the Spanish conquistadores in their early initiatives at exploring and conquering the Peninsula that now has that name.

2. It is important to appreciate that misunderstanding was a common occurrence when peoples of two cultures met suddenly. During the sixteenth century, European castaways in coastal Florida who came into contact with the Calusa were commonly ordered to dance and sing. The failure of the explorers to respond adequately was interpreted as meanness and rebelliousness—at which point they were slain by the Calusa. Until a captive Spaniard who had learned the Calusa language explained the problem, it had not occurred to the Calusa that the strangers did not understand their instructions (Fontaneda 1945: 33–34). Petitjean Roget (1997) gives a fascinating interpretation of the first contact between Columbus and the Taino Indians, in what is now known as the Caribbean. He explains that there were fundamental misunderstandings not only on the part of the Spanish adventurers; the Taino, using their basis of knowledge and cosmo vision, totally misunderstood the Spaniards: who they were, where they came from, what they wanted, what they were doing, and so on. Petitjean Roget refers to this as an "exchange" of misunderstanding.

3. Terms such as *postmodern*, *neoliberal*, and *globalization* have been part of the vernacular for years, and there is often a feeling that "we all know what we are talking about." However, these terms can be highly loaded, resulting in endless polemics to define them; too often, their meaning can vary considerably depending on context and the intent of the author. Precise meanings are not critical for the present study, for the terms are used simply as metaphors of argot, meant to express the high degrees of interconnectedness among diverse commercial, economic, political, and social systems throughout the world (see Sally 2002 and citations therein).

4. Remarkably, the negative social and environmental effects are not limited to the nations whose forests are being felled for exports to Japan, but also occur in Japan, further illustrating the complications of the impacts of globalized commerce and "shadow ecology" (Dauvergne 1997).

5. One reviewer recommended streamlining the citations included herein and providing just a key reference for major points instead of including a large number of citations for many points of discussion; however, because this chapter was written with a wide, multidisciplinary readership in mind and because one reviewer questioned the veracity of some of the claims contained herein, as they fly in the face of standard dogma, it was decided to provide a variety of diverse sources that support the various claims so that the reader can explore the generalities in greater detail and be alerted to sources from different disciplines.

6. Gonzalo Guerrero is regarded as the father of mestizos (the Spanish term for persons of mixed European and Native American ancestry) and a folk hero: he is reputed to have survived a shipwreck in 1511, during the early stages of Spanish exploration of the "New World," as well as enslavement by his Maya captors, and the threat of sacrifice, then to have married the daughter of a Maya noble with whom he had children (the first mestizos), and to have been an effective military commander for the Maya, notably in their battles against the Spanish (see Clendinnen 1987, 17–18; Restall 1998, 7).

7. Social sciences, excepting macro-level economics, have been underrepresented in conservation and development initiatives.

8. To these concerns must be added the fast-growing trend in many countries for governments to contract out certain functions that had been their sole responsibility, not only encouraging but mandating the private sector to take on roles that were previously the bailiwick of governments. As a result, lines of distinction between government and private sector are often blurred, and accountability and responsibility are lost (or hidden) in new, untried levels of bureaucracy.

9. The continued reference in this chapter to NGOs in Canada and the United States as examples is because of the relatively large body of information that exists on these organizations from these countries. This does not mean that similar problems do not exist elsewhere but rather that there is no systematic information. Indeed, a description of fundamental administrative, management, and ethical problems in World Wide Fund for Nature–India (Rai 2001) makes this point very clear.

10. This infatuation with compromise and patronization, with resulting superficiality and the courting of irrelevance, was driven home recently by the former Director of Program Development of a well-established, U.S.-based conservation organization that prides itself on having programs in forty-five nations. In reviewing a manuscript on animal conservation, this person censored the term *Third World* as "antiquated and pejorative" and stated, "Likewise, 'peasants' is offensive." It is difficult to imagine in

what language people like this communicate when they leave their plush offices in the United States. Will they disband the Third World Academy of Science and the Third World Network? Will they discredit some of the most forthright defenders of the peasants, biodiversity, and environment of the Third World, authors (e.g., Shiva 1992, 1993) who happen to have been born and live in the Third World, a term to which they constantly refer? What of the petitions for protection of the environment and indigenous denizens of the "Fourth World,"—too often regarded as inferior to peasants (e.g., Bodley 1988, 1990)? This is to say nothing of the fact that anyone who understands the etymology of the phrase *Third World* will appreciate that the term is in fact a sociopolitical satire (Isbister 1993, 14)!

11. It is important to clarify that although nowadays it is commonly assumed that the North is the source of information, initiative, and innovation in the environmental movement, the history of environmentalism is rich and diverse, easily predating the Christian era and encompassing a wide diversity of cultures and lands (Grove 1995), including Mexico (Simonian 1995). Hence, the flow of resources from donor to recipient is dictated more by contemporary discrepancies in financial resources and technical information than by a superior grasp of the problems and appropriate measures for their solution, or some deep-rooted cultural difference between donors and recipients. Grove (1995) provides a detailed analysis of "environmental imperialism."

12. The "elite" in the recipient society are not necessarily just people from the highest socioeconomic classes but can be people who have adopted what they consider to be a higher culture, such as through foreign schooling.

13. Indeed, the term *astroturf organizations* is now used to refer to those institutions that purport to represent the grassroots (i.e., marginalized members of society) but use this as a cynical deception to garner power, respect, and resources for their elite officers.

14. This is not to imply that complex, self-serving political maneuvers, intense sociopolitical competition, noncooperation (even if covered with a mantle of cordiality), and internecine combat are new to the Yucatan, or exist only because of foreign agencies; among Mexicans, the level of *grilla* in the Yucatan is renowned. Markowitz (2001, 43) knowingly asked, "How do we find a compass for sorting out the myriad internecine rivalries that characterize the NGO world?"

15. Although much of the literature available in the West focuses on relationships between Christianity and the environment, it is essential to understand that every religion has some position on the issue of environmental quality (Namgyal et al. 1986; ARC [2001?]), and non-Christian religions such as Buddhism (e.g., Batchelor and Brown 1992; Darlington 1998) and Islam (e.g., Agwan 1997; Kula 2001) also have "environmental movements" and environmental ethics.

16. An earlier, abbreviated version of this chapter was presented at the 1997 annual meeting of the American Anthropological Association (AAA), and since that time there has been growing involvement of some members of the anthropological community in issues of conservation biology. At the same 1997 meeting, the Anthropology and Environment Section (AES) of AAA was organized; at the 2000 annual meeting of AAA a special session for AES was held (Ragoné 2000, 79) and at the session on conservation anthropology the same year a new Internet discussion group, Conservation and Community (AECC-L@listserve.uga.edu), was initiated. A workshop of Conservation and Community was held during the 2001 annual meeting of the AAA (Brosius, in litt., October 24, 2001).

17. One of the editors of this volume, a founding member of the Society for Conservation Biology and a dedicated conservationist, commented, "We don't have the credentials to publish in *Conservation Biology*." This may be the attitude of some biologists who have the power of censorship, but it must not be tolerated if the field known as biological conservation is to be effective. It will be up to the social scientists to join the rebellion against this situation.

REFERENCES

Adam, B. 1998. *Timescapes of modernity: The environment and invisible hazards*. New York: Routledge.

Agwan, A. R., ed. 1997. *Islam and the environment*. Kuala Lumpur: Synergy Book International.

Allaby, M., ed. 1985. *The Oxford dictionary of natural history*. New York: Oxford University Press.

ARC (Alliance of Religions and Conservation). (2001?). *ARC: The first five years*. Manchester, UK.: World Wildlife Fund, ARC, MOA International; see also <http:// www.arcworld.org>.

Baskin, Y. 1997. *The work of nature: How the diversity of life sustains us*. Washington DC: Island Press.

Batchelor, M., and K. Brown, eds. 1992. *Buddhism and ecology*. London: Cassell & World Wide Fund for Nature.

Beck, U. 1992. Risikogesellschaft. *Risk society: Towards a new modernity*, translated by Mark A. Ritter. Newbury Park, CA: Sage Publications.

———. 1995. Politik in der Risikogesellschaft. *Ecological enlightenment: Essays on the politics of the risk society*, translated by Mark A. Ritter. Atlantic Highlands, NJ: Humanities Press.

Blase, M. G. 1986. *Institution building: A source book*. Rev. ed. Colombia: University of Missouri Press.

Bodley, J. H., ed. 1988. *Tribal peoples and development issues: A global overview*. Mountain View, CA: Mayfield.

———. 1990. *Victims of progress*. 3rd ed. Mountain View, CA: Mayfield.

Bookchin, M. 1994. *Which way the ecology movement?* San Francisco: A.K. Press.

Breslin, P. 1987. *Development and dignity*. Rosslyn, VA: Inter-American Foundation.

Brulle, R. 1994. Power, discourse, and social problems: Social problems from a rhetorical perspective. *Perspectives in Social Problems* 5: 95–121.

———. 1995a. Agency, democracy, and the environment: An examination of U. S. environmental movement organizations from the perspective of critical theory. Ph.D. diss., George Washington University, Washington, DC.

———. 1995b. Environmentalism and human emancipation. In *Social movements: Critiques, concepts, case-studies*, edited by S. Lyman. New York: Macmillan, 309–329.

———. 1995c. Foundations and environmental movement organizations: An examination of the role of foundations in facilitating social change for a sustainable society. Paper presented at the annual conference of the Society for the Advancement of Socio-Economics. April 1995; Washington D.C.

———. 1996. Environmental discourse and social movement organizations: A historical and rhetorical perspective on the development of U.S. environmental organizations. *Social Inquiry* 66 (1): 58–83.

———. 2000. *Agency, democracy and nature: The U. S. environmental movement from a critical theory perspective*. Cambridge, MA: MIT Press.

———, and R. H. Brown. 1997. Democratic science in practice: The experience of the environmental justice movement. In *Science as narration: Public knowledge and civic communication in a democratic society*, edited by R. H. Brown. New Haven, CT: Yale University Press.

Caldecott, J. O. 1996. *Designing conservation projects*. New York: Cambridge University Press.

Cernea, M. M. 1991. Knowledge from social science for development policies and projects. In *Putting people first: Sociological variables in rural development*, edited by M. M. Cernea. 2nd ed. Washington, DC: World Bank, Oxford University Press, 1–41.

Clark, J. 1990. *Democratizing development: The role of voluntary organizations*. West Hartford, CT: Kumarin Press.

Clendinnen, I. 1987. *Ambivalent conquests: Maya and Spaniard in Yucatan, 1517–1570*. New York: Cambridge University Press.

Cochrane, G. 1979. *The cultural appraisal of development projects*. New York: Praeger.

Crumley, C. L., ed. 1994. *Historical ecology: Cultural knowledge and changing landscapes*. Santa Fe, NM: School of American Research Press.

Czech, B. 2000. Economic growth as the limiting factor for wildlife conservation. *Wildlife Society Bulletin* 28 (1): 4–15.

Daily, G. C. 1997. *Nature's services: Societal dependence on natural ecosystems*. Washington, DC: Island Press.

Darlington, S. M. 1998. The ordination of a tree: The Buddhist ecology movement in Thailand. *Ethnology* 37 (1): 1–15.

Dauvergne, P. 1997. *Shadows in the forest: Japan and the politics of timber in Southeast Asia*. Cambridge, MA: MIT Press.

Decker, D. J., T. L. Brown, and W. F. Siemer. 2001. *Human dimensions of wildlife management in North America*. Bethesda, MA: The Wildlife Society.

Dewar, E. 1995. *Cloak of green: The links between environmental groups, government and big business*. Toronto: James Lorimer.

Dowie, M. 1995. *Losing ground: American environmentalism at the close of the twentieth century*. Cambridge, MA: MIT Press.

Durrenberger, P. 1990. Policy, power and science: The implementation of the turtle excluder device regulations in the U.S. Gulf of Mexico shrimp fishery. *Maritime Anthropological Studies* 3 (1): 69–86.

Dyer, C. L., and J. R. McGoodwin, eds. 1994. *Folk management in the world's fisheries: Lessons for modern fisheries management*. Niwot: University Press of Colorado.

Faust, B. B. 1998. *Mexican rural development and the plumed serpent: Technology and Maya cosmology in the tropical forest of Campeche, Mexico*. Westport, CT: Bergin and Garvey.

———. 2001. Maya environmental successes and failures in the Yucatan Peninsula. *Environmental Science & Policy* 4: 153–169.

———, and R. Bilsborrow. 2000. Maya culture, population and the environment on the Yucatán Peninsula. In *Population, development, and environment on the Yucatán Peninsula: From ancient Maya to 2030*, edited by Wolfgang Lutz, Leonel Prieto, and Warren Sanderson. Laxenburg, Austria: International Institute for Applied Systems Analysis, 73–107.

———, and R. C. Smardon. 2001. Introduction and overview: Environmental knowledge, rights, and ethics: Co-managing with communities. *Environmental Science & Policy* 4: 147–151.

Folen,W. J., B. Faust, W. Lutz, and J. D. Gunn. 2000. Social and environmental factors in the Classic Maya Collapse. In *Population, development, and environment on the Yucatán Peninsula: From ancient Maya to 2030*, edited by Wolfgang Lutz, Leonel Prieto, and Warren Sanderson. Laxenburg, Austria: International Institute for Applied Systems Analysis, 2–32.

Fontaneda, Hernando d'Escalante. 1945. Memoires of Do. d'Escalante Fontaneda respecting Florida. Written in Spain about the year 1575. Translated by Buckingham Smith, edited by David O. True. Coral Gables, FL: Glade House.

Frazier, J. 1990. International resource conservation: Thoughts on the challenge. *Transactions of the 55th North American Wildlife and Natural Resources Congress*. Washington D.C.: Wildlife Management Institute, 384–395.

———. 1994. Conserving sea turtles and other natural resources: What Ferdinand Marcos and Manuel Noriega can teach us. In *Proceedings of the thirteenth annual symposium on sea turtle biology and conservation*, compiled by B. A. Schroeder and B. E. Witherington. NOAA Tec. Mem. NMFS-SEFSC-341. Miami: U.S. Department of Commerce; National Oceanographic and Atmospheric Administration; National Marine Fisheries Service; Southeast Fisheries Center, 60–63.

———. 1997. Sustainable development: Modern elixir or sack dress? *Environmental Conservation* 24 (2): 182–193.

———. in press. Science, conservation and sea turtles: What's the connection? In *Proceedings of the twenty-first annual symposium on sea turtle biology and conservation*, compiled by M. Coyne. NOAA Tec. Mem. NMFS-SEFSC. Miami: U.S. Department of Commerce; National Oceanographic and Atmospheric Administration; National Marine Fisheries Service; Southeast Fisheries Center.

Friedmann, J., and H. Rangan, eds. 1993. *In defense of livelihood: Comparative studies on environmental action*. West Hartford, CT: United Nations Research Institute for Social Development and Kumarian Press.

García Q., A. 2000. El dilema de *Ah Kimsah K'ax*, "el que mata al monte": Significados del monte entre los mayas milperos de Yucatán. *Mesoamerica* 39: 255–285.

Garrett, W. E. 1989. La Ruta Maya. *National Geographic Magazine* 176 (4): 424–479.

Gates, W., trans. and ed. [1937] 1978. *Yucatan before and after the conquest. By Friar Diego de Landa. With other related documents, maps and illustrations*. Reprint, New York: Dover. [Originally published by the Maya Society of Baltimore, no. 20]

Gavin, M., and R. Hausmann. 1999. *Preventing crisis and contagion: Fiscal and financial dimensions*. Washington D.C.: Inter-American Development Bank, Office of the Chief Economist.

Ghimire, K. B., and M. P. Pimbert, eds. 1997. *Social change and conservation: National parks and protected areas*. London: United Nations Research Institute for Social Development (UNRISD) and Earthscan.

Grove, R. H. 1995. *Green imperialism: Colonial expansion, tropical island Edens and the origins of environmentalism, 1600–1860.* New Delhi: Oxford University Press.

Hackenbert, R. A., and N. Benequista. 2001. The future of an imagined community: Trailer parks, tree huggers, and trinational forces collide in the southern Arizona borderlands. *Human Organization* 60 (2): 153–158.

Hartasánchez H., I. E. 1997. Reporte de la revisión técnica a los proyectos de la microregión: Laguna de Términos/Central de Campache/Costa oriental de Yucatán/ Norte de Quintana Roo/Sur de Yucatán/Norte de Campeche; Organizaciones SSS Ixinha, Xch Keh Ghi SSS, y Lucero de Jesús; Proyecto Kinam/Milpa Ecológicamente Adecuada/Xunankab/Desarrollo de Estrategias de Cultivo de Arboles/Reforestación y Conservación de Biodiversidad/Proyectos Subsidiados por el PPS al CICY/Estudio Geohidrológico/Fundación Tun Ben Kin/Tinglado. Unpublished reports; Mérida, Mexico.

Heywood, V. H., ed. 1995. *Global biodiversty assessment.* New York: UNEP/Cambridge University Press.

Hymes, D., ed. 1964. *Language in culture and society: A reader in linguistics and anthropology.* New York: Harper and Row.

Isbister, J. 1993. *Promises not kept: The betrayal of social change in the Third World.* 2nd ed. West Hartford, CT: Kumarian Press.

Iwasaki, Y. 2001. Welcoming remarks by Yoshihiro Iwasaki, Director, Programs Department (West) Asian Development Bank; Conference on Preparing for Afghanistan's Reconstruction. Available at: http://lnweb18.worldbank.org/sar/sa.nsf/ Countries/Afghanistan/88D233B8A51B864385256B1100262A02?OpenDocument.

Johnston, B. R., ed. 1994. *Who pays the price? The sociocultural context of environmental crisis.* Washington, DC: Island Press.

Killingsworth, M. J., and J. S. Palmer. 1992. *Ecospeak: Rhetoric and environmental politics in America.* Carbondale: Southern Illinois University Press.

Kintz, E. R. 1990. *Life under the tropical canopy: Tradition and change among the Yucatec Maya.* Chicago: Holt, Rinehart and Winston.

Knight, J. 1997. A tale of two forests: Reforestation discourse in Japan and beyond. *Journal of Royal Anthropological Institute* 3(4): 711–730.

Korten, D. C. 1995. *When corporations rule the world.* West Hartford, CT: Kumarian Press.

Kula, E. 2001. Islam and environmental conservation. *Environmental Conservation* 28 (1): 1–9.

Lean, M. 1995. *Bread, bricks and belief: Communities in charge of their future.* West Hartford, CT: Kumarian Press.

Leff, E., ed. 2001. *Justicia ambiental: Construcción y defensa de los nuevos derechos ambientales culturales y colectivos en América Latina.* Mexico City: Programa de las Naciones Unidas para el Medio Ambiente, Oficina Regional para América Latina y el Caribe.

Lélé, S., and Norgaard, R. B. 1996. Sustainability and the scientist's burden. *Conservation Biology* 19 (2): 354–365.

Mangel, M., L. M. Talbot, G. K. Meffee, M. T. Agardy, D. L. Alverson, J. Barlow, D. B. Botkin, G. Budowski, T. Clark, J. Cooke, R. H. Crozier, P. K. Dayton, D. L. Elder, C. W. Fowler, S. Funtowicz, J. Giske, R. J. Hofman, S. J. Holt, S. R. Kellert,

L. A. Kimbal, D. Ludwig, K. Magnusson, B. S. Malayang III, C. Mann, E. A. Norse, S. P. Northridge, W. F. Perrin, C. Perrings, R. M. Peterman, G. B. Rabb, H. A. Regier, J. E. Reynolds III, K. Sherman, M. P. Sissenwine, T. D. Smith, A. Starfield, R. G. Taylor, M. F. Tillman, C. Toft, J. R. Twiss Jr., J. Wilen, and T. P. Young. 1996. Principles for the conservation of wild living resources. *Ecological Applications* 6 (2): 338–362.

Manno, J. P. 2000. *Privileged goods: Commoditization and its impact of environment and society*. New York: Lewis Publishers.

Margoluis, R., and N. Salafsky. 1998. *Measures of success: Designing, managing and monitoring conservation and development projects*. Washington, DC: Island Press.

Markowitz, L. 2001. Finding the field: Notes on the ethnography of NGOs. *Human Organization* 60 (2): 40–46.

McDonnell, M. J., and S. T. A. Pickett, eds. 1993. *Humans as components of ecosystems: The ecology of subtle human effects and populated areas*. New York: Springer.

Meffe, G. K., and S. Viederman. 1995. Combining science and policy in conservation biology. *Wildlife Society Bulletin* 23 (3): 327–323.

Metha, L. 2001. The World Bank and its emerging knowledge empire. *Human Organization* 60 (2): 189–196.

Meyer, C. A. 1992a. The irony of donor efforts to build institutions: A case study from the Dominican Republic. *Journal of Institutional and Theoretical Economics* 148: 628–644.

———. 1992b. A step back as donors shift institution building from the public to the "private" sector. *World Development* 20 (8): 1115–1126.

———. 1995. Opportunism and NGOs: Entrepreneurship and green North-South transfers. *World Development* 23 (8): 1277–1289.

———. 1996. NGOs and environmental public goods: Institutional alternatives to property rights. *Development and Change* 27: 453–474.

———. 1997a. The political economy of NGOs and information sharing. *World Development* 25 (7): 1127–1140.

———. 1997b. Public-nonprofit partnerships and North-South green finance. *Journal of Environment and Development* 6 (2): 123–146.

———. 1999. *The economics and politics of NGOs in Latin America*. Westport, CT: Praeger.

Miller, B., R. Reading, C. Conway, J. A. Jackson, M. Hutchins, N. Snyder, S. Forrest, J. Frazier, and S. Derrickson. 1994. A model for improving endangered species recovery programs. *Environmental Management* 18 (5): 637–645.

Moran, E. F., ed. 1990. *The ecosystem approach in anthropology: From concept to practice*. Ann Arbor: University of Michigan Press.

Myers, N. 1992. *The primary source: Tropical forests and our future*. London: W. W. Norton.

Nader, L., ed. 1996. *Naked science: Anthropological inquiry into boundaries, power, and knowledge*. New York: Routledge.

Namgyal, Rinpoche Lungrig, Lanfranco Serrini, Karan Singh, Abdullah Omar Nasseef, Arthur Hertzberg. 1986. *The Assisi Declarations: Messages on man and nature from Buddhism, Christianity, Hinduism, Islam and Judaism*. London: World Wildlife Fund.

Norton, B. G. 1988. What is a conservation biologist? *Conservation Biology* 2 (3): 237–238.

Noss, R. F., M. A. O'Connell, and D. D. Murphy. 1997. *The science of conservation planning*. Covelo, CA: Island Press.

Orbach, M. K., and J. C. Johnson. 1988. The transformation of fishing communities: A public policy perspective. Paper presented at Marine Resource Utilization: A Conference on Social Science Issues, Mobile, Alabama, May.

Painter, M., and W. H. Durham, eds. 1995. *The social causes of environmental destruction in Latin America*. Ann Arbor: University of Michigan Press.

Pell, E. 1990. Buying in: How corporations keep an eye on environmental groups that oppose them—by giving big wads of money. *Mother Jones*, April/May, 23, 25, 27.

Petitjean Roget, H. 1997. The Taino vision: A study in the exchange of misunderstanding. In *The indigenous people of the Caribbean*, edited by S. M. Wilson. Gainesville: University Press of Florida, 169–175.

Place, S. E. 1998. Society and nature: Recent trends in the study of Latin American environments. *Latin American Research Review* 33 (2): 221–236.

Praden, A. R., trans. and ed. 1975. *The Maya: Diego de Landa's account of the affairs of Yucatán*. Chicago: J. Philip O'Hara.

Primack, R. B. 1998. *Essentials of conservation biology*. 2nd ed. Sunderland, MA: Sinauer Associates.

Pye-Smith, C., G. B. Feyerabend, and R. Sandbrook. 1994. *The wealth of communities: Stories of success in local environmental management*. West Hartford, CT: Kumarian Press.

Ragoné, H., program chair. 2000. *Program. American Anthropological Association; 99th annual meeting, San Francisco, CA; The public face of anthropology; November 15–19, 2000*. Arlington, VA: American Anthropological Association.

Rai, U. 2001. Endangered organisation. *Hindu Newspaper*, August 26.

Raven, P. H. 1987. Conservation biology: A new discipline comes of age. Review of *Conservation biology: The science of scarcity and diversity*. *Conservation Biology* 1 (1): 82–83.

Reed, N. 1964. *The Caste War of Yucatan*. Stanford, CA: Stanford University Press.

Restall, M. 1998. *Maya conquistador*. Boston: Beacon Press.

Richy-Vance, M. 1993. *The art of association: NGOs and civil society in Colombia*. 2nd ed. Arlington, VA: Inter-American Foundation.

Sally, R. 2002. Rival visions of globalisation—anything new after September 11th? *Frankfurter Allgemeine Sonntagszeitung*. Available at http://www.globaldimensions.net/articles/sally/rivals.html.

Salvatierra, M. 1995a. Bioma y los delfines. *Profauna* (Ministerio del Ambiente y Recursos Naturales Renovables, Caracas) 3 (5): 2–3.

———. 1995b. Delfines en conflicto. *Profauna* (Ministerio del Ambiente y Recursos Naturales Renovables, Caracas) 3 (5): 15–26.

SCB (Society for Conservation Biology). 1987. Goals and objectives of the Society for Conservation Biology. *Conservation Biology*. 1 (1): 91.

Scott, J. C. 1998. *Seeing like a state: How certain schemes to improve the human condition have failed*. New Haven, CT: Yale University Press.

Shiva, V. 1992. *Decolonizing the North*. The INTACH Environmental Series, 18. New Delhi: Indian Trust for Art and Cultural Heritage, 1–16.

———. 1993. *Monocultures of the mind: Perspectives on biodiversity and biotechnology*. Dehra Dun, India: Natraj.

Simonian, L. 1995. *Defending the land of the jaguar: A history of conservation in Mexico*. Austin: University of Texas Press.

Smith, K. 1994. *The human farm: A tale of changing lives and changing lands*. West Hartford, CT: Kumarian Press.

Struhsaker, T. H. 1998. A biologist's perspective on the role of sustainable harvest in conservation. *Conservation Biology* 12 (4): 930–932.

Terán, S., and C. Rasmussen. 1994. *La milpa de los Mayas*. Mérida, Mexico: DANIDA and Government of Yucatán.

Thornton, A., and D. Currey. 1991. *To save an elephant: The undercover investigation into the illegal ivory trade*. New York: Doubleday.

Utting, P., D. Ghai, C. Hewitt de Alcántara, Y. Bangura, and J. Vivian. P. Stalker, eds. 1995. *States in disarray: The social effects of globalization*. London: United Nations Research Institute for Social Development and Banson.

Vandermeer, J., and I. Perfecto. 1995. *Breakfast of biodiversity: The truth about rain forest destruction*. Oakland, CA: Food First, The Institute for Food and Development Policy.

Wedel, Janine R. 1998. *Collision and collusion: The strange case of western aid to Eastern Europe 1989–1998*. New York: St. Martin's Press.

Winthrop, R. 1997. The real world: Participation and other fictions of development. *Practicing Anthropology* 19 (4): 35–36.

World Bank. 2001. Poverty to rise in wake of terrorist attacks in US: Millions more people condemned to poverty in 2002. The World Bank Group, Washington, DC. News Release no. 2002/093/S. Available at http://lnweb18.worldbank.org/news/pressrelease.nsf/.

———. 2002. World Bank anticipates global upturn, urges increased help to poor countries. World Bank, Washington, DC. News Release no. 2002/225/S. Available at http://lnweb18.worldbank.org/news/pressrelease.nsf/.

Wright, A. 1990. *The death of Ramón González: The modern agricultural dilemma*. Austin: University of Texas Press.

13

Conclusion: Reflections on Rights, Resources, and Responsibilities in Participatory Research

Mary Lindsay Elmendorf

Commenting on the previous chapters in this book has been both a privilege and an honor. The research findings presented confirm my long-held belief that many of the negative impacts of change, in both human and environmental dimensions, can be mitigated by full participation of the community—women, men, and children.

One of the chief themes throughout sixty-five of my eighty-five years of life as wife, mother, social worker, teacher, anthropologist, and activist has been the importance of women within their homes and communities—and now globally—as agents of change. As gatekeepers of values and resources, as educators and decision makers in daily life, women are critically needed as full partners in working toward sustainable development.

At the invitation of the authors, I am sharing personal reflections from my life experiences that are related to the cultural and environmental challenges so vividly portrayed in the previous chapters. I also suggest applications of lessons learned. In so doing, I have taken the liberty of casting a wide net, drawing in examples and illustrations at times linked by theme and memories. In many ways, the same challenges confronting the Yucatan villages documented in these chapters—environmental disasters, population growth and migration, pollution, resource depletion, and violence—are the same ones buffeting our global community today.

Before trying to reach any conclusions, I must say how amazed I was in the session that led to this book at the many lessons we have learned from each other and the Maya, lessons we can share with others. I would also like to add some very personal comments on the chapter by Alicia Re Cruz, in which she emphasized how I was ready to listen to and learn from the wisdom of Maya women. It was thanks to two things that I was able to use this approach.

First, I probably would not be here and certainly would not have had the freedom to do research in the Yucatan had it not been for my first husband, John Elmendorf. Over the forty-three year mosaic of our marriage, we lived life fully and cooperatively, interested in each other's careers, sharing the housework, and raising our two children. We were ahead of our time in that. These were not the accepted things for either of us to do. Nevertheless, in 1968 he, as president of New College, encouraged me to "escape" from some of the expected wifely roles and duties to complete the degree in anthropology that I had started in 1945. In 1972, twenty-seven years after beginning it, I finished! My graduate studies had been interrupted first by World War II and then by the births of our son and daughter, by ten years in Mexico and five in Rhode Island, where John Elmendorf was vice president of Brown University. Later we moved to New College in Sarasota with our son and daughter. While complicating the completion of my degree, these years of interruptions enriched my life enormously, particularly my volunteer work designing and supervising pilot projects in off-campus experiential learning at Brown and New College.

Thanks to the Union Graduate School (UGS, now the Union Institute and University), I was able to complete my doctoral work (begun earlier at the University of North Carolina at Chapel Hill). My previous thirty-plus years of experience as a social worker and administrator in the rural South, in the slums of New England, war-torn Europe, and Mexico, and as a consultant for the Peace Corps, League of Women Voters, and the Department of State all enriched my doctoral studies. UGS allowed me to design my own research program and to choose my Ph.D. committee: Alfonso Villa Rojas, Dorothy Lee, Fernando Cámara, and Erich Fromm (later replaced by Michael Maccoby).

After John Elmendorf's death, I found it difficult to continue my interests in Latin America and started accepting assignments in Asia, perhaps intrigued by the cultures and religions there—perhaps reincarnation. In 1981, I married widower anthropologist John Landgraf, an Asia specialist. We have enjoyed discussing our very different anthropological learning and experiences, visiting our respective sites, and helping each other on various projects. His long-standing interests in ecological anthropology became a central and most relevant point of reference in my growing concern for quality of life and change—not only for the Maya but for all peoples, everywhere. His interests in ecological anthropology had derived originally from 1940s fieldwork with the Navajo, as an assistant to Ralph Linton. This research was published by Harvard in a Peabody Museum monograph, *Land Use in the Ramah Area of New Mexico* (Landgraf 1954).

I wanted to add research data to document my anecdotal evidence. I wanted to understand how women felt about their lives and to fill in what I perceived as a gap about women in the accepted ethnologies. There was a scarcity of research about their worlds. I was particularly interested in the roles women play in traditional peasant societies facing change and modernization. Alfonso Villa Rojas, with whom I had worked in the 1950s, suggested Chan Kom. "Go to Chan Kom; maybe that's the best village for you. I'll give you a letter of introduction to don Eustaquio Ceme, the mayor of the village." I focused intensively on the women of Chan Kom, building my research on the extensive studies of Redfield, Villa Rojas himself, and others, starting with the Carnegie group in the 1930s.

I had never before done research in the Yucatan, but during my eight years in Mexico as director of CARE (Cooperative for American Relief Everywhere), I had supervised development programs extending technical assistance to existing agencies throughout Mexico. As noted in chapter 3, one of the first large donations was for medical equipment for a new hospital in Pisté to be used by José Estrella Ek, the first Maya to graduate from medical school. Little did I know then of the past relationships to Chan Kom, let alone future ones!

By the late 1960s, when I began the fieldwork for my graduate degree, nearly two decades had passed. Mérida, Yucatan, was only an hour away from Tampa, Florida, which was a twenty-minute flight from Sarasota, so I could quickly slip away from my various duties at New College in order to do my own graduate work and then get back when needed—to pour tea, host visitors, or meet with students.

Some people in Sarasota were shocked by my leaving for weeks at a time, eating in thatched huts, sleeping in hammocks, and then waiting for buses or walking for six hours through the jungle. John Elmendorf understood why I was doing these things. So did the New College students I helped with independent studies, some in the Yucatan. And so did my son and daughter, both of whom later visited the village, he to take the photographs used in my book (*La Mujer Maya y el Cambio*, Elmendorf 1973, which became *Nine Mayan Women: A Village Faces Change*, Elmendorf 1976) and she to interview the young women.

During my research in Chan Kom, with the encouragement of Margaret Mead, Timothy Asch, and Alan Lomax and with the help of David Pini, a New College student, I started using Super-8 film to document the daily life of the Mayan women and their various ceremonies, to supplement my taped recordings of our conversations and field notes. My attempt to convince my committee to allow me to use this as the "project of excellence" for my Ph.D. was rejected by Michael Maccoby, who insisted that I finish writing up my research. Disappointed that my film had not been considered adequate, I completed the manuscript for *Mayan Women and Change*, which not only was accepted by all, but within a few weeks was published by Ivan Illich as a case study for a seminar at the Mexican Intercultural Documentation Center (Centro Intercultural de Documentación, CIDOC), conducted with Paolo Freire, called "Quality of Life." Alfonso Villa Rojas shared his copy of the CIDOC manuscript with Gonzalo Aguirre Beltrán, who was then in the Mexican Ministry of

Education. Aguirre Beltrán translated and published *La Mujer Maya y el Cambio* (Elmendorf 1973) as a part of his SEP-70 series (published by Mexico's Department of Education, the Secretaria de Educación Publica in the 1970s) for distribution in the public schools throughout Mexico. So, within a year after I completed the draft, it was published twice.

Just a few months later, a second edition of *La Mujer Maya* was given as Mexico's gift to the OAS (Organization of American States) Seminar on Indian women leaders held in Guatemala, which I was invited to attend as an observer. At that time I was doing research with a Ford Foundation grant, Changing Roles of Women in Latin America, and became involved in a three-week workshop on research about women, the first ever. It was sponsored by the Social Science Research Council, and its purpose was to discuss various techniques with young scholars. I worked closely with June Nash, Larissa Lomnitz, Carmen Elu Lenero, and others, sharing our different research approaches. Many of those young scholars have gone on to make outstanding contributions to women's studies and gender issues. One of them, Marianne Schmink, of the University of Florida, has received prestigious awards for her continuing environmental work in Brazil.

But I want to say at this point how delighted I was when Alicia Re Cruz, then a young Spanish graduate student at SUNY-Albany (State University of New York), expressed sincere interest in knowing more about my research in Chan Kom. I invited her to spend time in my home, where together we went through my field notes and unpublished papers. We reviewed the early photos and film, some of which were already on their way to the archives at the University of Florida in Gainesville, thanks to a grant from the Wenner-Gren Foundation. Re Cruz, who was particularly interested in migration from Chan Kom to Cancún, was able to extract valuable baseline data, starting with the late 1960s, which she incorporated into her perceptive book, *The Two Milpas of Chan Kom* (Re Cruz 1996). Additional information came from the research Deborah Merrill-Sands had helped me with during the mid-1970s, when we studied the socioeconomic impact of development in Chan Kom for the World Bank (Elmendorf and Merrill-Sands 1976).

The few subsequent return trips that Re Cruz and I made together to Chan Kom, including the ones we coordinated for the Society for Applied Anthropology in 1994 and 2001, have been special. Re Cruz helped me in annotating and editing my research films with Jake Homiak, director of the Human Studies Film Archives at the Smithsonian. She has also filmed follow-up footage to the ethnographic film taken before the road was built, which together we shared with the people in Chan Kom. Her documentation of their reactions to the "before and after" films is extremely interesting.

In this book are numerous references to NGOs (nongovernmental organizations) and also local community-based groups, sometimes called CBGs. Over the last decade there has been increasing recognition of the need for local participation, or popular participation, or what in the 1930s, 1940s, and 1950s was called community development and self-help. At last, binational organizations such as the United States Agency for International Development (USAID) and international organizations such as the United Nations Development Program (UNDP) and the World Bank have realized that they must reach the grassroots for their projects and programs to be successful and sustainable. As we provided careful case studies, large agencies began to understand the need for working with NGOs, CBGs, and even anthropologists in order to understand and reach the people—women, men, and children. At last, most agencies understood that an official exchange with community leaders is not enough. "Talking with the chief" is *not* community participation.

Joann Andrews's chapter, "Shifts of Strategies and Focus of the Conservation Efforts of Pronatura on the Yucatan Peninsula: A Personal History," will, I am sure, become a classic example of how to build a truly collaborative cooperative agency. In 1968, when I started my field work in the Yucatan, I had not met Andrews but knew of her. Alfonso Villa Rojas, who admired both her and her late husband E. Willys Andrews IV very much, he for his truly pioneering interpretation of the archaeological findings of Dzibilchaltún and she for the extraordinary way she had kept the library of Tulane's Middle-America Research Institute, MARI, not only up to top academic standards but open to scholars, both Mexicans and outsiders. Villa Rojas had urged me to get in touch with Joann, but I was shy. My old friend Stirling Dickinson, of San Miguel de Allende and ex-president of the International Orchid Society, who in 1940 turned my focus south to Mexico, said to me in 1968 on hearing that I was starting research in the Yucatan, "Do give my best to Joann, my favorite orchid collector, a person you will certainly enjoy."

When I finally called Joann Andrews, she and MARI became for me, as she has for so many others, a valuable resource for research. For me she also was a vital link with home and civilization—mailing address, telephone connection, later a parking space for my old Volkswagen Beetle—as well as an admired friend. When in the early 1980s she focused her energies and knowledge on getting a conservation group started in the peninsula, I knew it would work and that she would get great satisfaction in making it a success. In many ways, she and I were both very lucky wives who had worked closely with our husbands, sharing contacts, efforts, and long-term residences in a specific culture.

In 1996, as a founding president of the board of Pronatura Península de Yucatán, Andrews reported on a training course, in cooperation with The Nature Conservancy, for men and women living within the Calakmul Reserve and its environs, to manage and conserve the resources of this important ecosystem. One of the strategies of the workshops was to combine the knowledge of the Maya with the existing data on the flora and fauna of the region. At the same time, efforts for sustainable use of the land, promotion of ecotourism, and sale of forest products are underway with local participation in Calakmul. Similar activities have started in other reserves: Celestún, Río Lagartos, Punta Laguna, and Sian Ka'an.

Jenny Ericson's chapter, "Population and Land Use: An Essential Link for Conservation in the Calakmul Biosphere" is an excellent example of the human dimensions of the problems being faced by both insiders and outsiders, by new and old residents in the numerous ejidos and small landowner settlements "encircling and infiltrating the vast tropical forests of the Calakmul Biosphere Reserve" in what she calls "a challenging ecological and social environment." Covering 723,185 hectares, Calukmul is the largest tract of protected tropical forest in Mexico and was accepted into the UNESCO network of biosphere reserves in 1993, making it, as Ericson notes, an "important element in larger system on the Yucatan Peninsula, forming an ecological corridor of over 2 million hectares. . . ." Beneath the Calakmul forests lie the ruins of the Classic Maya civilization, which prospered with high population densities from 200 to 800 C.E.

Before doing the fieldwork for the case studies with a six-member team trained in participatory research methods, Ericson had coordinated a macro-level assessment of population dynamics and social, political, and environmental changes taking place within and around the biosphere. In collaboration with Pronatura, she organized a team of local people in three communities, all with rapid population growth rates from 1990 to 1995 and with unique mixtures of indigenous and nonindigenous groups, for discussion of local-level population dynamics. In an interesting example of a participatory exercise designed to determine historical changes in one of the isolated communities, a mixed group of residents was asked to fill in a matrix based on criteria they themselves established, using piles of beans. Through the matrix and the ensuing discussion, these individuals were given the opportunity to reflect on how quickly the population of their community had grown over the years and, by looking toward the future, determine where this growth would lead if it continued at the same pace. As a result of this exercise, the ejidatarios in this group decided to introduce a policy change to limit the number of newcomers admitted during the coming years, thus allowing time for current *pobladores* to become accepted as ejidatarios.

After completing the macro study and the three case studies, the team conducted an analysis that, according to Ericson, led them to believe that "stakeholder debate and collaboration can generate a foundation for formulation of appropriate policy responses to migration, and they can shape an approach to land use planning in the region at large, as well as the individual communities within and around the reserve."

Another factor to consider that Ericson pointed out is fertility rates of present residents, which hopefully may change following the 1996 reproductive health project initiated by Pronatura with assistance from USAID, which, among other things, attempted to "integrate reproductive health issues into community-level discussions." Understanding fertility determinants is complicated by the diverse cultural backgrounds of the residents.

In 1992, while a consultant for CEDPA (Center for Development and Population Activities), I attended the 1992 United Nations Conference on

Environment and Development (UNCED) and the Earth Summit in Rio de Janeiro as well as the Forum in Curitiba. Extremely relevant to this book was the preparatory workshop Discovering Linkages Between Environment and Population in Mexico, sponsored by InterAction, a coalition of 150 agencies, where my assignment involved working with representatives of agencies involved in population or environment to develop together an environmental strategy. Most were completely unaware of the programs of the others. In fact, defining a shared vocabulary was necessary before we could develop a background paper. In her evaluation of the InterAction workshop, Rose Marie Vidal, one of the outstanding participants working on environmental problems in the Maya area then and now, summed up one of the greatest problems in our society today: "Traditionally, when dealing with the South, northern agencies and funding sources have promoted very specialized project work along sectoral lines. Any approach to integration will thus require a fundamental paradigm shift for northern organizations, ironically one that will move them closer to ancient indigenous views on the inter-relatedness of individuals, society and nature." Later she urged "regional approaches in combining population and environmental issues" (personal communication, 1997). I agreed then, and I still do, that we can no longer keep them separate because both the ecological and sociocultural dimensions must be incorporated into planning and action on a global scale as well as on regional and local bases.

In fact, environmental issues raised by representatives from the population groups were forceful in Rio de Janeiro, and later at the United Nations Population and Health Conference (UNPHA) in Cairo. This international meeting, with a large representation from NGOs, greatly revived interest and concern for linkages between environmental and population issues. Urbanization, migration, movement of refugees, and displacement of people because of environmental disruption are examples of realities that are directly linked to population, as we find in the Yucatan today.

E. N. Anderson's chapter, "Valuing the Maya Forests," gives us hope for the Yucatan Peninsula with its description of a sustainable management system: Valuing forests is a difficult matter. As he surmises, "On the one hand, they have cash value—normally understood to be the cash value of the products of the forest, but now including the value of the forests themselves as ecotourism destinations and genetic preserves. On the other hand, they have spiritual, aesthetic, and 'heritage' value placed on them by the people who live in or near them—and, today, often by people who live far from them. In Quintana Roo, all these concerns are matters of intense debate. The Yucatec Maya almost all desire to save their forests."

As Anderson notes, I had in my study *Nine Mayan Women* in the late 1960s pointed out the importance of understanding the specific social situation in the local community and its dialogue with the outside world. Now, as he says so well, "Government agencies, nongovernmental organizations, and Maya communities are interacting in a complex and shifting political and economic universe. This is not a dialogue; it is a polylogue." Some local communities have cooperated with development agencies in devising new management systems and economic opportunities as the encroaching commercialized outside world presses on a fragile base. Over the years, the Maya, using their traditional slash-and-burn agricultural system which Anderson carefully describes, have been serious and caring managers who "have been farming their land for 3,000 years." Even though we do not yet have all the answers, the collapse of the Classic Maya civilization is widely believed to be due in part to overdraft on the environment. Other theories postulated prolonged drought, lowering water tables, and/or pollution, disease, and of course war, all of which might well be related to population pressure. But the present-day Maya are survivors, still struggling to survive.

In one of Anderson's descriptions of the important role of Maya moral traditions in Quintana Roo, he relates a story one of his informants told. "After he had killed five pacas one night, he heard a whistling from above, which he knew to be the Lords of the Forest." These are the same Lords of the Forest whom Bob Laughlin's theater group, Jtz'ibajol (Tzotzil for "Writers' House"), from San Cristóbal de las Casas portrayed in 1994, at the conference Wisdom of the Maya, as being distressed by the violence and the destruction of the forests and the animals in Chiapas. There is a Mundo Maya. But as Anderson notes, the moral rules are not enough with modern technologies, roads, trucks, chainsaws, rifles, and shotguns. More stringent rules are necessary—in short, economic forces and an ethic of "use and destroy" are changing the Maya world in a most unfortunate way.

The Plan Forestal, started as a pilot project nearly ten years ago, was so successful that it has become a government program. With help from German government biologists, a wildlife dimension was added. The details worked out by the Maya in their polylogue with outsiders are clearly outlined in Anderson's chapter and are relevant to similar situations. As he says so well, "Top-down control means policy is set by people who do not know the conditions on the ground, and who invariably make mistakes, for which they are not accountable and from which they do not suffer.... Other plans ... have failed because of the contrary case: so much was left to the local communities that they were overwhelmed with the responsibility and had no help with things that were beyond their ability to manage (e.g., marketing products in distant markets). Anarchy is, alas, not a viable alternative to top-down authoritarianism. The only viable alternative is good management, of the sort idealized in the relevant literature."

Betty Faust in her chapter, "The End of Innocence in a Modernizing Maya Community: 'There Is No More Timber; Let's Sell the Land!'" adds new information to the issues she raised during our 1997 panel, based on recent events in the town of Pich, Campeche. She had previously described a history of community involvement in commercial cutting of tropical hardwoods. In the 1960s, this had escalated in response to both the improved access provided by a paved road and competition from outsiders also using the road, to become a "lumber fiesta." Along with overexploitation of community forests, she noted that the consequences included increases in political conflicts, economic insecurity, emigration, looting of archaeological sites, wife beating, hunger, and "sadness." Some of Pich's young people began to distrust modernization, asking, "Is it too late to go back?" Faust thought then that perhaps the young, as they surveyed the damage done by their parents, might return to some of the environmental wisdom that guided their grandparents. They would need to incorporate new information and new practices, and invent as well as adopt forms of social organization adequate to the new challenges. However, before those changes could evolve, a new source of income suddenly became available in January 2001: the community was offered an attractive price for some of their ejido land and the opportunity to be employed by the buyer, a new agribusiness firm. They decided to sell a large part of their forestry commons. Faust examines some of the factors involved in this decision, concluding that with the changes in federal law allowing for sale, communal land is likely to be relinquished for cash unless some economic benefits can be found that are compatible with resource conservation, which in turn requires new forms of social organization to limit potential overuse.

Grace Bascopé's chapter, "Economic, Political, and Environmental Changes in a Yucatec Maya Village," based on research in Yaxunah, the community at the other end of the *sac be* from Cobá, is an excellent example of how important it is to relate to local communities and the individual families, starting with the local authorities but not stopping there. We have to be sure that we involve traditional as well as elected leaders and outside authorities. What is so exciting about her chapter is that during ten years of investigation by anthropologists from Southern Methodist University, local people were hired on a rotational basis each week, ensuring that all members who wished to work on the dig uncovering the ancient Maya ruins in Yaxunah received equal employment and equal pay. The anthropologists and archaeologists also involved them in doing the research while combining local myths and oral history with published materials. They trained local Maya in restoration and management skills. When the government of Mexico decided to include Yaxunah on La Ruta Maya in 1996, the permit for SMU was denied and archaeologists from the National Institute of Anthropology and History (Instituto Nacional de Antropología e Historia, INAH) were sent to manage the consolidation of the site. They brought in stonemasons from outside and did hire some local laborers, but not on a rotational basis. The local people were able to confront the INAH project director, explaining that they were all equally needy and that, "as ejidatarios, they shared certain democratic values and should have equal access to work on a site located on their own lands." Community efforts, facilitated by help from the anthropologists, resulted in the people from Yaxunah convincing the INAH archaeologists to accept their rotational system and not import outside workers, but they have not yet become the managers of their archaeological treasure.

Most important, Bascopé carefully describes the years of coping strategies of the families as they faced disaster after disaster. Many innovative but for various reasons unsatisfactory or unsustainable survival techniques—cattle, chickens, charcoal, bees, and wood carving—were attempted. The proud Maya of Yaxunah have their archaeological site and the possibility of ecotourism to supplement their milpas and other income-producing activities. Hopefully they can continue living in their beloved but environmentally degraded community with pride and dignity.

There is much we can learn from Bascopé's insightful chapter. We can see what we as anthropologists can do and facilitate. Should it be just the representatives from each community who have to go to the national authorities in Mexico City (such as to INAH in the case of Yaxunah) to get permission for local people to be hired to help excavate, run, and manage an archaeological site and to become the Maya guides? They need knowledge, advocates, or at the very least facilitators to accomplish this. There are many positive things anthropologists can do to help people help themselves while giving them more dignity, empowering them by involving them in participatory research and sharing our findings.

As anthropologists, we have long been aware of the treasures that might be in Yaxunah, the terminal point of the *sacbe* that runs from Cobá. In 1970, when the governor of Yucatan came to inaugurate the opening of the new road into Chan Kom from the Mérida-Valladolid highway, I wondered whether access to the *sacbe* in Yaxunah would intrigue the thousands of tourists who visit Chichén Itzá annually. After testing the idea out on the governor's assistant, I sat down beside the governor, who was eating breakfast in Chan Kom with the local authorities, schoolteachers, and me. I asked him if he knew that the *sac be* from Cobá to Yaxunah was less than two miles away. The governor looked excited and interested and asked specific questions. I quoted my sources and showed him the *National Geographic* with maps I had highlighted. The mayor of Chan Kom proudly assured him that the trail from the plaza went right to the *sacbe* but was not passable for cars. Without further hesitation the governor rose and said, "I just had a great idea!" He turned back to thank me, speaking over his shoulder, as he gave loud orders for the road to be continued the few miles to the *sac be*. There was great applause. But few tourists have come to Chan Kom or to the *sac be*.

Now, twenty-seven years after the governor's celebration, will the newly excavated site in Yaxunah (now on La Ruta Maya) be interesting enough to become a real tourist attraction? Will the new road to Yaxunah bring tourists to the ruins? Will the still-passable narrow paved road that runs straight from the old Mérida-Valladolid highway through Chan Kom bring tourists into the plaza that was so proudly built in the early 1970s and after the 1970s around the cenote? Some of the elders ask today, "Why did we cut down our beautiful mahogany trees to get money to build a plaza no one goes to? Why did we build the little kiosks to show our embroidery? Where are the tourists?" Everyone who comes to Chan Kom is an anthropologist! Perhaps Cobá and Yaxunah and Chan Kom can profit from ecotourism and La Ruta Maya, but I still wonder.

As I said nearly three decades ago, "Were the tourists going to be more polluting than helpful?" I just did not know. If they came to visit the *sac be*, a monument to the skill and ingenuity of the ancient Maya, would the tourists somehow share in the sense of this heritage? And would this somehow cause less of a loss of pride in the past and respect for their own way of life in Chan Kom? I wished I could be sure. I left the village the last time with ambivalent feelings.

> From the huge Coca-Cola advertisement painted on the multicolored store, once all white, to the miniskirted schoolteachers, it was obvious that Chan Kom was turning away from its past and beginning a new stage of existence. Would the men and women remain contented with their lives, each believing the other's job to be equally important? Or would the exposure that the road might bring make them feel backward, underdeveloped, and soon discontented to live the lives in which they once had such joy and pride? Would the road bring in the many benefits of modernization, or would the changes overwhelm them, sending those still able to leave back up into the peaceful bush country? Would the women have any part in making the choices that would affect their lives, as many had before? Would they see their roles as women and particularly as mothers minimized as they came into contact with a civilization to which their past experience did not relate? (Elmendorf 1975, 126)

Last year, thanks to the Wenner-Gren Foundation, I returned on a brief visit to Chan Kom to deliver a copy of the ethnographic video based on the 8 mm film I had taken in Chan Kom in the 1970s. The video was prepared by Jake Homiak, director of the Human Film Archives at the Smithsonian Institution, where the original film and background data are available. Alicia Re Cruz had videotaped the reaction of the people to the showing of the film during a village fiesta in 1996. The mayor, son of one of the *Nine Mayan Women*, was delighted to get a copy of the film and proudly introduced me to the visiting regional supervisor of schools. He explained to the supervisor what the film portrayed, the *ch'a chaak* and early ceremonies and even shots of older people in their traditional Maya clothes and of the mayor as a small boy! The supervisor, much to my surprise, said that he would like to make copies of the video to show in other schools in the area "so that all our children can be proud of their Maya heritage and not feel like second-class citizens in their own land. Our young people are ashamed of being Maya." Maybe the film, which was not acceptable for my Ph.D., is at least useful to the Maya. Let me urge all of you who are doing fieldwork, as Margaret Mead did earlier, where any of the traditional ceremonies are still being performed, take your cameras and record them with and for the people. Get the sound of their voices, both old and young, explaining their interpretation of the meaning. All of us will benefit: anthropologists, other societies, and Maya.

David Forrest's chapter, "The Landscape of Santa Rita Komchen: The Role of Metaphor in the Ecological History of a Yucatecan Rural Estate," evoking images of Maya beliefs, was beautifully read at the 1997 American Anthropological Association (AAA) panel. His careful research and excellent presentation, both written and visual, are examples of how anthropologists can use the economic history of a village such as Santa Rita Komchen to explain to planners and engineers the human–environmental interactions within their cultural settings. He carefully guided us through the details of daily living while relating those activities to their pre-Columbian heritage in the shapes and forms of geographical arrangements. When he presented his paper, "A Solar-Based Water: Transition from Traditional to Modern Systems of Water Management in Yucatan," during my panel at the 1996 Society for Applied Anthropology meeting, I urged him to get in touch and work with the engineers and planners, some of whom are eager to find anthropologists like him to help them. I also suggested that he send his paper to the International Reference Center in Holland, which publishes an annual abstract, *Journal on Women, Water, and Sanitation.* This publication grew out of the PROWESS project I designed in 1983 for the World Bank at the request of John Kalbermattan, Promotion and Support of Women in Water Supply and Sanitation, for the UNDP. The project was approved with additional support coming from NORAD, the Norwegian bilateral aid organization, and later from USAID. Soon it became a budgeted part of the worldwide UNDP/World Bank Water Supply and Sanitation Program.

During the twenty years I have been a part-time consultant to the World Bank, I have enjoyed working with the engineers more than with the economists! Engineers are proud of their skills and upset when the systems they designed do not work or are not used. Everyone can see the broken systems of engineers, while economic mistakes are not as obvious to the general public. Engineers, not only at the World Bank but around the world, are intrigued when given specific strategies and techniques for involving people in the planning, operation, and maintenance of improved water supply and sanitation services. As one of them said to me after he returned from an international symposium in Singapore, where at my request he had presented my paper "Water, Waste and Women: The Hidden Dimension" at the panel on community involvement—with key points carefully underlined in yellow—"Mary, you know I like women, but I've never understood before why we should involve them, but now I know. It's because we want our projects to be successful—to continue to be operated and effectively used—to work." From then on, there were funds in the budgets for "software" as well as "hardware," as the engineers were inclined to say.

In their chapter, "The Transformation of 'Paradise': Deep, Social, and Political Ecology Among the Yucatec Maya of Cobá, Quintana Roo, Mexico—a Dialogical Approach," Ellen Kintz and Amanda Ritchie have documented the beauty of Yucatec Maya women through photographs and descriptions. They describe how the Maya have maintained their communities located in dense tropical forests, retaining their indigenous religious, kinship, economic, and political systems. Cobá is near the beautiful archaeological site by the same name, which is inland from Tuluum, the Mayan ruin overlooking the blue waters of the Caribbean. It is the starting point of the *sacbe*, which the *National Geographic* described as the longest extant Mayan highway—a beautiful, broad road, fifteen feet wide, raised high and built in a perfectly straight line through the jungle between the two well-known classic ruins Cobá and Yaxunah, where Bascopé and Kintz work respectively. The *sacbe* passes through the jungle only a few miles from "my village" of Chan Kom—as we anthropologists say.

Kintz and Ritchie have observed how the outside world is disturbing the balance that the Maya have created over the years. They describe that rather than trying to resist the expansion of the invading commercial system, the Maya have interfaced with it while maintaining their traditional lives. They seem to feel that the families in Cobá have been able to survive in spite of the pressures of the outside world, but one wonders how much longer they can continue as the outside world threatens and pressures them. A key to their economic survival may be their refusal to abandon their traditions even though they have entered the international commodity market.

In her chapter, "Women's Human Rights Issues Among the Maya," Susannah Glusker shares her concern about the violence against the Maya, all the Maya but especially those suffering in Chiapas, against them as an ethnic group and against their political and cultural ecology. Glusker fits into this in two ways. Her book, *Anita Brenner: A Mind of Her Own* (Glusker 1998), is an extremely interesting study of a brilliant, eccentric woman, her mother. Anita Brenner knew the art worlds of both New York and Mexico better than most people knew either one. She also understood both the Spanish and indigenous cultures of Mexico. Her amazing book, *Idols Behind Altars* (Brenner 1929), was published before she had completed her doctorate in anthropology at Columbia, during the days of Franz Boas. Brenner's book was my constant reference as I tried to understand the many Mexicos I was encountering in the 1950s, particularly in my work with the Instituto Nacional Indigenista (INI, National Indian Institute). Along with many of us, Brenner suffered discrimination during the Joe McCarthy days, but she kept meticulous diaries during her lifetime, including the details of her early work as a participant in the original Carnegie explorations at Chichén Itzá. At the 1994 meeting of the Society for Applied Anthropology, in a panel to honor Alfonso Villa Rojas, Glusker presented a paper covering these events. She also used Brenner's diaries as part of an outstanding research project for the Union Graduate School. As a member of her committee, I watched Glusker do an extraordinary job of analyzing the wealth of material and converting it into the biography of her mother—at once a scholarly and deeply personal publication. Both she and her mother added greatly to my understanding of Mexico from the 1940s until the present.

In chapter 11, Glusker starts with a reference to Mariclaire Acosta, the brave Mexican woman who has been so active in human rights in Mexico during the last three decades. Since the New Year's 1994 Zapatista rebellion in Chiapas, in which of course many Maya were involved, Acosta has devoted much of her time to helping them defend their rights to their homes, their culture, and their land.

In spite of concern for her safety and against the advice of her father, a retired Mexican Foreign Service officer, and other worried friends, in the early 1990s Acosta turned down an invitation by the president of Mexico to become a founding member of Mexico's first official human rights agency. As she said, "Mexico needs a nongovernmental organization, a private agency." She did not want to jeopardize the credibility of the Mexican Academy of Human Rights, which she had founded in 1984 after ten years as coordinator of Amnesty International's efforts to abolish torture in Mexico. As Glusker points out, in 2001, Vicente Fox, as the new president, appointed Mariclaire Acosta as ambassador of human rights in the Office of Foreign Affairs. She accepted, but was not confirmed by the Mexican Congress. President Fox thereupon appointed her undersecretary for human rights. She and many of us have new hopes for the rights of Mayas and others in Mexico.

It was at the 1970 conference Man in the Americas, sponsored by CONACYT and the AAAS (American Association for the Advancement of Science), that I first realized the power of Mariclaire Acosta when she and I were on the same panel addressing women's rights. She challenged the pervasive and demeaning stereotypes of Mexican women in the media, both in Mexico and abroad. Now, as Glusker points out in her chapter, Acosta is not only an internationally recognized human rights activist, but never forgets that human rights are women's rights and that the rights of Maya men and women are human rights. Glusker's exhaustive documentation broadens the scope of her chapter to include other human rights issues that Mexico is facing today.

This leads to the penultimate chapter, which is another synthesis, with a theme that pervades most of the chapters. I think we need to look for linkages to other disciplines, to build on the research, on the existing database, and on all of the lessons that have been learned. Then we need to add new lessons and new interpretations of them. As anthropologists we should do more, as John Frazier says so well in his provocative chapter, "The 'Yucatan Syndrome': Its Relevance to Biological Conservation and Anthropological Activities." The Yucatan Peninsula is a classic case of persistent imperialism: the geographic name is a corruption of the Mayan "*Ma t'aan, ku y u'ub a t'aan*" (or similar form meaning "I do not understand you"). "We—humanity, at least thinking, responsible people—cannot afford to continue with the same mentality.... Not understanding or appreciating the people and culture that they had come to discover... the Spanish Conquistadores totally misconstrued the simple (and expected) response to their question, "Where do you live?"

As a biologist deeply concerned about the degradation of the people and the environment of the Yucatan, self-professed outsider John Frazier challenges anthropologists to share our knowledge about the Maya, their local customs and culture, political structure, and environment. He notes, "Over the decades I have come to realize that biological conservation, often described as a vibrant new discipline, is not ultimately about biology. . . . It must deal with human customs, human relations, human motivations, political institutions, and political structures that drive the interactions between people and their environment."

Frazier not only challenges us; he eggs us on as professionals, trained in the identification, description, and evaluation of human social processes, political structures, social interactions, and cultural matrices, the sources of human motivation that illuminate the questions of why people do what they do, to "go to meetings about biological conservation; let them know the relevance of your work. Publish in their journals. Publicly debate their conclusions, illuminate their unspoken assumptions, test their taboos, shake their intellectual structure."

Those of us who were working in the Yucatan when Cancún was still being planned by people using computers in offices in Mexico City felt that no consideration was given to how that investment could have been made for and with the Maya, who were even then facing catastrophic economic disaster as henequen fields and factories were being abandoned. Many of us felt then, and still do, that when tourist hotels in places like Cancún are built, local people should be part of the planning, implementation, and management. Now even the World Bank has anthropologists working with their tourism section. After the research paper Deborah Merrill-Sands and I prepared for the World Bank, "Socio-Economic Impact of Development Projects on Chan Kom 1971–1976" (Elmendorf and Merrill-Sands 1976), was circulated, the bank's training division used parts of it, especially the section on community participation and the impact of the road, in training seminars at the bank. Chris Willoughby, then director of that section, later invited me to help arrange a conference on tourism for the bank, but I was so overcommitted that I could not do it. However, he did accept my suggestions for participants, including I Gusti Bagos, who was the head of the Department of Anthropology at Udayan University in Denpasar, Bali. Bagos was later invited to continue as a consultant for the World Bank in designing of a new hotel in Bali, one of the first to be designed in cooperation with local people, using textiles, lamps, furnishings, and decorations made or purchased locally. Even though I could not do this in the Yucatan, I was able to help the Balinese preserve some of their culture. The World Bank published the tourism conference proceedings and has since invited other anthropologists to advise them on similar projects. Within our own AAA is a small section on tourism with panels where we can present our findings. Some of my colleagues, including Betty Faust, have done this and are in communication with the tourist industry. More should be. How many of us attended the events in Mérida giving publicity to La Ruta Maya? How many of the people of the villages we are working in have ideas to share with the planners? How many of them were there? How many are eligible for jobs or income-producing activities? Are we anthropologists in touch with the planners and managers of La Ruta Maya, helping villagers seek out opportunities to seize?

Some of us feel that developers should be required to protect the environment by using appropriate technologies—saving water, using solar energy, recycling. We wonder why they do not use thatch, bamboo, and other beautiful and effective indigenous building techniques, as was done in Bali. Why were lovely Mayan thatched buildings, like the dining room at Akumal, overlooked as models? Cancún could have been a demonstration to others of what can be done and could have made the Maya proud of what they have and know.

Such demonstrations should not stop in the hotel zone but should extend to the barrios, where most of the workers live in slumlike conditions without safe water, electricity, or adequate sanitation. In Chan Kom, the cement block rooms were symbols of wealth, but many of the families continued using their thatched huts as kitchens, bathing areas, and often for sleeping since they were so much cooler and better ventilated. Undoubtedly they would be considered unsafe in Cancún, but safe, affordable housing with appropriate sanitation could be designed. Perhaps this is the time and place to considered reintroducing into the Yucatan the SIRDO (Sistema Integral de Reciclamiento de Desechos Organicos, or Integrated System for Recycling Organic Waste), which is being used successfully in suburban Mexico City today.

During the meetings in Curitiba preceding the 1992 UN Conference on the Environment, the king of Sweden presented *Eco-Cycles*, an amazing publication containing case studies that highlighted the importance of linking rural areas surrounding cities and towns into the urban cycle of productivity and planning as a basis for sustainable development. The careful collection in Curitiba—called "the eco-capital of the world" by Jaime Lerner, then their mayor—of organic waste ("garbage that is not garbage") by unemployed young people for composting to use in community gardens was also an interesting project. In exchange for "garbage," householders were given fresh vegetables to supplement their diets. A number of the young people had learned gardening skills that they were using as income-producing activities—some as gardeners at hotels, others selling their produce. Could this idea be adapted to the Yucatan? What variations would the Maya suggest?

But what we as anthropologists can do is ensure that those who "develop" not only the real estate but also the natural resources of the forests and the waters include the Maya themselves in the planning and management of all resources under consideration: the archaeological sites, the vanishing jungles, the fertile waters, and the lovely beaches. We as anthropologists could have a voice in some of that planning, or at least amplify the voices of the Maya, as Bascopé and others in these chapters have pointed out.

At the November 1997 AAA meeting, I circulated copies of the May–August 1997 issue of *Kambul*, titled "Women in Conservation," (Vol. VII[2]) to all of the members of our panel (Biodiversity II). It had just been announced that a Yucatec Maya, Norma Poot, the environmental education coordinator in Calakmul, had been awarded a fellowship from the MacArthur Foundation. The current chair of the National Fish and Wildlife Foundation (affiliated with the U.S. Fish and Wildlife Service), Magalen O. Bryant, who has made significant contributions to the environment in the United States, Africa, and Latin America, understands the cultural sensitivities involved in merging interests of the environment and commerce and is committed to bringing together individuals from all perspectives in a search for solutions. To this end, Ms. Bryant is famous for her potluck suppers; however, she does not ask that you bring a dish but rather that you dish out ideas to the many guests in an effort to come up with creative solutions to today's global environmental problems.

A related point that came up several times during the Earth Summit was the importance of avoiding distinctions between rural and urban areas, since cities and towns are fringed with marginal, underserved communities made up of rural people looking for a better life and livelihood, as has been noted in all the chapters.

During the 1977 AAA meeting, E. N. Anderson and I were at the panel in honor of Roy Rappaport. A student spoke up from the audience, saying that few people care about the environment. In a survey he had just completed, "only 11 percent of the young people thought that the environment was of a high priority." The Kyoto conference was coming up soon. As you know, the U.S. position was dreadful. We in the United States are the greatest polluters. The other overdeveloped countries are not far behind. We as anthropologists should take a lead in this debate and work with other groups cooperatively, both as academics and activists, in both the developed and developing world.

One way for anthropologists to reach out is to work here in the United States with the AAAS (American Association for the Advancement of Science), which has one of the highest profiles. AAAS encourages interdisciplinary projects. Work up a panel of people from other disciplines and present your papers at their meetings. Discuss your findings with people from other disciplines. Talk to the press. In fact, Louise Lamphere, president of the AAA, in a mailing to all members, urged us to attend workshops to learn how better to get our messages to the media.

Along with the destruction of the environment and the ecological and military violence that have been alluded to by many people, there is another kind of violence: terrorism. Violence around the world has in the past been due to conflict over limited resources available to growing populations. Boutros Boutros-Ghali said some time ago at the United Nations that the next war would be about water. In fact, not all but certainly some of the many reasons for today's conflicts are, and in the past probably were, due to the lack of basic needs for all families. Infant mortality caused by lack of safe drinking water and sanitation is violence against children. It is violence against women who have to carry water for miles, use polluted water, or care for dying children. Both polluting and not sharing our resources are forms of violence.

SUMMARY

Among the lessons learned over the years are that the studies referred to in these chapters are urgently needed by development planners to complement their studies. Small in-depth case studies using participatory techniques such as Erickson's and Bascopé's are being put to immediate use.

For example, the people of Chan Kom refused introduction of *achiote*, a berry used as a condiment, as a cash crop for them to sell on the open market. Their question was, "What would we do with the extra *achiote* if people stopped buying it? If we could have citrus like that other village, we and our children could eat it, and whatever is left over we could preserve or sell locally. What would we do with extra *achiote*?" Very logical thinking!

In 1997 the American Friends Service Committee (AFSC) held a Peace Gathering in Philadelphia to commemorate the fiftieth anniversary of the Nobel Peace Prize award to the American and British Friends Service Committees. I would like to quote the 1947 words of Henry Cadbury, the wonderful American chairman of the Board of Directors of the American Friends Service Committee (AFSC), words as true today as they were when first spoken at the acceptance ceremony: "Common folk, not statesmen nor generals nor great men of affairs, but just simple men and women, if they devote themselves, can do something to build a better, peaceful world."

Among those attending the 1997 Peace Gathering was former Costa Rican President Oscar Arias Sanchez, who challenged world leaders and ordinary citizens to end world poverty, hunger, oppression, and the global arms trade. "We can no longer say business is business and turn a blind eye to the poverty and oppression caused by arms transfers. Just like slavery and the drug trade, the arms trade reaps profits tainted with blood.... The 20th Century has been marked by cynicism, selfishness, greed and the desire to please all without changing the status quo. In the 21st century, we must resurrect solidarity and compassion." At about this time, thirteen other Nobel laureates joined with Arias and the AFSC to promote an international code of conduct on arms sales (personal communication from Jack Sutter of the AFSC archives, Philadelphia, December 9, 2003). But that is not enough.

The annual AAA meeting in November 2001 included a full day of panels in honor of Margaret Mead on the centennial of her birth, by fourteen of the past recipients of the Margaret Mead Award who had received the honor for "applying the principles of anthropology to the resolution of issues of contemporary human concern and influencing a concerned public outside of anthropology." My paper, "Women and Technological Change in Developing Countries: Lessons from the Maya," focused on two human rights or basic needs—knowledge of contraceptives as well as safe water and environmental sanitation.

The year 1982, when I was honored with the award, was the first in which it was presented jointly by the AAA and the Society for Applied Anthropology. This was a sign of the coming of age of practicing development anthropologists. At the Margaret Mead panel her daughter, Mary Catherine Bateson, did not want people merely to provide eulogies, because she knew her mother would not like that. Instead, she hoped that anthropologists would remember her often-quoted statement: "Never doubt that a small group of thoughtful, committed citizens can change the world." And that is what I would like to leave with you—that we use the lessons we have learned from each other and the Maya as together we try to bring about a peaceful and sustainable world.

This is a world in which, as Mikhail Gorbachev (2001, 4) noted in an editorial in *World Watch*, "Nature Will Not Wait": "The impact and forecasts of global warning are worsening; desertification is advancing; deforestation and pollution are endangering our ecosystems; and more than 1.2 billion people do not have access to drinking water. . . . This is a very grave situation. It is critical that we find a way to bring about rapid, sweeping change of human consciousness worldwide—something that enables us to provoke a large scale shift, of course, in very short time."

Perhaps the shock of September 11, 2001, the overwhelming fear of terrorists and terrorism has helped us as members of the AAA and as citizens of the United States think of ways we can help in proactive, peaceful ways. As Paul and Anne Ehrlich (2001, 5) have reminded us, "The United States is responsible for more environmental devastation than any other nation, as indicated by America's use of about a quarter of the world's energy and its accounting for almost the same proportion of CO_2 emissions."

During the 2001 AAA meeting, an even more important event took place—the unanimous passage of a resolution in support of the Earth Charter on November 29 at the business meeting. At my suggestion, with support from Barbara Johnston of AAA's Committee on Human Rights, the resolution was prepared and presented by Pete Brosius, chair of the section on anthropology and environment, "affirming AAA support for the Earth Charter principles, urging anthropologists to address Earth Charter concerns in the dissememination of their research findings where possible, and urging the U.S. government to use Earth Charter principles to guide foreign and domestic policy in the future." Among the original thirty signers of the resolution are several authors contributing to this book.

"The Earth Charter Community Summit has been the most open and participatory consultation process ever conducted in connection with an international document," says Steven Rockefeller, chair of the drafting committee. The leading spirits internationally in negotiation at all levels have been Maurice Strong, the outstanding Canadian who chaired Habitat I in Vancouver in 1976 and then the UNCED Conference, and Mikhail Gorbachev, president of Green Cross International. Or as Jan Roberts (2001, 1), a remarkable volunteer who coordinated the satellite-linked Earth Charter Summits in ten cities across the United States, said, "While the world's governments and corporations are jockeying for a greater share of the earth's largesse, citizens worldwide are drafting and signing a declaration of responsibility and interdependency."

Since September 11, I have listened to all the reports of violence and plans for retaliation. I wonder how the voices of those who want peace for all in a shared world can be heard. I was pleased to get this message from a friend in Mexico:

"This could be a time to advocate and promote the fundamental principles for building a just, sustainable, and peaceful global society." These words are from the Earth Charter (see the Web site, http://www.earthcharter.org).

The new millennium is the time for ordinary people to make an unprecedented difference. Let's not waste our time, our energy, our money, the lives of our young on missiles and missile defense, on armies and weapons of destruction. Let's not turn to terrorism to get even with terrorists. Let's look for new alternatives—peaceful solutions, caring solutions. May those concerned with the environment—the air, the water, all living creatures—join those seeking peace, not war, as a way to face the enormous challenges confronting the world.

At the first Earth Charter Community Summit, September 29, 2001, ten communities across the United States were linked by satellite, primarily on college campuses, to present their concerns and activities while viewing presentations given by national leaders such as David Korten and Steven Rockefeller of Rockefeller Brothers. Individual signatures on a "Declaration of Interdependence" became the beginning of a united effort for peace worldwide. The first international professional association to support the charter was the World Federation of Engineering Associations, with 8 million members in eighty nations. The engineers have developed a code of ethics based on the Earth Charter. In the 1970s, as the first anthropologist working in the Office of Environment and Health at the World Bank, I noticed how much more concerned the engineers were about the impact of their projects on the environment and the users than were the economists and most of the social scientists.

Collectively, we have written this book as a wake-up call. With our biologist guest, John Frazier, we challenge our anthropological colleagues to get involved. This is a crisis, and we need all the anthropological expertise we can bring to bear on the environmental crisis that confronts us in the Yucatan and elsewhere. As anthropologists once helped with the crisis of World War II, we can again serve by bringing our knowledge to bear on problems of critical importance for the future, not just of our own families or research sites, but of our species and others that share the biosphere with us.

REFERENCES

Brenner, Anita. 1929. *Idols behind altars*. New York: Harcourt, Brace.

Cadbury, Henry E. 1947. Quakers and peace. In *Nobel Lectures: Peace, 1926–1950*, edited by Frederick W. Haberman. Amsterdam: Elsevier, 392–398.

Ehrlich, Paul, and Anne Ehrlich. 2001. The United States: Another leap back. *World Watch*, March/April, 5.

Elmendorf, Mary. 1973. *La mujer Maya y el cambio*. Translated into Spanish by G. Aguirre Beltran. Mexico: Secretaria de Educación Publica.

———. 1975. The Mayan woman and change. In *Women cross-culturally: Change and challenge*, edited by R. Rohrlich-Leavitt. The Hague: Mouton, 111–127.

———. 1976. *Nine Mayan Women: A Village Faces Change*. New York: Schenkman.

———, and Deborah Merrill-Sands. 1976. Socio-economic impact of development projects on Chan Kom 1971–1976. Executive summary and full report to Department of Development Economics, The World Bank, Washington, DC.

Glusker, Susannah. 1998. *Anita Brenner: A mind of her own.* Austin: University of Texas Press.

Gorbachov, Mikhail. 2001. The world: Nature will not wait. *World Watch*, March/April, 4.

Landgraf, John. 1954. *Land use in the Ramah area of New Mexico.* Cambridge, MA: Peabody Museum of Harvard University.

Re Cruz, Alicia. 1996. *The two milpas of Chan Kom.* Albany: State University of New York Press.

Roberts, Jan. 2001. Earth Charter. *Yes: A Journal of Possible Futures*, 1–3.

INDEX

ABOUT THE CONTRIBUTORS

E. N. ANDERSON is Professor of Anthropology at the University of California at Riverside.

JOANN M. ANDREWS is founder and honorary President of Pronatura Península de Yucatán, A.C., and lives in Mérida, Yucatán, Mexico.

GRACE BASCOPÉ is Adjunct Professor in the Anthropology Department of Texas Christian University, Fort Worth.

MARY LINDSAY ELMENDORF is a consultant and development anthropologist, and lives in Sarasota, Florida.

JENNY ERICSON is an independent consultant, and lives in Arlington, Virginia.

BETTY B. FAUST is Researcher and Professor in the Departamento de Ecología Humana, Centro de Investigación y Estudios Avanzados del IPN, Mérida, Yucatán, Mexico.

DAVID W. FORREST is Research Associate at the Institute for Health and Social Policy at the University of Akron.

JOHN G. FRAZIER is Research Associate for the Conservation and Research Center at the Smithsonian Institution, Washington, DC.

SUSANNAH GLUSKER is a historian, journalist, and Adjunct Professor at Universidad Iberoamericana, Mexico City, Mexico.

ELLEN R. KINTZ is Professor and Chair of the Department of Anthropology at the State University of New York at Geneseo.

ALICIA RE CRUZ is Professor in the Department of Anthropology at the University of North Texas, Denton.

AMANDA S. RITCHIE is Resource Coordinator in the Department of Anthropology at the University of Maryland, College Park.